AF326954

STP 1180

Masonry: Design and Construction, Problems and Repair

John M. Melander and Lynn R. Lauersdorf, editors

ASTM Publication Code Number (PCN)
04-011800-60

ASTM
1916 Race Street
Philadelphia, PA 19103

Library of Congress Cataloging-in-Publication Data

Masonry : design and construction, problems and repair / John M.
  Melander and Lynn R. Lauersdorf, editors.
        (STP ; 1180)
    Includes bibliographical references and index.
    ISBN 0-8031-1492-3
    1. Masonry--Congresses. I. Melander, John M., 1949-   .
II. Lauersdorf, Lynn R., 1937-   . III. Series: ASTM special
technical publication ; 1180.
TA670.M343 1993
693'.1--dc20   93-1147
                                                             CIP

Copyright © 1993 AMERICAN SOCIETY FOR TESTING AND MATERIALS, Philadelphia, PA. All rights reserved. This material may not be reproduced or copied, in whole or in part, in any printed, mechanical, electronic, film, or other distribution and storage media, without the written consent of the publisher.

Photocopy Rights

Authorization to photocopy items for internal or personal use, or the internal or personal use of specific clients, is granted by the AMERICAN SOCIETY FOR TESTING AND MATERIALS for users registered with the Copyright Clearance Center (CCC) Transactional Reporting Service, provided that the base fee of $2.50 per copy, plus $0.50 per page is paid directly to CCC, 27 Congress St., Salem, MA 01970; (508) 744-3350. For those organizations that have been granted a photocopy license by CCC, a separate system of payment has been arranged. The fee code for users of the Transactional Reporting Service is 0-8031-1492-3/93 $2.50 + .50.

Peer Review Policy

Each paper published in this volume was evaluated by three peer reviewers. The authors addressed all of the reviewers' comments to the satisfaction of both the technical editor(s) and the ASTM Committee on Publications.

The quality of the papers in this publication reflects not only the obvious efforts of the authors and the technical editor(s), but also the work of these peer reviewers. The ASTM Committee on Publications acknowledges with appreciation their dedication and contribution to time and effort on behalf of ASTM.

To make technical information available as quickly as possible, the peer-reviewed papers in this publication were printed "camera-ready" as submitted by the authors.

Printed in Fredricksburg, VA
May 1993

Foreword

This publication, *Masonry: Design and Construction, Problems and Repair*, contains papers presented at the symposium of the same name held in Miami, FL on 8 Dec. 1992. The symposium was sponsored by ASTM Committees C-1 on Cement, C-7 on Lime, C-12 on Mortars for Unit Masonry, and C-15 on Manufactured Masonry Units. Lynn R. Lauersdorf, State of Wisconsin, and John M. Melander, Portland Cement Association, presided as symposium co-chairmen, and were editors of this publication.

Contents

Overview

These proceedings are the seventh in a series of ASTM symposia on masonry that began in 1974. Committee C-1 on Cement formally joined with Committees C-7 on Lime, C-12 on Mortars for Unit Masonry, and C-15 on Manufactured Masonry Units in sponsoring this symposium. Eighteen papers were presented orally at the symposium and the other ten were presented in poster sessions. The symposium continued to provide a forum for the dissemination and exchange of information and experiences related to all facets of masonry.

Special Technical Publications (STP) containing papers presented at five of the six preceeding masonry symposia were also published by ASTM. The list of these symposia follows.

- *STP 589—Masonry: Past and Present*, from the symposium held June 25, 1974, in Washington, D.C., was the first in this series. It provided a basis for future symposia.

The second symposium in this series was held June 29, 1976, in Chicago, IL. Twelve papers were presented, but an STP was not published from this symposium. Several of the papers appeared in ASTM's *Journal of Testing and Evaluation*.

- *STP 778—Masonry: Materials, Properties, and Performance*, from the symposium held Dec. 9, 1980, in Orlando, FL, covered the third in the series.
- *STP 871—Masonry: Research, Application, and Problems*, from the symposium held Dec. 6, 1983, in Bal Harbour, FL, covered the fourth in the series. It was dedicated to J. Ivan Davison.
- *STP 992—Masonry: Materials, Design, Construction, and Maintenance*, from the symposium held Dec. 2, 1986, in New Orleans, LA, covered the fifth publication in the series.
- *STP 1063—Masonry: Components to Assemblages*, from the symposium held Dec. 5, 1989, in Orlando, FL, covered the sixth in the series. It was dedicated to Alan H. Yorkdale.
- *STP 1180—Masonry: Design and Construction, Problems and Repair*, from the symposium held Dec. 1992, in Miami, FL, covered the seventh and latest in the series that continues.

Russell H. Brown, Clemson University, John T. Conway, Holnam, Inc., Kenneth A. Gutschick, National Lime Association, Harry Harris, Ash Grove Cement Co., George Judd, consultant, and John H. Matthys, University of Texas at Arlington, served as the symposium steering committee. Russell H. Brown, George Judd, Harry A. Harris, and John H. Matthys each chaired the respective oral presentation sessions titled: Design and Detail, Installation and Materials, Testing and Evaluation, and Strategies and Techniques. The 28 papers presented at the symposium and published in this STP were peer-reviewed by 90 ASTM committee members from C-1, C-7, C-12, and C-15.

Numerous ASTM staff members provided needed guidance. Thanks are extended to these as well as the authors and all others who made the symposium and proceeding publication a successful reality.

The symposium co-chairmen, session chairmen, and presenters are pictured on the following pages.

We look forward to the next ASTM masonry symposium scheduled for Dec. 5, 1995, in Norfolk, VA.

John M. Melander

Portland Cement Association, symposium co-chairman and editor.

Lynn R. Lauersdorf

State of Wisconsin, symposium co-chairman and editor.

Lynn R. Lauersdorf   John M. Melander

Symposium Co-Chairmen

R.H. Brown

I.R. Chin

R.S. Piper

S. Foltz

N.V. Krogstad

M.I. Hammons

G. Judd

A.T. Krauklis

R.E. Klingner

B.E. Trimble

W.M. McGinley

M.E. Driscoll

M. Vickers

J.E. Lovatt

J.H. Matthys

A.R. Whitlock

M.P. Schuller

P.L. Kelley

B.L. Kaskel

R.J. Kenney

Design and Detail

Ian R. Chin[1] and Lee Petry[2]

**DESIGN AND TESTING TO REDUCE EFFLORESCENCE
POTENTIAL IN NEW BRICK MASONRY WALLS**

REFERENCE: Chin, I.R., and Petry, L., **"Design and Testing to Reduce Efflorescence Potential in New Brick Masonry Walls,"** Masonry: Design and Construction, Problems and Repair, ASTM STP 1180, J. M. Melander and L. R. Lauersdorf, Eds., American Society for Testing and Materials, Philadelphia, 1993.

ABSTRACT: Efflorescence is a deposit of substances leached from the masonry either onto the surface or into the pores of brick masonry walls. These deposits may be the substances themselves or secondary reaction products of them with the atmosphere. Investigations of efflorescence on brick masonry walls on dozens of structures throughout the United States by the authors and their colleagues have revealed that the efflorescence on approximately 50 percent of the 43 samples studied consists of water soluble sulfate compounds and that the efflorescence on approximately 40 percent of the samples studied consist of calcium carbonate (carbonated calcium hydroxide) that is not soluble in water. Whenever sulfate compounds are predominant in efflorescence on brick masonry, the source of the efflorescence is usually the brick. Calcium carbonate "efflorescence" originates from the mortar. These data strongly suggest that water soluble efflorescence on brick masonry is most likely a sulfate compound which originated from the brick.

To reduce the potential for efflorescence development in brick masonry walls, the bricks manufactured for a specific building should be tested in accordance with ASTM C 67, Method of Sampling and Testing Brick and Structural Clay Tile. In addition, they should be chemically analyzed for water-soluble constituents to assess the efflorescence potential of the brick.

The potential for efflorescence development in brick masonry walls can also be reduced through the use of drainage type walls with a proper flashing and weep system and through the use of good construction practices and proper material selection.

KEYWORDS: alkali, brick, calcium carbonate, cement, copings, design, efflorescence, flashing, mortar, pavements, preconstruction testing, sills, sulfates, water penetration, water-soluble salts.

Efflorescence has been defined as a deposit of water-soluble salts either on the surface or within the pores of brick masonry walls [1]. In the vast majority of cases where efflorescence has occurred, it has been white in color, as shown in Fig. 1. In a few cases, certain vanadium and

---

[1]Principal and Chicago Unit Manager, Wiss, Janney, Elstner Associates, Inc. (WJE), Chicago, IL

[2]Consultant, Erlin Hime Associates (EHA) Division of WJE, Northbrook, IL

molybdenum components present in some brick units can produce noticeable green efflorescence on the surface of white brick masonry walls.

The "brown stains" on brick masonry walls resulting from the presence of either iron or manganese compounds in the brick units are technically not efflorescence because these compounds are not water-soluble and efflorescence by definition is a deposit of water-soluble salts.

Fig. 1 -   View of efflorescence on brick masonry wall

The presence of efflorescence on the surface of brick masonry walls is normally unsightly and, therefore, adversely affects the aesthetics of the wall and building.   Efflorescence on the surface of brick masonry walls normally does not, per se, affect the strength or durability of the wall.   However, when efflorescence is deposited behind the surface of the wall within pores in the bricks, forces produced from confinement of continuing efflorescence deposits in the pores can lead to cracking and spalling of bricks which adversely affect the strength and durability, as well as the appearance of the wall.

Confinement of continuing efflorescence deposits by the glaze on glazed brick units in brick masonry walls or by "water proofing" and "water repellent" coatings applied on the surface of brick masonry walls, can produce forces that can cause spalling of bricks, as shown in Figs. 2 and 3.   Spalling of the type pictured can also be caused by forces resulting from freezing of water trapped inside the brick by the water-impermeable glaze or coating.

Fig. 2 -   Spalling of glazed brick due to efflorescence
           build-up behind glaze

Fig. 3 -   Spalling of bricks due to efflorescence build-up
            behind "water repellant" coating

In order for efflorescence to occur on brick masonry walls, the Brick Institute of America (BIA) [2] suggests that the following three conditions must simultaneously exist in the wall:

1.   Water-soluble salts must be present within or in contact with the brick masonry wall.
2.   Water must be able to gain access into the wall in sufficient amounts and come in contact with the water-soluble salts for a sufficient time to permit the salts to dissolve.
3.   The wall environment must be conducive to evaporation of water that penetrates into the wall.

Based upon the above conditions, the potential for the development of efflorescence on brick masonry walls can be eliminated if none of the materials used to construct the brickwork contains water-soluble substances, or if no water is permitted to penetrate into the wall after it is constructed. These conditions are not possible in brick masonry walls exposed to the weather because water-soluble substances cannot be practically eliminated totally from all materials used in masonry construction and because brick masonry walls are not impermeable to water.

Therefore, from a practical standpoint, the potential for efflorescence cannot be totally eliminated from brick masonry walls. However, the potential for efflorescence development on brick masonry walls on a building can be reduced by the following:

    1.   Preconstruction testing of the brick manufactured for the specific building to determine the potential of the brick to cause efflorescence.
    2.   Utilization of details that reduce water penetration into brick masonry walls.
    3.   Utilization of proper construction methodologies that reduce water penetration into brick masonry walls.

TESTING

The authors and their colleagues have performed dozens of investigations of efflorescence on brick masonry walls. The investigations have included the following:

    1.   Analyses of efflorescence to determine the compound(s) present in the efflorescence using x-ray diffractometry, chemical analysis, and petrography.
    2.   Analysis of mortar and brick specimens using x-ray diffractometry, chemical testing, petrography, and ASTM C 67 efflorescence testing of unused brick samples, if available, to determine the source of the compound(s) that caused the efflorescence.
    3.   Water penetration testing of the effloresced wall to determine path(s) of water entry into brickwork.

ANALYSIS OF EFFLORESCENCE SPECIMENS

The principal efflorescence compounds found on brickwork in 43 different samples from 24 separate projects investigated by the authors and their colleagues throughout the United Stated are as follows, in order of frequency of occurrence:

    1.   Calcium carbonate, $CaCO_3$
    2.   Sodium sulfate, Na_2SO_4
    3.   Potassium sodium sulfate, $K_3Na(SO_4)_2$
    4.   Calcium sulfate dihydrate, $CaSO_4 \cdot 2H_2O$
    5.   Magnesium sulfate hydrates, $MgSO_4 \cdot 6H_2O$, $MgSO_4 \cdot 7H_2O$ and $MgSO_4 \cdot 4H_2O$
    6.   Potassium chloride, KCl
    7.   Sodium chloride, $NaCl$
    8.   Sodium hydrogen carbonate hydrate, $Na_3H(CO_3)_2 \cdot 2H_2O$

    The most common efflorescence compounds found by other researchers are:

    1.   "Alkali (sodium and potassium) sulfates and carbonates; and alkali-earth (calcium, magnesium and aluminum) sulfates and, to a lesser degree, carbonates" [3].
    2.   "Sulfates of calcium, magnesium, aluminum, sodium and potassium. Chlorides are never formed, but in rare cases carbonates of calcium, sodium and potassium appear" [4].

    The efflorescence compounds found on the brickwork of the 43 samples investigated by the authors and their colleagues were also found by the other researchers. However, aluminum carbonate, magnesium carbonate, potassium carbonate, sodium carbonate and aluminum sulfate which were found by other researchers were not found on the effloresced buildings investigated by the authors and their colleagues. We did find other substances, including chlorides, that were not mentioned by those researchers.

<u>Efflorescence Consisting of Calcium Carbonate</u>

Calcium carbonate was found to occur by itself in the "efflorescence" on 15 (approximately 35 percent) of the efflorescent samples investigated and in combination with other efflorescence components in two of the other samples (Table 1). Calcium carbonate was, therefore, found in the "efflorescence" in 17 (approximately 40 percent) of the 43 samples.

TABLE 1--Efflorescence Compounds Identified on Brickwork in 43 Samples

Efflorescence Compound(s) or Mixtures	Number of Samples	Presumed or Determined Source of Efflorescence
Calcium carbonate	15	Mortar
Sodium sulfate/Potassium sodium sulfate	7	Brick
Calcium sulfate dihydrate	3	Brick
Magnesium sulfate hexahydrate	3	Brick
Magnesium sulfate heptahydrate magnesium sulfate hexahydrate	2	Brick
Sodium sulfate/potassium sodium sulfate/ calcium sulfate dihydrate	2	Brick
Potassium chloride/sodium chloride	2	Muriatic Acid/Chloride Accelerator
Potassium chloride/sodium chloride/ potassium sodium sulfate	1	Muriatic Acid/ Chloride Accelerator/Brick
Sodium sulfate	1	Brick
Magnesium sulfate tetrahydrate	1	Brick
Sodium sulfate/sodium carbonate sulfate/ sodium carbonate disulfate	1	Brick
Potassium sodium sulfate	1	Brick
Sodium hydrogren carbonate hydrate/ calcium carbonate	1	Mortar
Sodium hydrogen carbonate hydrate/ sodium sulfate/sodium carbonate decahydrate/calcium carbonate	1	Mortar
Sodium sulfate/potassium sodium sulfate/ calcium sulfate dihydrate/ potassium calcium sulfate hydrate	1	Brick
Calcium glycolate/calcium glycolate hydrate/calcium sulfate dihydrate	1	Anti-freeze (ethylene glycol)

Calcium carbonate is technically not efflorescence (by the previous definition) because it is not water soluble and did not originate as a salt in the masonry. When the "efflorescence" on brickwork consists primarily of calcium carbonate, the source of the efflorescence is usually the mortar. The formation mechanism of calcium carbonate "efflorescence" on brickwork is as follows:

   1.   As the portland cement in the mortar hydrates, it forms calcium hydroxide which is water soluble.
   2.   Calcium hydroxide is also present in any hydrated lime used to make mortar.
   3.   When exposed to significant amounts of water, the calcium hydroxide is dissolved by the water, and as the water evaporates it brings the calcium hydroxide to the face of the brickwork. When exposed to carbon dioxide in the air at the face of the brickwork, the calcium hydroxide carbonates to form calcium carbonate which is not water soluble and will not be washed off the brickwork by rain.

   Because calcium carbonate "efflorescence" originates from portland cement, masonry cement, and lime, every brick masonry wall has the potential of forming calcium carbonate "efflorescence" when significant amounts of water are able to penetrate the brickwork. Calcium carbonate "efflorescence", therefore, cannot be controlled by preconstruction testing and screening of mortar material ingredients. However, the development of calcium carbonate "efflorescence" can be minimized and perhaps eliminated with the use of proper design details and good construction practices that result in a wall that does not permit significant amounts of water to penetrate it.

   Other researchers have suggested that in order to minimize the development of alkali (sodium and potassium) based efflorescence, the "free alkali" solutions of sodium and potassium hydroxide in cements used to make mortar should be "specified as low as possible" [5]. These hydroxides undergo carbonation and form "new building bloom." However, since these compounds are water soluble and of limited quantity (typically less than half a percent in portland cement), they are soon washed away by rainwater. This condition explains why alkali (sodium and potassium) carbonates rarely chronically remain as efflorescence on brick masonry.

<u>Efflorescence Consisting of Sulfate Compounds</u>

   A combination of sulfate compounds was found in efflorescence on 20 (approximately 45 percent) of the 43 samples investigated. Magnesium sulfate tetrahydrate was found by itself in the efflorescence on one of the samples and sodium sulfate was found by itself in the efflorescence on another one of the samples investigated. Combinations of sulfate compounds were, therefore, found in the efflorescence on 22 (approximately 50 percent) of the 43 samples investigated.
   Whenever sulfate compounds are predominant in efflorescence on brick masonry, the source of the efflorescence is usually the brick [6]. The predominance of calcium sulfate in efflorescence indicates that the clay raw materials used to make the brick contain this compound [7]. When potassium or sodium sulfate is predominant in efflorescence, "their development will be found in the firing process" [8].
   The sulfate compound in cements used to make mortars is introduced into cement during manufacture as gypsum which is interground with the clinker. When cement hydrates during usage, calcium sulfoaluminates are produced by the action of calcium sulfate on calcium aluminate to produce trisulfoaluminate (ettringite) and monosulfoaluminate, both of which are only slightly soluble in aqueous solutions resulting from rainwater that penetrates masonry. As a consequence, these sulfate compounds in cement do not cause significant efflorescence.
   Sulfate efflorescence compounds are water soluble, and except for calcium sulfate, which has limited solubility, they are normally washed off the wall by rain water. Since the most common mortar-originated efflorescence, calcium carbonate, is not water soluble, the presence of a water soluble efflorescence on brick masonry indicates that the

efflorescence is most likely a sulfate compound and originated from the brick.

To reduce the potential of brick-originated efflorescence "changes must be made in the plant operation to correct the cause. This may mean additives to the raw materials, higher firing temperatures, more uniform firing distribution, different firing schedule, alterations to the flow of waste-heat gases, or changes to the firing atmosphere" [9].

To reduce the potential of brick-originated efflorescence in completed brick masonry walls, the following preconstruction testing of bricks specifically manufactured for a particular building should be performed in the laboratory to assess the efflorescence potential of the brick and to aid in the final approval and acceptance of the brick:

1.   Water extraction of pulverized brick specimens for identification of water soluble compounds, and analyses of them for sulfate, alkalies, magnesium, and calcium.

2.   ASTM C 67 testing
     a.   Efflorescence
     b.   Initial Rate of Absorption (IRA)
     c.   Saturation coefficient

3.   Identification of efflorescence compounds, if any, from the ASTM C 67 efflorescence test.

Additionally, where possible the performance history of the bricks in similar environments should be obtained.

Efflorescence Consisting of Chlorides

The efflorescence on three (approximately 7 percent) of the 43 samples investigated was found to consist primarily of a combination of potassium chloride and sodium chloride. Potassium and sodium chloride efflorescence has several sources:

1.   Muriatic acid (hydrochloric acid) used to clean the brickwork
2.   Accelerant ("anti-freeze" compounds) in mortar
3.   Sea water either in the mortar (usually from sand washed in sea water) or deposited on the building in an ocean-side environment.

To reduce the potential of chloride-based efflorescence compounds, the use of muriatic acid types of cleaning agents and chloride-based mortar additives should be avoided.

Efflorescence Consisting of Glycolate

The efflorescence in one (approximately 3 percent) of the 43 samples investigated was found to contain calcium glycolate compounds which probably originated from an anti-freeze (ethylene glycol) additive used in the mortar during cold weather placement.

DESIGN AND CONSTRUCTION

The sources of water that penetrate brick masonry walls include rain water, ground water, and condensation from the interior of the building. Of these three sources, rain water is the primary source of water for the formation of efflorescence in brick masonry walls.

Water penetration tests performed on uncracked brick masonry walls during investigations of efflorescence and of water leakage conditions by the authors and their colleagues have revealed that the vast majority of the rain water that penetrates the brick masonry walls tested enters through the mortar joints, primarily at the interface between the mortar and brick units.

Reduction in the amount of rain water that is able to penetrate into a brick wall will reduce the potential for efflorescence development. The following conditions will significantly improve the water tightness of mortar joints in brick walls:

1.   Use of proper mortar with compatible bricks.
2.   Good bond between mortar and brick.
3.   Full contact between mortar and brick
4.   Joints that are completely filled with mortar.
5.   Properly tooled concave and v-groove mortar joints.
6.   Mortar that did not freeze during cold weather construction.

All of these 6 items can be achieved in brick walls with good specifications and normal, proper workmanship practices.

Tests performed by Brown [10] revealed that, under the conditions of his tests, "walls constructed with portland cement/lime mortars are more resistant to water permeance than those constructed with masonry cement mortars."   Tests performed by Matthys [11] revealed that under the conditions of his tests, "the masonry cement/mortar walls leaked significantly more than the portland cement/lime mortar walls".   Our experience is similar. Tests performed by other researchers [12], [13], and [14] have indicated that under the conditions of their tests, there was no significant difference in water penetration resistance of masonry assemblies constructed with portland cement/lime mortars and those constructed with masonry cement mortar. The more watertight a wall is the less potential it has to effloresce.

To reduce the potential for efflorescence from water that does penetrate brick walls, the water should be collected and drained out of the wall as quickly as possible. This condition is best achieved with the use of drainage type walls with a proper flashing and weephole system located at all of the following strategic locations:

1.   Base of wall. The flashing at this location will prevent ground water from contacting the brickwork.
2.   Above all openings (doors, windows, HVAC, etc.) in the wall.
3.   Above shelf angles.
4.   Below window sills.
5.   Below copings.

To be most effective, the front edge of the flashing should be extended beyond the exterior face of the brickwork and be turned downward to form a drip, and the joint directly below the flashing should be sealed with sealant. In addition, end dams should be designed and installed at the discontinuous ends of the flashing to prevent water collected by the flashing from flowing off its ends into the wall and building. Flashing is usually not effective in collecting and diverting water out of the wall when its front edge is recessed in from the front face of the brickwork or when end dams are not installed.

The use of drainage type walls and the use of flashing below copings and sills and at the base of the walls further reduces the potential for development of efflorescence, because the exterior brickwork wythe in the wall is separated from dissimilar materials such as concrete block, reinforced concrete, and stone that may contain soluble salts that could contribute to efflorescence on the brickwork.

Design elements that enhance the potential for development of efflorescence and that should be avoided include:

1.   Brickwork sills, as typically shown in Fig. 4.
2.   Brickwork copings, as typically shown in Fig. 5.
3.   Brickwork pavements with mortar joints, as typically shown in Fig. 6.
4.   Brickwork planter boxes, as typically shown in Fig. 7.

These brickwork design elements, especially sills, copings, and pavements, enhance the potential for development of efflorescence because they are more susceptible to rain water penetration due to their mortar joints being horizontally exposed.

Fig. 4 -   View of efflorescence below brickwork sill

Fig. 5 -   View of efflorescence below brickwork coping

Fig. 6 -   View of efflorescence on brickwork pavement

Fig. 7 - View of efflorescence on brickwork planter box

CONCLUSIONS

1.   The results of investigations of efflorescence on brickwork performed on 43 samples removed from 24 buildings throughout the United States by the authors and their colleagues revealed the following:

   a.   The efflorescence on 22 (approximately 50 percent) of 43 samples investigated was a sulfate compound or a combination of sulfate compounds.

   b.   The "efflorescence" on 17 (approximately 40 percent) of the 43 samples investigated was calcium carbonate. Calcium carbonate is technically not efflorescence because it is not water soluble nor originated as a salt, and efflorescence by definition is a deposit of water soluble salts leached from the masonry onto the face of a wall. However, it is a carbonation product of a water-soluble base.

   c.   The efflorescence on 3 of the remaining 4 samples was a combination of chloride compounds, and the efflorescence on the remaining samples consisted of calcium glycolate compounds.

2.   Other researchers have concluded that whenever sulfate compounds are predominant in the efflorescence on brick masonry, the source of the efflorescence is usually the brick.

3.   Whenever the "efflorescence" on brickwork consists primarily of calcium carbonate, the source of the efflorescence is usually the portland cement, masonry cement, and/or the lime used to make the mortar.

4.   The potassium and sodium chloride efflorescence found may have originated from muriatic acid (hydrochloric acid) used to clean the brickwork, chloride-based additives, or sea water in the mortar or deposited on the building in an ocean-side environment.

5.   The glycolate compound efflorescence found was caused by an anti-freeze additive that was added to the mortar.

6.   The sulfate efflorescence compounds that were found in approximately 50 percent of the samples investigated are water soluble and according to other researchers, the source of sulfate efflorescence compounds on brick masonry is usually the brick. The calcium carbonate "efflorescence" compound that was found in approximately 40 percent of the samples investigated is not water soluble and originated from the mortar used to construct the brickwork walls. These data strongly suggest that water soluble efflorescence on brick masonry is most likely a sulfate compound that originated from the brick.

7.   Efflorescence on the face of brick masonry walls adversely affects the appearance of the wall but does not usually adversely affect the strength and durability of the wall. However, forces from continuing efflorescence deposits in brick pores and from confinement of efflorescence deposits on the face of bricks by "waterproofing" and water "repellent" coatings applied to the brickwork and by the glaze on the glaze bricks can cause spalling and deterioration of the bricks.

8.   From a practical standpoint, the potential for efflorescence cannot be totally eliminated from brick masonry walls exposed to the weather because efflorescence causing salts cannot be totally eliminated from the materials used to construct brick masonry walls and because brick masonry walls are not impermeable to water.

9.   The potential for efflorescence development on brick masonry walls can be reduced by the following:

a.   Preconstruction testing of brick manufactured for the specific building to evaluate the efflorescence potential of the brick. The preconstruction testing should include the ASTM C 67 efflorescence test, the ASTM C 67 absorption tests used to determine brick saturation coefficient, the ASTM C 67 IRA test, and water extraction of pulverized brick for analysis for sulfate, alkalis, calcium, and magnesium.

b.   Utilization of details that reduce water penetration into the brickwork and details that collect and drain water that has penetrated the brickwork out of the wall as quickly as possible. These details include the use of drainage type walls with a proper flashing and weep system at base of wall, above openings, above shelf angles, and below window sills and copings.

c.   Proper construction of brickwork walls that results in minimal water penetration. Such a wall should have good bond and full contact between brick and mortar, joints that are completely filled with mortar, properly tooled concave or v-groove joints, and mortar that did not freeze during cold weather construction. To enhance bond between bricks and mortar, bricks with IRA greater than 30 gallons/minute per 30 square in. (30 gallons/minute per 194 square cm) should be wetted prior to laying.

10. Contrary to tests performed by other researchers, tests performed by Brown and Matthys and our experience have indicated that the use of portland cement-lime mortar will produce masonry walls that are perhaps more watertight than walls constructed with masonry cement mortars. The more watertight a wall is, the less potential it has to effloresce.

11. Design elements that enhance the potential for development of efflorescence and that should be avoided include brickwork sills, brickwork copings, brick pavements with mortared joints, and brickwork planter boxes.

REFERENCES

[1]   Plummer, Harry C. _Brick and Tile Engineering_, Structural Clay Products Institute, Washington, DC, 1950, pp 48.

[2]   "Efflorescence Causes and Mechanisms, Part I of II," Technical Note 23 Revised, Brick Institute of America, Reston, VA, May 1985, pp 1.

[3]   Plummer, Harry C. _Brick and Tile Engineering_, Structural Clay Products Institute, Washington, DC, 1950, pp 48.

[4]   Brownell, Wayne E. "The Causes and Control of Efflorescence on Brickwork" Research Report Number 15, Structural Clay Products Institute, McLean, VA, August 1969, pp 2.

[5]   Brownell, Wayne E. "The Causes and Control of Efflorescence on Brickwork" Research Report Number 15, Structural Clay Products Institute, McLean, VA, August 1969, pp 6.

[6]   Brownell, Wayne E. "The Causes and Control of Efflorescence on Brickwork" Research Report Number 15, Structural Clay Products Institute, McLean, VA, August 1969, pp 9.

[7]   Brownell, Wayne E. "The Causes and Control of Efflorescence on Brickwork" Research Report Number 15, Structural Clay Products Institute, McLean, VA, August 1969, pp 10.

[8]   Brownell, Wayne E. "The Causes and Control of Efflorescence on Brickwork" Research Report Number 15, Structural Clay Products Institute, McLean, VA, August 1969, pp 10.

[9]   Brownell, Wayne E. "The Causes and Control of Efflorescence on Brickwork" Research Report Number 15, Structural Clay Products Institute, McLean, VA, August 1969, pp 17.

[10] Brown, Russell H. "Effect of Mortar on Water Permeance of Masonry" _Proceedings of North American Masonry Conference_, University of Colorado, Boulder CO, August 1978, pp 115-8.

[11] Matthys, John H. "_Conventional Masonry Mortar Investigation_", University of Texas, Arlington, TX, 1988. National Lime Association, Arlington, VA, pp 7.

[12] Gillam, Kenneth "Effect of Sand on Water Permeance of Masonry" _Proceedings of the North American Masonry Conference_, University of Colorado, Boulder, CO, August 1978, pp 118-1.

[13] Ribar, J. W., "Water Permeance of Masonry: A Laboratory Study" _Masonry: Materials, Properties, and Performance_, ASTM STP 778, J. G. Borchelt, Ed., American Society for Testing and Materials, Philadelphia, PA, 1982, pp 200.

[14] Ritchie, T. and Davison J. I. "Factors Affecting Bond Strength and Resistance to Moisture Penetration of Brick Masonry" _Symposium On Masonry Testing_, ASTM special technical publication No. 320, American Society for Testing and Materials, Philadelphia, PA, 1962, pp 16.

DISCUSSION

J. Carrier[1] *and R. Evans*[2] *(written discussion)*--In an attempt to discern the cause of efflorescence, potential sources of soluble salts are often overlooked. There is a tendency to consider brick and/or mortar as the only contributors to efflorescence. Statements such as, "Calcium carbonate 'efflorescence' originates from the mortar." tend to exemplify this tendency. Other elements of a building wall may contribute part, if not all of the soluble salt deposited on or within the masonry. Concrete products, including materials used to form mortar and concrete block, may contain two to seven times as much soluble material [1]. This material includes sulfates, as well as calcium hydroxide, that may be deposited on the surface of the brickwork. While sulfates from gypsum used during the manufacture of cement may be chemically altered during hydration, other materials used to make cementitious products, including slag, pumice, and other aggregates, may contribute to the sulfate content of the brick wall system [2]. These sulfates are water soluble. Portland cement-lime mortars themselves may contain relatively appreciable quantities of sulfates, even after hydration [3].

While manufacturing research has effectively reduced the efflorescence potential of brick, it is virtually impossible to eliminate all soluble salts from mortar. Unfortunately, brick appears to be a victim of its own industry's research. The efflorescence test in ASTM C67 has been used for decades to determine the efflorescence potential of brick itself, while further research has provided other tests to determine even the slightest potential for brick originated efflorescence. The results of these tests are often used as a basis for accepting or rejecting brick. Meanwhile, there is no standard efflorescence test for other products, such as mortar, block, or even exterior wall board that are used in the same wall system because each of the products inherently produce soluble salts, including sulfates.

The relatively high percentage of sulfate (50%) found in the 43 samples appears to give foundation to the reasoning for added brick testing. However, the 43 samples were taken from 24 buildings, which means that more than one sample was taken from individual buildings. Multiple sampling for the purposes of determining the cause of efflorescence is appropriate when trying to determine the cause of efflorescence on a specific building; however, duplication of samples must be considered when determining the overall efflorescence potential of one type of soluble salt vs. another. Without such consideration (even if the samples were from different areas of the building), such double sampling (or duplication) may significantly effect the final percentage of sulfates in relation to other salts, while incorrectly concluding that sulfates form the brick are the major cause of efflorescence.

---

[1]Design advisor, Glen-Gery Corporation, Baltimore, MD.
[2]Quality assurance manager, Glen-Gery Corporation, Shoemakersville, PA.

In new brick masonry, brick and mortar are only two parts of a wall system. In a true brick wall, one that is constructed solely of brick and mortar, efflorescence due to the deposit of sulfates may well be caused by the brick or mortar, but a true brick wall is relatively uncommon in today's construction. Sulfates, carbonates, and even chlorides, may appear on the face of new brickwork from many sources. In order to determine the actual source of efflorescence salts, each element of the system, and it's potential to supply soluble salts and/or moisture to the rest of the wall system must be considered.

[1]   Young, J., "Backup Materials as a Source of Efflorescence," *Journal of the American Ceramic Society*, Vol. 40, No. 7, July 1957, pp. 240-243.

[2]   Morrish, C. F. and Johnston, R., "A Survey of Recent Research Into the Control of Efflorescence in Concrete Masonry," *Technical Report 52*, Washington, Nov. 1980.

[3]   Ritchie, T., "Study of Efflorescence Produced on Ceramic Wicks by Masonry Mortars," *Journal of the American Ceramic Society*, Vol. 38, No. 10, Oct. 1955, pp. 362-366.

Chin, I. R. and Petry, L. (Author's closure)-- In the 24 projects investigated and reported on in this paper, wall elements other than brick and mortar were either not present in the wall or were found not to have an influence on the formation of efflorescence. The comment by Carrier and Evans that elements, other than brick and mortar, "of a building wall may contribute in part, if not all, of the soluble salt deposited on or within the masonry" was, therefore, not substantiated by our data.

With respect to sulfates in portland cement, hydration causes their incorporation in the mortar primarily as ettringite, which is only slightly water-soluble. Consequently, sulfates in portland cement generally produce no noticeable sulfate efflorescence.

Carrier and Evans imply that we found 50 percent sulfate in our samples. More correctly, we found that almost 50 percent of the 43 samples contain some sulfate form of efflorescence.

We did take more than one efflorescence sample from some of the projects. Often this was done because the efflorescence in different areas of the building is visually not the same texture and appearance or is located at different locations on the wall (below bed joint, on face of brick, etc.). The efflorescence of the set of samples removed from each of the buildings where multiple samples were removed was found to be either calcium carbonate only or sulfate only, or one sample of the set is calcium carbonate, one is sulfate, and one is chloride or glycolate. On a building by building basis, the efflorescence we examined was determined to be as follows:

1.   Calcium carbonate "efflorescence" was found by itself in 31 percent of the buildings investigated.

2.   Sulfate efflorescence was found by itself in 52 percent of the buildings investigated.

The efflorescence samples removed from the remaining 17 percent of the buildings investigated were found to be calcium carbonate, sulfate, chloride, or glycolate based, with different efflorescence appearing at different locations on each of the buildings.

In our case studies, the overall percentage of sulfate based efflorescence found is not significantly different when it is reported as a percentage of samples or when it is reported as percentage of buildings.

DISCUSSION

C. T. Grimm[1] *(written discussion)*--A map indicating the location of the projects would be a helpful addition to the paper. The addition of the following citation to the references may be useful:

Grimm, Clayford, T.: "Water Permeance of Masonry Walls: A Review of the Literature," *Masonry: Materials, Properties, and Performance, ASTM STP 778*, American Society for Testing and Materials, Philadelphia, 1982, pp. 178-199.

Chin, I. R. and Petry, L. (Author's closure)

States Where Efflorescence Samples Were Obtained

---

[1]Consulting architectural engineer, 1904 Wooten Drive, Austin, TX 78757.

Richard S. Piper[1] and Russell J. Kenney[1]

BRICK VENEER WALLS - PROPOSED DETAILS TO ADDRESS COMMON AIR AND WATER
PENETRATION PROBLEMS

REFERENCE: Piper, Richard S., and Kenney, Russell J., **"Brick Veneer Walls - Proposed Details to Address Common Air and Water Penetration Problems,"** _Masonry: Design and Construction, Problems and Repair, ASTM STP 1180_, John M. Melander and Lynn R. Lauersdorf, Eds., American Society for Testing and Materials, Philadelphia, 1993.

ABSTRACT: Common problems with brick veneer walls include inadequately sealed flashing, poorly draining cavities, excessive air infiltration, and large thermal bridges through the back-up walls. This paper proposes brick veneer wall details that directly address these problems and which provide a durable wall assembly that is expected to retain its weather resistance and thermal properties for the life of the wall. It is proposed that a properly sealed air barrier membrane be applied at the back of the cavity and that glass fiber drainage insulation be installed in the cavity, the full depth of the cavity.

KEYWORDS: air barrier, air infiltration, brick veneer, flashing, pressure equalizing, rain screen, thermal bridge, water penetration.

Brick veneer walls have been used extensively for many years, and much has been written on their proper design and construction. The literature emphasizes the need for careful detailing and high quality workmanship if the walls are to perform well and provide the durability that is expected of exterior masonry. The Brick Institute of America notes that brick veneer walls "may not tolerate even minor errors in detailing, material selection, and construction." [1] The authors' inspection of masonry work during construction and investigation of distressed masonry walls indicates that poor workmanship is common and may be the major cause of defective brick veneer walls. Other investigators have questioned the long term durability of brick veneer walls with steel stud back-up. This paper proposes materials and details that address the most common deficiencies in these walls and that reduce the reliance currently placed on high quality workmanship.

---

[1]Architect and President, respectively, R. J. Kenney Associates, Inc., P.O. Box 1748, Plainville, MA 02762.

COMMON DEFICIENCIES

The problem the authors are most often asked to investigate on brick veneer walls is water penetration through the masonry and into the building interior. Excessive air infiltration, efflorescence, corrosion of metal components, and poor thermal performance are often found during the investigation of water penetration problems. Some of the most common deficiencies observed in the field and their consequences are:

•   Flashing that is not properly sealed at laps or to the horizontal surface supporting it allows water to bypass the flashing and enter the back-up wall.

•   Flashing that is not panned at the ends or otherwise sealed to prevent water from moving laterally allows water to enter the wall. Flashing details are almost always drawn as vertical sections, with little thought given to how the ends of the flashing are terminated. Flashing that is otherwise well installed is often ineffective because water can flow off the ends and into abutting construction.

•   Flashing that is not secured to the back of the cavity allows water that has bridged the cavity above to run down the face of the sheathing and enter behind the flashing.

•   Mortar droppings that fill the cavity can block weep holes and cause water to back-up over the top of the flashing or through poorly sealed laps. Mortar droppings several inches deep are often found in veneer walls.

•   Damaged, deteriorated, or improperly installed gypsum sheathing and voids and cracks in concrete block work allow air movement into and out of the back-up wall. Warm, moist air exfiltrating the building can condense considerable amounts of moisture in the cavity.

•   Holes and unsealed penetrations through the sheathing and sheathing that is not continuous over columns, beams, floor slabs, and intersecting walls allow water and air penetration.

•   Sheathing that is not sealed to windows, doors, air conditioners, and similar penetrations allows water and air penetration. Field pressure tests of window air infiltration often show that several times as much air infiltrates between the window frame and abutting construction than infiltrates through the window system itself.

•   Misplaced or dislodged rigid insulation allows cold air in the cavity to circulate around the insulation, negating its thermal value. This can also cause a bridge for moisture to cross the cavity.

These are all common problems seen by the authors and reported by other investigators. These problems allow excessive water and air movement through the walls which can cause damage to interior materials, deterioration of structural members, loss of thermal insulating value, masonry deterioration, efflorescence, and mildew. Because these problems remain common even though the need for careful workmanship and the consequences of poor workmanship are well known, new details and materials that reduce the reliance on high quality workmanship must be used. The proposed details are intended to allow a greater margin for below average workmanship and still provide weathertight, durable walls.

DESIGN ISSUES

Air Barriers

The importance of an air barrier to the proper functioning of an exterior wall has been known to researchers for many years [2, 3], but has not generally been included in the standard masonry reference literature and recommended details in the United States. The National Building Code of Canada has required air barriers in exterior walls since 1985. Air barriers perform distinctly different functions do than vapor retarders, and the two should not be confused [4].

A vapor retarder is intended to reduce the rate of water vapor movement into exterior walls by vapor diffusion. It does not have to be continuous, and small gaps are not harmful because the rate of vapor transfer is directly proportional to the percent of the area covered by the vapor retarder.

An air barrier, however, is intended to limit the mass flow of air through exterior walls and must be continuous and fully sealed, because the rate of air flow is proportional to the square root of the pressure difference across the opening. The air barrier is by far the more critical of the two because as much as 10 to 100 times or more moisture can pass through a typical exterior wall due to air flow than by vapor diffusion [4]. In buildings in cold climates and in buildings that are maintained under a positive pressure by mechanical ventilation, large amounts of moisture can be carried into the exterior wall by air flow. Much of this moisture can condense in the exterior wall causing "leaks" that are often blamed on the masonry, parapet, or roofing. A proper air barrier must be continuous across the full extent of the wall and be sealed at all openings and penetrations. It must have sufficient strength and rigidity to resist the strong positive and negative pressures from gusting wind and the lower but sustained pressures from the stack effect and mechanical ventilation. The air barrier must also be durable and should have the same expected service life as other wall materials that are not accessible for maintenance.

A good air barrier at the sheathing line has several significant advantages. The rain screen principle of the brick veneer is most effective when the inside face of the cavity is several times more air tight than the brick veneer. This allows the cavity pressure to equalize with the exterior thus reducing the pressure differential across the brick veneer. Without a pressure drop across the veneer, there will be very little water penetration through the veneer because the pressure differential is the force that moves water through hairline cracks and other openings.

A few additional details are necessary to achieve a true pressure equalizing cavity. There must be sufficient openings in the veneer to allow enough air flow to quickly pressurize the cavity. Open head joints at 40 cm on center at the bottom of the cavity are thought to be a minimum requirement for this purpose. At the building corners, the cavity must be reasonably well air sealed to prevent air flow around the corner, preventing pressure equalization. A pre-compressed expanding foam sealant or a piece of the air barrier membrane brought out to the inside plane of the brick veneer before the brick is installed are two of several ways to seal the corners. Intermediate seals should also be provided, because the external wind pressures are not uniform across the wall. Pressures are higher at the corners and tops of walls, and the separations should be more closely spaced in these areas. Further work is necessary to establish required limits on the horizontal and vertical spacing of cavity seals.

With proper material selection, the air barrier can also serve as a secondary moisture barrier at the back of the cavity. If it has been designed as an air barrier with good seals and continuity, it will be a far more effective moisture barrier than the 15# felt or house wrap commonly used over gypsum sheathing.

The need for an air barrier with stud walls and gypsum sheathing is apparent. Less apparent, but no less necessary, is the need for an air barrier with masonry back-up walls. Concrete block masonry back-up walls may have voids in the mortar joints and often shrink away from or are not fully grouted to columns and beams. The resulting cracks can allow significant air flow and moisture transfer.

<u>Thermal Bridges</u>

The structural frame of a building causes many thermal bridges through the insulation when the insulation is in the stud walls or on the inside face of masonry back-up. Floor slabs, columns, spandrel beams, and abutting partitions and bearing walls can all interrupt the insulation. Steel studs themselves are a major thermal bridge that can reduce the effective R value of stud cavity insulation by 40% or more for 150 mm walls [5]. Thermal bridges not only increase a building's operating cost but can also cause "ghosting" of studs on interior finish surfaces and, more importantly, condensation within the walls. The fasteners for the veneer anchors penetrate the sheathing and steel studs and are potential points of condensation. With is thermal insulation exterior to the sheathing, the stud cavity will be at a higher temperature and, therefore, have a lower relative humidity and dew point. This will reduce the potential for corrosion of steel studs at veneer anchors, one of the long-term durability concerns with brick veneer steel stud walls.

Thermal insulation in the masonry cavity can be continuous over the entire opaque wall area, insulating all major thermal bridges. Rigid foam board insulation is often used for this in masonry cavities. A typical configuration is a 25 mm board in a 50-75 mm wide cavity. A disadvantage of rigid insulation is the necessity for the insulation to be in full contact with the back-up if it is to function properly. It is seldom possible to achieve full contact on all surfaces because of the irregular and misaligned surfaces typical in veneer cavities. The insulation boards most often fit between the brick ties with masonry back-up and are cut out around the ties on stud back-up walls. This allows cold air in the cavity to easily circulate behind the insulation and negate its insulation value.

Another disadvantage of rigid foam insulation is the need to increase the depth of the cavity to accommodate the insulation and maintain an adequate air space. The Brick Institute of America recommends a minimum 50 mm air space [1]. When 25-50 mm of insulation is added, the depth becomes 75-100 mm. There is not always enough room in the overall wall dimension for a cavity this deep. The added depth can also cause difficulties with brick veneer anchor stiffness and windows and door details. A 50 mm air space with 50 mm of glass fiber insulation provides greater thermal value than does a 88 mm cavity with 38 mm of polystyrene insulation.

PROPOSED DETAILS

<u>Air Barrier</u>

A fully adhered sheet membrane applied over the full extent of the sheathing provides a continuous air barrier that is also water resistant and a good secondary moisture barrier. Several different products are marketed for this purpose by a number of manufacturers. The product used on the building shown in Fig. 1 was a self-adhering

FIG. 1--Lafayette Place, Worcester Housing Authority, Worcester MA
Johnson Olney Associates, Inc., Architects, Boston MA

membrane composed of 0.8 mm (0.032 in.) of rubberized asphalt on a 0.2 available from several manufacturers with different combinations of film and rubberized asphalt thicknesses. However, membranes less than 0.6 or 0.7 mm thick may not provide adequate thickness of rubberized asphalt to make a permanent seal around the screws used to fasten the brick ties. The membrane must be elastic enough to tolerate minor movement in the back-up wall and also be strong enough to bridge over joints and voids in the sheathing. These are sometimes fairly wide due to unforeseen construction problems such as the sheathing not being continuous over the outside face of a column or floor slab.

     Other acceptable products include single-ply modified bitumen roofing or waterproofing membranes. These can be torch applied or adhesive applied. They are very durable membranes with excellent puncture resistance and are able to span larger gaps in the sheathing than are the polyethylene backed membranes.

     Hot and cold liquid-applied membranes and mastics are acceptable on masonry back-up walls but do not always provide adequate elongation without rupturing over joints between abutting materials such as the top of a non-bearing masonry wall and a beam or over the edge of floor slabs. They are also more difficult to seal at penetrations, around windows, and over expansion joints. When applied at less than the recommended thickness, they have very little ability to bridge cracks. They can be used successfully in combination with compatible sheet membranes when the liquid-applied or mastic product is used over solid areas and the sheet membrane is used to span voids and joints and to seal the air barrier to flashing, lintels, and penetrations.

     The heavier sheet membranes can also be used as flashing. They are installed continuously from the back-up block or stud wall over the lintel and across the cavity to within approximately 12 mm of the exterior brick face. The bituminous membranes should not be exposed to the exterior to avoid possible heating by direct sunlight and bleeding

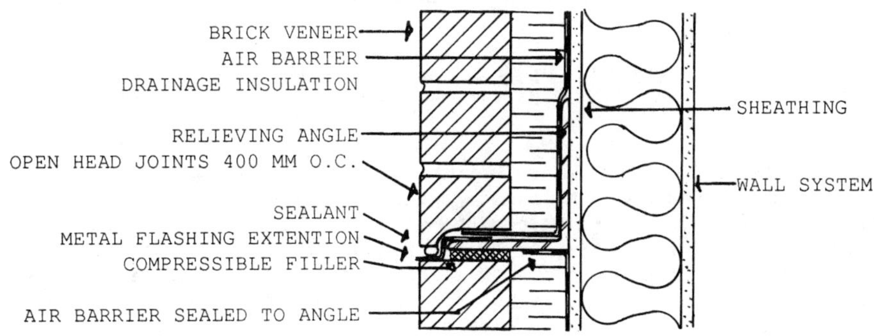

FIG. 2--Relieving angle.

of the bitumen or deterioration due to exposure to ultraviolet light. The through-wall flashing should extend to the exterior brick face, preferably, project out and form a drip. An approximately 50 mm wide strip of metal flashing or other durable flashing material is used for this, and the membrane flashing is sealed to it (Fig. 2). The self-adhering sheet membranes are easily sealed to themselves, to lintels and similar supporting surfaces, and to metal or other compatible flashing materials. The ease of sealing these membranes increases the likelihood of watertight flashings as compared to rigid flashings requiring joints or dry membranes requiring mastic seals.

Fig. 3--Jamb detail at window. The air barrier membrane is sealed to the window surround.

The air barrier membrane shown in Fig. 3 was applied to the sheathing, turned into the window opening, and sealed to the framing. ahead of the brick work. After the windows were installed, sealant was applied to seal the perimeter of the window frame to the air barrier. Pre-compressed, expanding foam seal tapes or foamed-in-place urethane are also acceptable products for this purpose. The compressed foam sealant tapes work best with smooth, regular surfaces, while the urethane foam is preferred with irregular or rough textured surfaces. When the windows are installed before the brick veneer, the air barrier membrane can be sealed directly to the window frame (Fig. 4). Doors, ducts, air conditioning sleeves, and other penetrations must also be sealed to the air barrier.

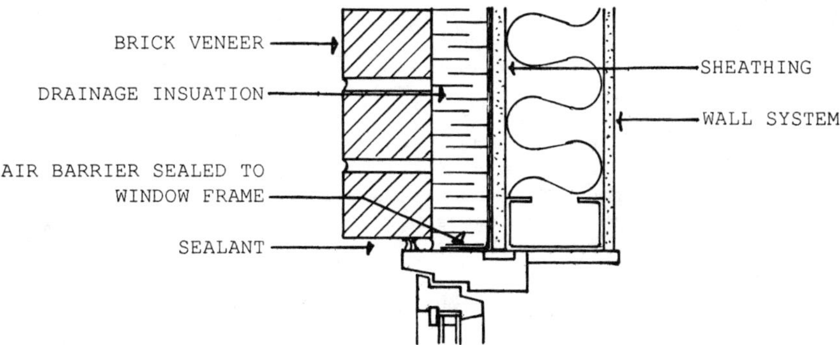

This detail is drawn for window installation prior to masonry
work. For windows installed after masonry work is complete,
the air barrier should be turned into the window opening and
sealant applied between the air barrier and window frame.

FIG. 4--Window jamb.

The wall air barrier must be continuous with the air barrier at
the roof, soffit, and similar changes in plane. Vertical joints between
brick veneer and abutting construction need to be sealed at the air
barrier to prevent horizontal movement of water and air from the cavity
into the adjacent wall construction. This is especially important when
the abutting wall contains a cavity that could allow air and moisture
flow to or from the interior or when it is a moisture sensitive material
such as an exterior insulation finish system.

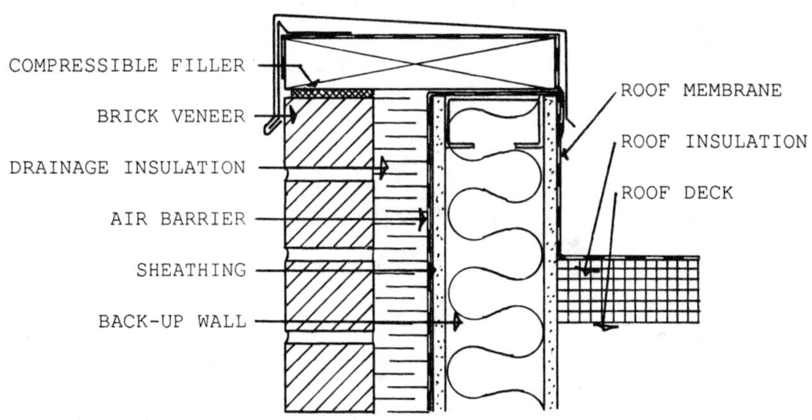

Wall air barrier is continuous over top of framing and
connects with roof air barrier, which is generally but not
always the roof membrane.

FIG. 5--Parapet detail.

Cavity Insulation

One of the most difficult (and often unsuccessful) construction quality control efforts is to assure a clean cavity with unobstructed flow to the flashing and weeps. If the cavity insulation provides drainage and fills the full depth of the cavity, a clean, mortar free cavity is assured. Three pound density (50 kg/m^3) semi-rigid glass fiber insulation provides good drainage (the glass fibers are all oriented vertically and moisture will not move perpendicular to the face of the wall) and air flow through the insulation for pressure equalization and venting. This product has been used for below grade foundation insulation and drainage for approximately ten years in the U.S. and Canada. The masonry cavity is a much less severe exposure because there is less water and no soil pressure on the insulation.

The insulation is installed between the brick anchors ahead of the brick work. It is easily cut to fit over an anchor where additional ties are used, such as around window openings. The glass fiber insulation is more flexible and easier to install with tight joints than are rigid foam boards. Because the insulation is the full thickness of the cavity, mortar cannot fall into the cavity, and the base of the cavity will remain clean and free draining. A clean cavity is assured without any special effort by the masons. Pull boards, brick removal to clean out the cavity, pea gravel, and other special work and inspections are eliminated. Once the insulation is installed on the flashing, the weeps cannot become blocked by mortar dropping into the cavity.

The insulation cannot become displaced and allow air circulation between it and the back-up wall because it is the full thickness of the cavity. Because the insulation is compressible, it conforms well to the typical irregularities of stud or masonry back-up walls and to variations in cavity width. A 50 mm cavity will provide a thermal value of RSI 1.5 (R 8.4) over the full extent of the opaque wall. The 20% lower R value per unit thickness than extruded polystyrene is more than compensated for by the greater thickness used and fewer gaps between and behind boards. No special care is required to assure full contact with the back-up wall or to adhere the board to the back-up. The boards are easily and quickly installed, and average workmanship will result in an application equal to or better than that achieved by careful, quality work with rigid foam boards.

CONCLUSION

Because of the continuing problem of poor workmanship and defective brick veneer walls, details are proposed that provide reduced water penetration through the brick veneer and good flashing seals. A positive air barrier at the back of the cavity and air seals at building corners reduce the water penetration through the brick veneer due to the reduced pressure differential across the veneer. A self-adhering, waterproof air barrier membrane also provides a positive secondary moisture barrier at the back of the cavity and can be used as flashing or easily sealed to the flashing. Semi-rigid glass fiber insulation completely filling the cavity and installed prior to the brick veneer assures a clean, free draining cavity with no possibility of mortar droppings blocking the weeps. The entire opaque wall is insulated, greatly reducing the many thermal bridges in typical exterior masonry walls. The possibility of condensation at the sheathing is greatly reduced because of the thermal insulation on the outside of the sheathing and positive air barrier. The major benefit of the full depth glass fiber insulation is that clean cavities are assured independent of the masonry workmanship.

More investigation and testing is needed of the details and spacing of air seals in masonry cavities necessary to achieve pressure equalization.

REFERENCES

[1]   "Brick Veneer Steel Stud Panel Walls," <u>Technical Notes on Brick Construction No. 28B</u>, Brick Institute of America, Reston, VA, February 1987.

[2]   Garden, G.K., "Rain Penetration and Its Control," <u>Canadian Building Digest No. 40</u>, National Research Council Canada, Division of Building Research, 1963.

[3]   Latta, J.K., "Walls, Windows and Roofs for the Canadian Climate," <u>Special Technical publication No. 1</u>, National Research Council Canada, Division of Building Research, 1973.

[4]   Quiroutte, R.L., "The Difference Between a Vapor Barrier and an Air Barrier," <u>Building Practice Note No. 54</u>, National Research Council Canada, Division of Building Research, 1985.

[5]   Strzepek, W.R., "Thermal Resistances of Metal Frame Wall Constructions Incorporating Various Combinations of Insulating Materials," <u>Insulation Materials, Testing, and Applications, ASTM STP 1030</u>, D.L. McElroy and J.F. Kimpflen, Eds., American Society for Testing and Materials, Philadelphia, 1990.

Stuart Foltz and Charles W.C. Yancey

THE INFLUENCE OF HORIZONTAL REINFORCEMENT ON THE SHEAR PERFORMANCE OF CONCRETE MASONRY WALLS

REFERENCE: Foltz, S., Yancey, C.W.C., "The Influence of Horizontal Reinforcement on the Shear Performance of Concrete Masonry Walls," _Masonry: Design and Construction, Problems and Repair_, _ASTM STP 1180_, John M. Melander and Lynn R. Lauersdorf, Eds., American Society for Testing and Materials, Philadelphia, 1993.

ABSTRACT: This report documents one experimental investigation which is part of an overall program of research on the properties of concrete masonry. The primary objective of this testing was to evaluate the effect of the quantity and the distribution of horizontal reinforcement on masonry walls under cyclic in-plane loading which fail by a shear mode (primarily diagonal cracking). The ten wall specimens in the current investigation were subjected to an average of over forty loading cycles. Nine to eighteen of these cycles were applied before yield and loss of load capacity. The walls were subjected to cycles of increasing displacements separated by "degradation cycles."

All ten walls were seven courses high (56 inches or 1.42 m) and three blocks wide (48 inches or 1.22 m) and made from similar concrete block and mortar. The concrete blocks used were hollow block having a unit compressive strength of 1900 psi (13.1 MPa) based on the gross area. The mortar was proportioned as type S. Reinforcement was provided by two types of steel. Rebar steel was placed in grouted bond beams. As an investigation of an alternative method, 9-gage ladder type reinforcement was placed in the bed joint of some walls. Axial load was 39.3 kips (174.9 kN), resulting in a uniformly distributed pressure on the net cross section of 200 psi (1.38 MPa).

KEYWORDS: concrete block masonry, in-plane loading, shear behavior, cyclic loading, vertical reinforcement, horizontal reinforcement, bond beams, bed joint reinforcement

---

Masonry testing has been conducted on a continuing basis at the National Institute of Science and Technology (NIST) since the completion of a tri-directional testing facility in 1984. This report documents one experimental investigation which is part of an overall program of research on the properties of concrete masonry. Previously published reports investigated the influence of vertical compressive strength, block and mortar strength, and aspect ratio on shear resistance of concrete block masonry walls. Testing conducted as part of previous investigations was on the behavior of unreinforced masonry and generally included only monotonic lateral loading to failure and some limited post-elastic cycling. The ten wall specimens in the current investigation were the first to contain reinforcement and were subjected to an average of over forty loading cycles.

SPECIMENS

<u>Material Properties</u>

Every effort was made to hold material properties constant for this series of tests.

<u>Concrete masonry blocks</u>--Two concrete masonry unit shapes were used in the construction of the walls.
1. 8 in. by 8 in. by 16 in. (0.2 m by 0.2 m by 0.4 m), 2 core hollow stretcher block.
2. 8 in. by 8 in. by 16 in. (0.2 m by 0.2 m by 0.4 m), 2 core hollow kerfed corner block with a steel sash groove in one end.

The dimensions represent nominal sizes. The half blocks at each end of alternating wall courses are made by sawing kerfed corner blocks in half through the kerf. Both halves produced by this procedure are used in the wall panels.

Stretcher blocks conform to ASTM Specification for Hollow Load-Bearing Concrete Masonry Units (C90) and have a compressive strength of approximately 1900 psi based on gross area of block. All of the concrete masonry units used in the wall panels and prisms were manufactured by the same manufacturer in one day. The mixture proportions were:

    1950 lbs (882 kg) lightweight expanded shale aggregate
    1250 lbs (566 kg) sand
     260 lbs (118 kg) portland cement
     190 lbs (86 kg) NewCem

NewCem is the proprietary name for a very finely ground water granulated blast furnace slag manufactured by Atlantic Cement Co., Inc. and is a partial replacement for portland cement. It meets the requirements of ASTM Standard Specification for Blended Hydraulic Cements (C595).

<u>Mortar</u>--Type S mortar was used in the construction of the wall panels according to the specifications of ASTM Standard Specification for Mortar for Unit Masonry (C270). Each batch of portland cement-lime mortar was mixed to the same proportions. The materials used in the mortar were:
1. Sand - a natural bank sand that was dug locally with its primary use being for masonry mortar and meeting the specifications of ASTM Standard Specification for Aggregate for Masonry Mortar (C144).
2. Portland cement - a commercially available, bagged, 94 lbs (43 kg) per bag, Type I portland cement identified as meeting the specifications of ASTM Standard Specification for Portland Cement (C150).
3. Lime - a commercially available, bagged, 50 lbs (23 kg) per bag, hydrated lime, Type S, identified as meeting the specifications of ASTM Standard Specification for Hydrated Lime for Masonry Purposes (C207).

These materials were proportioned 1:3/8:4 by volume with one part cement, 3/8 part lime, and 4 parts sand. Dry materials were mixed in a motorized mortar mixer with most of the required amount of water. Finally, small amounts of water were added to produce mortar of a consistency acceptable to the mason. Re-tempering of the mortar, if required, was permitted only once per batch. Unfortunately, mortar cubes made to verify the consistency of formulation between batches were damaged.

<u>Grout</u>--Grout for bond beams was used in the construction of the wall panels according to the specifications of ASTM Standard Specification for Grout for Masonry (C476). Each batch of grout was mixed to the same proportions. Grout was proportioned to have

approximately the same strength as blocks, based on trial mixes. The
materials used in the mortar were:
1.   Sand - same fine aggregate used for mortar.
2.   Pea gravel - coarse aggregate graded by size according to the
      specifications of ASTM Aggregates for Masonry Grout (C404).
3.   Cement - same type I portland cement used for mortar.
The aggregate, cement, and water were proportioned 8.07:1:1.56 by
volume. Aggregate was 38 percent sand and 62 percent pea gravel by
volume. ASTM specifies a slump of eight to ten inches. Our
measurements resulted in an average slump of nine and one half
inches. Grout blocks made during wall construction were also damaged
and could not be tested. Cubes were cut from the bond beams after
wall panel testing. Strengths of tested cubes range from 1210 to
2310 psi (8.34 to 15.93 MPa).

Steel reinforcement--Two types of steel reinforcement were used
in the walls tested. Deformed concrete reinforcement bars were used
in bond beams and ladder type bed joint reinforcements were used in
bed joints. Rebars used included #3, #4, and #5. The bed joint
reinforcement was nine gage steel made with two pieces which lay in
the faceshell bed joint held together by perpendicular cross pieces
welded every 16 inches (0.4 m).

Wall Construction

Test walls were built using running bond with 50% overlap. The
nominal size of the walls was 56 inch (1.44 m) high by 48 inch
(1.22 m) wide by 8 inch (0.2 m) wide. Walls were constructed by an
experienced mason using construction techniques representative of
good workmanship. Fabrication was done in a controlled environment
laboratory and materials were stored in the same area for at least
thirty days.
The walls were three blocks wide and seven courses high with
four block wide isolation courses at the top and bottom of the walls.
These two courses were reinforced with reinforcement and grout
similar to the bond beams described subsequently.

Bond beams--Bond beam construction was not a simple process.
Regular stretcher blocks were used instead of special bond beam
blocks with a solid bottom and partially removed webs. To create a
floor for the concrete, two materials were used. The main component
was rectangular sections cut from concrete blocks. Small pieces of
foam insulation board were used to fill in the cracks around the
block pieces. A mason's chisel was used to remove most of the web of
the blocks. Low slump grout was used to fill in the bond beam after
the reinforcement had been positioned.
Reinforcement for the bond beams was cut longer than the
placement area and was bent 180 degrees at each end to improve load
transfer. The diameter of the bend was dependant on the bar size and
was according to ACI 318 minimum requirements. Final length of the
bent bars were approximately 1/2 inch (1.3 cm) less than the opening
length.

Individual specimen details--Specific details of the type and
quantity of reinforcement and bond beams are provided in Table 1.
Provided below is some explanatory information and specific notes.
Explanatory information includes the reinforcement methods used and
the walls built by that design.
Specimen R1 is an unreinforced wall.
Specimens R2, R3 and R4 contain 9-gage bed joint reinforcement.
(Specimen R3 was damaged early in the test by failure of testing
machine computer controls. This specimen is not further discussed in
this paper.)

TABLE 1--RWall reinforcement configuration and percentages

Wall	As% (%)	Rebar (%)	Bond Beams	Rebar (all)	Bedjt. (%)	Bedjt sp. (in.)
R1	0	0	0	None	0	...
R2	0.024	0	0	None	0.024	16 (0.4 m)
R4	0.049	0	0	None	0.049	8 (0.2 m)
R5	0.094	0.094	1	2 #4	0	...
R6	0.218	0.218	1	4 #5	0	...
R7	0.145	0.145	2	2 #5	0	...
R8	0.218	0.218	1	4 #5	0	...
R9	0.076	0.029	1	2 #3	0.024	16 (0.4 m)
R10	0.215	0.092	1	2 #4 & 1 #5	0.049	8 (0.2 m)
R11	0.145	0.145	2	2 #5	0	...

Specimens R5, R6, and R8 each have one grouted bond beam with
deformed reinforcement. (Specimen R8 was tested under axial load
control instead of displacement control.)
Specimens R7 and R11 each have two bond beams with deformed
reinforcement. The construction of these walls is similar to
placement of a #5 bar every 32 inches (0.8 m) used in standard
practice.
Specimens R9 and R10 each have a grouted bond beam with deformed
reinforcement and 9-gage bed joint reinforcement.

Testing apparatus--The NIST tri-directional testing facility
was used for loading the wall panels. This machine is a computer-
controlled force/displacement system with six servo-controlled
hydraulic actuators connecting the loading frame consisting of two
plus sign shaped (+) steel crossheads and two reaction buttresses.
The lower crosshead is rigidly attached to the structural tie-down
floor and the force/displacement in all six degrees of freedom is
applied through the upper crosshead. See Figure 1.

Testing Procedure

Horizontal displacements were cyclically applied in both in-
plane directions according to a loading pattern designed by the third
meeting of the Joint Technical Coordinating Committee on Masonry
Research for TCCMAR testing [1]. It is very similar to the loading
pattern finally used for TCCMAR testing by P. Shing at Colorado
University at Boulder [2,3,4]. The pattern included cycles of
increasingly larger peak displacements each followed by many smaller
displacement cycles named degradation cycles. Peak displacements are
based on multiples of the "first major event" (FME), which is based
on the occurrence of a load drop and/or substantial cracking.
Axial load on the masonry wall was targeted to result in a
pressure of 200 psi (1.38 MPa) on the net cross sectional area. The
axial load which resulted in this pressure was 39.3 (174.9 kN) kips.
Ideally, this force, which simulates a constant floor load, would be
automatically maintained throughout the test. Unfortunately, the
computer-control system used was not powerful enough to adequately
adjust the vertical movements as horizontal movements changed the
vertical load. This was largely due to the nonlinearity and
unpredictability in the relationship between horizontal translation
and vertical load. For this reason, vertical displacement was held
constant during cycling and vertical load was manually adjusted
between cycles on nine of the ten walls tested. One wall built with
the same reinforcement content as another wall was tested using
vertical load control to evaluate its potential use in future test

series.

After every increase in total deflection and at any time after significant events, horizontal loading was interrupted to examine the condition of the wall more closely and to take pictures. All cracks were highlighted with felt tip markers so the crack pattern would be visible on photos.

<u>Instrumentation</u>

Recorded forces and displacements included those measured by the hydraulic actuators, linear voltage displacement transducers (LVDTs), and strain gages. Data recorded from the actuators included three displacements and three forces in the two horizontal and one vertical direction and three moments about those same directions. The LVDTs measured the horizontal in-plane displacement of the wall at its top and bottom and the displacements occurring across the diagonals of the wall surface. Strain gages were placed on the bond beam reinforcement bars and the bed joint reinforcement.

TEST RESULTS

All walls were tested using an initial displacement of 0.02 inch (0.051 cm) and the displacement was incremented by 0.02 in (0.051 cm) each direction until the occurrence of the first major event (FME). All subsequent displacements were based on the FME displacement and the model displacement pattern in Figure 2. Actual displacement patterns varied among tests.

Ductility calculations were made by comparing the deflection at which the FME occurred and the ultimate deflection which maintained the FME load. This ductility calculation will be called DuctTest since it is based on information readily available during the test. Ductilities used in this report will be based on these deflections in the first direction of loading.

<u>Wall R1</u>

The first major event (FME) was attained after a displacement of 0.10 inch (0.254 cm) and 22.0 kips (98.1 kN). Ultimate strength of 23.9 kips (106.4 kN) occurred at a deflection of 0.095 inch (0.241 cm). DuctTest was only 1.0. This lack of ductility is to be expected for an unreinforced wall which has endured 6 inelastic loading cycles before a larger displacement is applied.

This wall was tested to provide information on the effect of the masonry on the performance of reinforced walls. The failure was extremely brittle and the confidence in the wall's behavior being average is lower than for the reinforced walls. As the wall degraded, the upper crosshead was not lowered to maintain the 39.3 kips (174.9 kN) and axial load was greatly reduced for much of the test. This may have been a factor in the inability of the wall to carry any shear load greater than the FME load. The failure was by the shear mode.

Degradation was mainly due to enlargement of singular block cracks although there was an area on the wall which had predominant mortar cracks. Failure resulted from separation of two triangular shaped areas which left an hour-glass shaped section in the center. See Figure 3.

<u>Wall R2</u>

The FME was attained after a displacement of 0.10 inch (0.254 cm) and a load of 26.8 kips (119.2 kN). Ultimate strength of 35.2 kips (156.6 kN) occurred at a deflection of 0.3 inch (0.762 cm).

Figure 1--Tri-axial testing facility

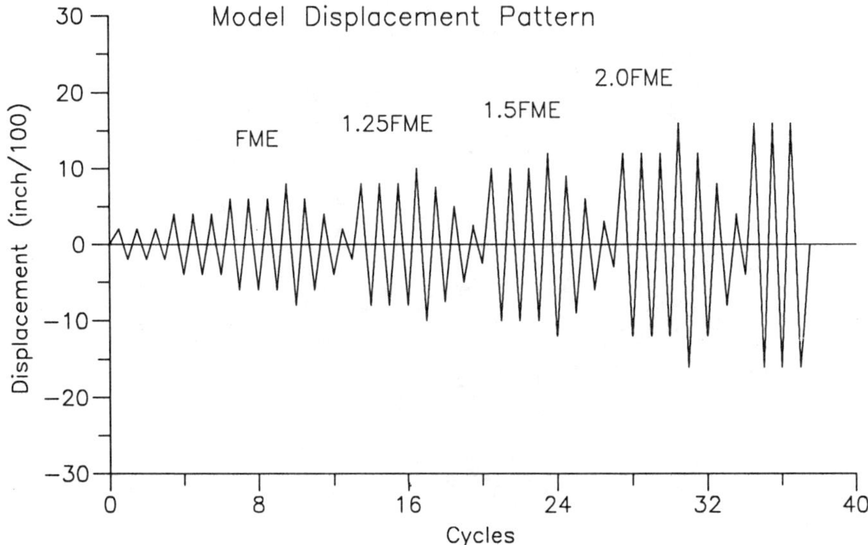

Figure 2--Model cycle displacement pattern

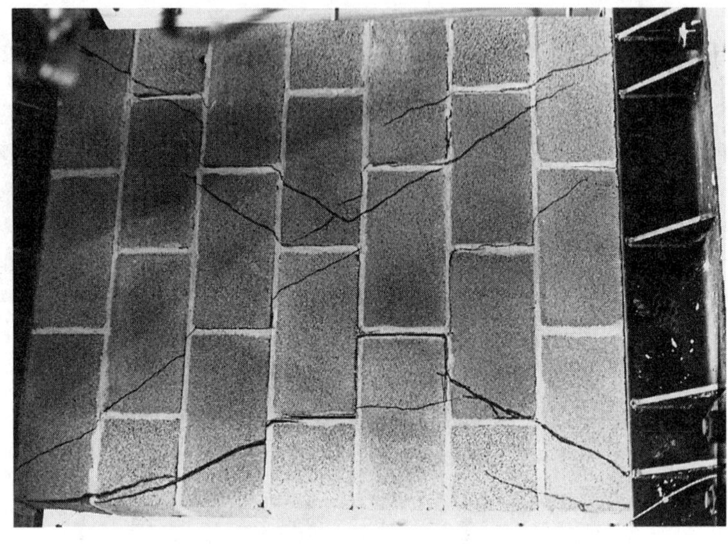

(b) South side

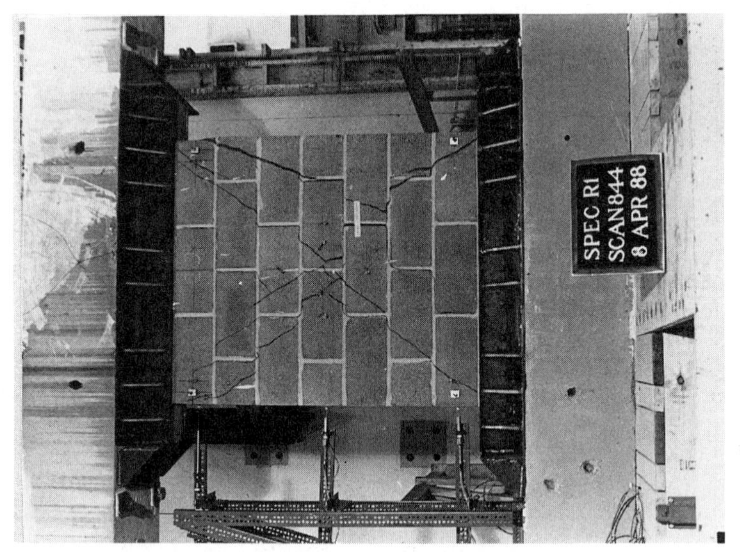

(a) North side

Figure 3--Pictures of RWall 1

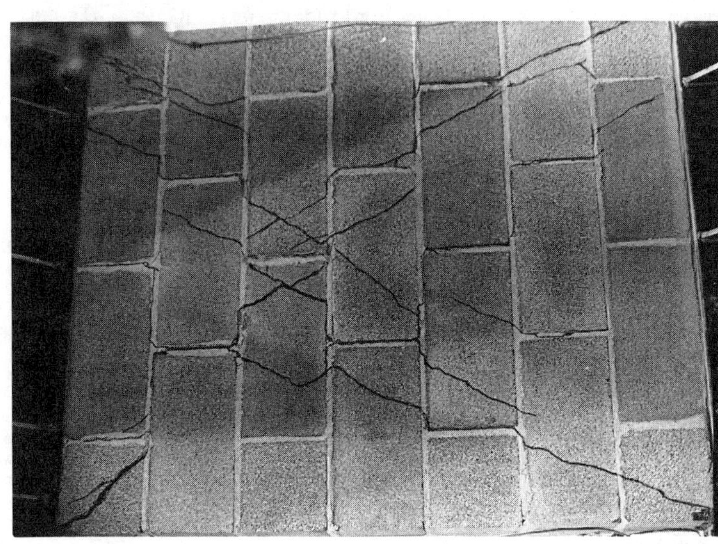

(b) South side

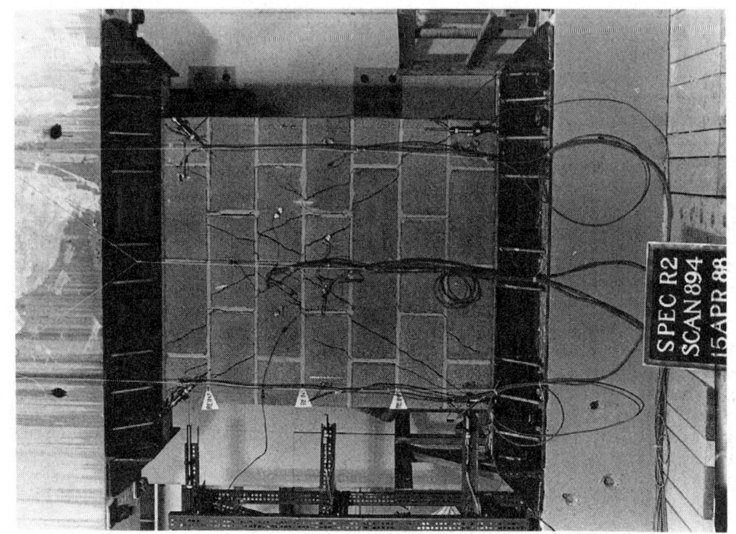

(a) North side

Figure 4--Pictures of RWall 2

DuctTest was 3.0. Axial load applied to wall R2 was also allowed to diminish, reducing the ductility. The failure was by the shear mode with some flexural distress.

Figure 4 of wall R2 shows multiple cracks of four blocks in the upper center portion of the wall. The cracks were prone to remain closed. This is attributable to the bed joint reinforcement in every other course. The development of multiple cracks improves performance under cyclic loading.

Wall R4

The FME was attained after a displacement of 0.12 inch (0.305 cm) and a load of 25.6 kips (114.1 kN). Ultimate strength of 32.8 kips (146.0 kN) occurred at a deflection of 0.36 inch (0.914 cm). DuctTest was 3.0. The failure was by the shear mode with moderate flexural distress.

Figure 5 of wall R4 shows the development of multiple cracks similar to or slightly more numerous than wall R2. The appearance of walls R2 and R4 do not justify the addition of bed joint reinforcement in every course instead of every other course. The ultimate deflection which maintained the maximum load under elastic loading was 0.3 inch (0.762 cm) for both wall R2 and R4. Therefore, deflection and load data supports the visual observation that wall R4, with bed joint reinforcement in every course, performed essentially the same as wall R2, with bed joint reinforcement in every other course.

Wall R5

The FME was attained after a displacement of 0.12 inch (0.305 cm) and a load of 29.5 kips (131.4 kN). Ultimate strength of 44.6 kips (198.5 kN) occurred at a deflection of 0.30 inch (0.762 cm). DuctTest was 2.5. Failure was by a combined shear/flexural mode with the shear distress dominating.

Wall R5 was the only wall with a performance which was significantly inconsistent with the other walls. Although it only had one lightly reinforced bond beam and no bed joint reinforcement, its ultimate shear load was equalled only by wall R10. Only one other wall without bed joint reinforcement (R8) had an ultimate displacement of 0.30 inch (0.762 cm). Wall R8 reached this displacement with substantially reduced strength. Additionally, wall R8 was tested under axial load control which may result in larger deflections than axial displacement control.

It is interesting that the course of blocks immediately below the bond beam had the most extensive cracking. The bond beam prevented the cracks from propagating to the top half of the wall until the second cycle at the FME displacement. The stiffness reduction caused by the existing cracks was adequate to delay the origination of new cracks above the bond beam. The wall failed due to a large crack, which separated a triangular section, and failure of the compression corners.

Axial load on wall R5 was allowed to diminish during the last two cycles. This reduced the lateral resistance and made the lateral stiffness appear less than it was. Despite this, it is doubtful that the wall could have withstood further loading cycles with full axial load restored. This judgement is based on visual inspection of the wall. Figure 6 shows severe deterioration of the top left and bottom right corners. It also shows the separation of a triangular section on the left side of the wall which also greatly weakened the wall.

Wall R6

The FME was attained after a displacement of 0.12 inch

Figure 5--Picture of RWall 4

(a) South side

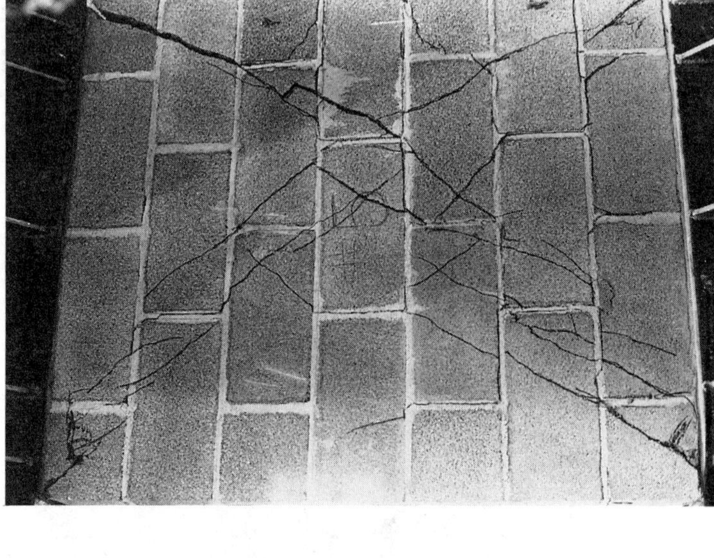

(b) South side

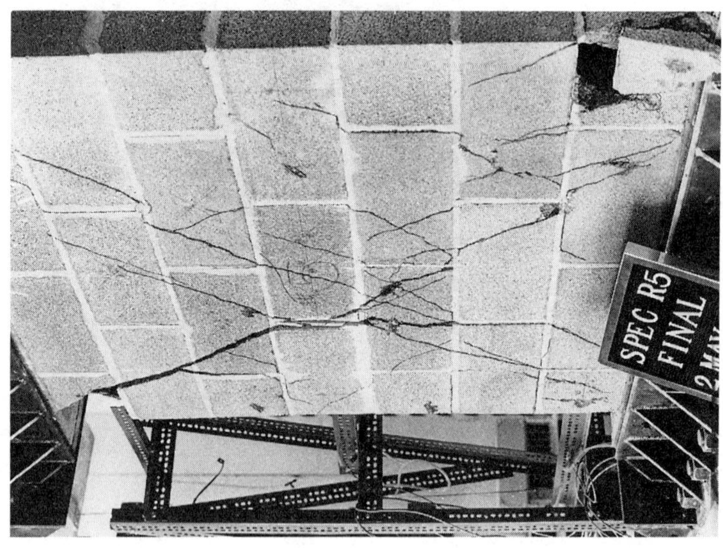

(a) North side

Figure 6--Pictures of RWall 5

(0.305 cm) and a load of 26.0 kips (115.9 kN). Ultimate strength of
33.3 kips (148.2 kN) occurred at a deflection of 0.24 inch (0.700
cm). DuctTest was 2.0. Failure was by a combined flexural/shear
mode with the flexural distress dominate.

Although R6 was more heavily reinforced than R5, it did not
perform as well. Test notes indicate a potentially contributory
factor. The bed joints between the first three courses on the north
side were cracked before the start of the test.

The final picture (Figure 7) shows the results of flexural
failure. The compression corners are severely damaged and were
obviously involved in the loss of load resistance. The center part
of the wall is relatively undamaged. The extensive damage to the
compression corners may be caused by the stiffness of the wall which
is indicated by the few, small cracks. Comparison of this wall and
R5 raises a significant question. Can increased shear stiffness
cause flexural failure to occur at a smaller load? Although not
specifically mentioning this question, M.J.N. Preistley has stated
that performance under combined flexural/shear failure mode is not
well understood.

Wall R7

The FME was attained after a displacement of 0.18 inch
(0.457 cm) and a load of 30.2 kips (134.4 kN). Ultimate strength of
35.6 kips (158.4 kN) occurred at a deflection of 0.27 inch (0.686
cm). DuctTest was 1.5. Failure was by a combined shear/flexural
mode with the shear distress dominating.

Wall R7 had moderately heavy shear cracking. The cracking was
confined to between the bond beams until near the end of the test
when the cracks began extending into the bond beams. After the first
cycle at the FME displacement, all four corner blocks had diagonal
faceshell cracks and there was spalling along the vertical edges.
Failure, after three more cycles at ultimate deflection, resulted
from separation of two slender triangular shaped areas which left an
hour-glass shaped section in the center. See Figure 8.

Wall R8

The FME was attained after a displacement of 0.12 inch
(0.305 cm) and a load of 23.5 kips (104.6 kN). Ultimate strength of
26.6 kips (118.4 kN) occurred at a deflection of 0.36 inch (0.914
cm). DuctTest was 2.5. Failure was by a combined shear/flexural
mode.

Although this wall was built with the same reinforcement as
wall R6, ultimate deflection was larger. It was also larger than for
R7 which had two bond beams. Its ultimate deflection was the same as
for the less reinforced wall R5. The shear load on wall R8 also
differed from other walls. At any chosen deflection, the load was
much less. Ultimate load was also much less. The difference in load
and ultimate deflection can both be explained by the loading method.
It was by load control instead of deflection control. This
eliminated the large increases in vertical load as the lateral
deflection increased and the reduction in axial load when at zero
deflection due to working of the cracks. Although vertical load
increases ultimate shear load, it reduces ductility and seismic
performance.

Shear cracking was mild. The blocks were able to slide along
the mortar joints until cycles at ultimate displacement were applied.
Flexural failure of the compression corners was heavy. All blocks
along both outside edges are broken or separated from the center of
the wall. Although the wall does not look the same in pictures, the
failure is very similar to the walls with the final hour glass shape.
The diagonal shear cracks originating in the corners are apparent in

Figure 7--Picture of RWall 6

(a) North side

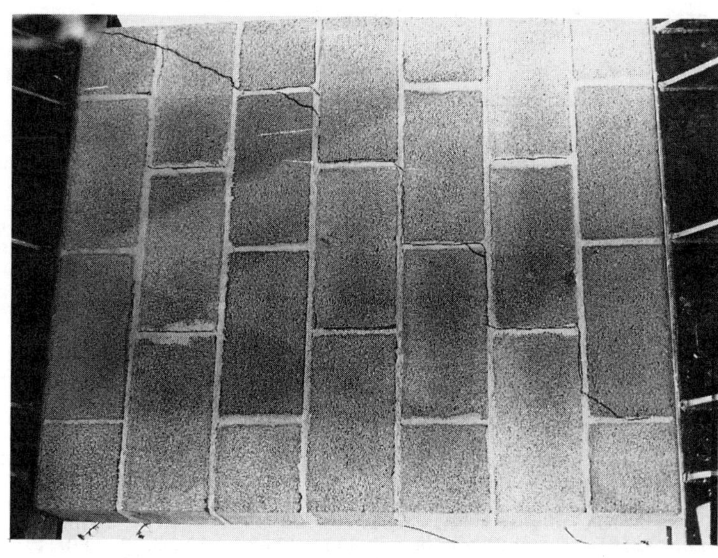

(b) South side

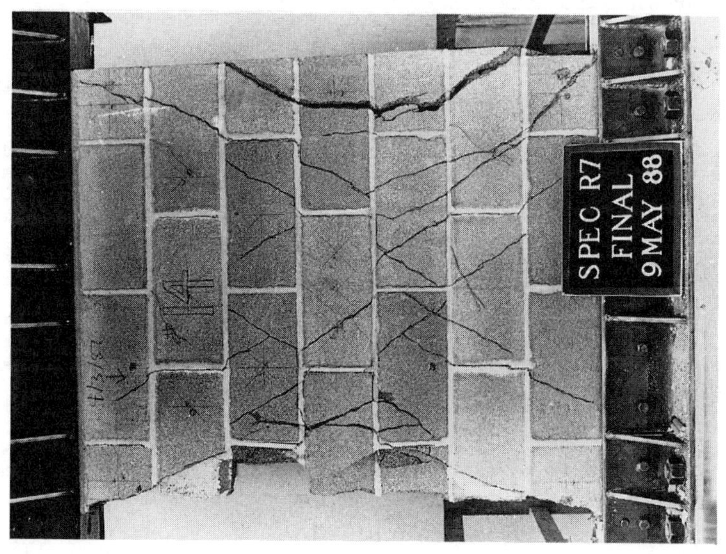

(a) North side

Figure 8--Pictures of RWall 7

the bottom two courses. See Figure 9.

<u>Wall R9</u>

The FME was attained after a displacement of 0.15 inch
(0.381 cm) and a load of 21.6 kips (96.1 kN). Ultimate strength of
39.5 kips (175.8 kN) occurred at a deflection of 0.375 inch (0.953
cm). DuctTest was 2.5. Failure was by a combined shear/flexural
mode with the shear distress dominate.

Shear damage to wall R9 was extensive. See Figure 10. The
compression corners were moderately damaged. Although shear failure
is not desirable, the uniform distribution of numerous small shear
cracks is the best possible scenario if the shear mode dominates.
The results of the test of this wall, as well as the others, strongly
indicates that the addition of bed joint reinforcement results in
improved crack distribution and improved performance.

<u>Wall R10</u>

The FME was attained after a displacement of 0.24 inch
(0.610 cm) and a load of 36.7 kips (163.3 kN). Ultimate strength of
45.3 kips (201.6 kN) occurred at a deflection of 0.3 inch (0.762 cm).
The FME and the ultimate deflections and strengths were large
relative to the other walls. DuctTest was 2.5.

Wall R10 was the only reinforced wall with a significantly
different ultimate deflection. All other walls had an ultimate
deflection between 0.24 and 0.375 inch (0.610 cm and 0.953 cm) versus
0.6 inch (1.524 cm) for wall R10. See Figure 11.

<u>Wall R11</u>

The FME was attained after a displacement of 0.18 inch
(0.457 cm) and a load of 34.2 kips (152.2 kN). Ultimate strength of
38.5 kips (171.3 kN) occurred at a deflection of 0.3 inch (0.762 cm).
DuctTest was 1.5. See Figure 12.

DISCUSSION OF RESULTS

<u>Crack Patterns</u>

The walls in this test showed a predominance to initially
crack along mortar joints. This may be attributable to the relative
strengths of the block and mortar or the bond between them. Once the
mortar joints were all cracked, the only way to further reduce strain
is to work the cracks or develop cracks through the blocks. The
unreinforced wall mainly worked the cracks, but walls with even the
smallest amount of reinforcement showed substantial ability to
develop more cracks through the blocks.

The advantage of distribution, regardless of quantity, of
steel was visually evident in all walls. Figures 3-12 show that
walls without bedjoint reinforcement had more pronounced block
separation than the walls with bed joint reinforcement.

Axial load had a large effect on the formation of block
cracks. Axial load on wall R8 was limited to the initial 39.3 kips
(174.9 kN) and this wall did not develop block cracks as quickly or
extensively. Energy was released predominately through working of
the mortar joints.

<u>Behavior of Unreinforced Masonry</u>

It would have been beneficial to have tested a second
unreinforced wall. It is not certain the one unreinforced wall

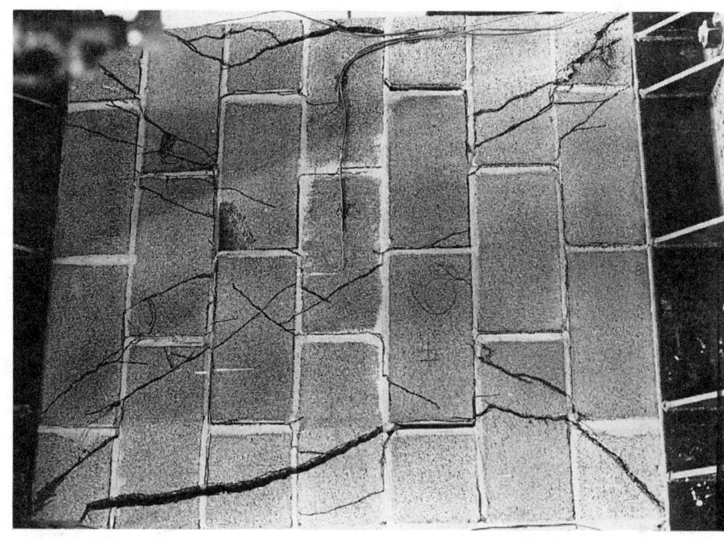

(b) South side

(a) North side

Figure 9--Pictures of RWall 8

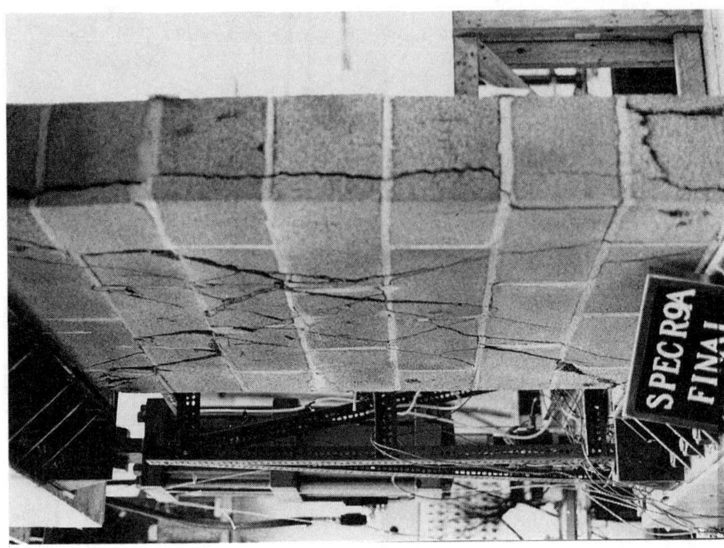

(b) North and west sides

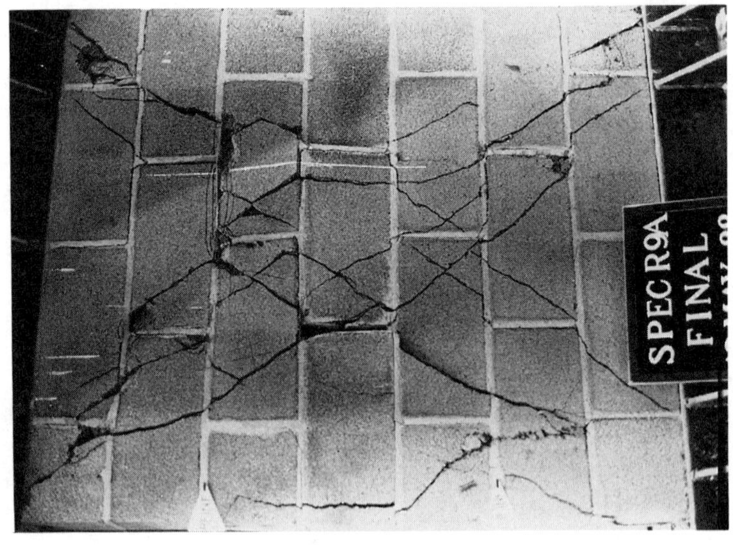

Figure 10--Pictures of RWall 9

(a) North side

(b) North and west sides

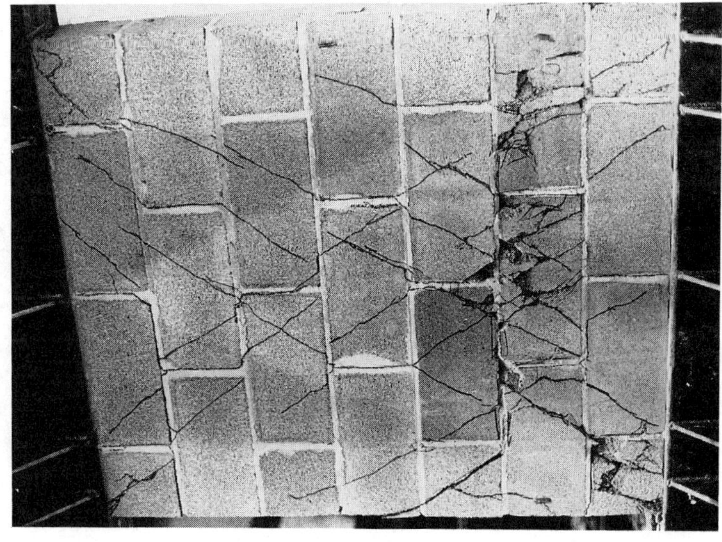

(a) South side

Figure 11--Pictures of RWall 10

(b) North side

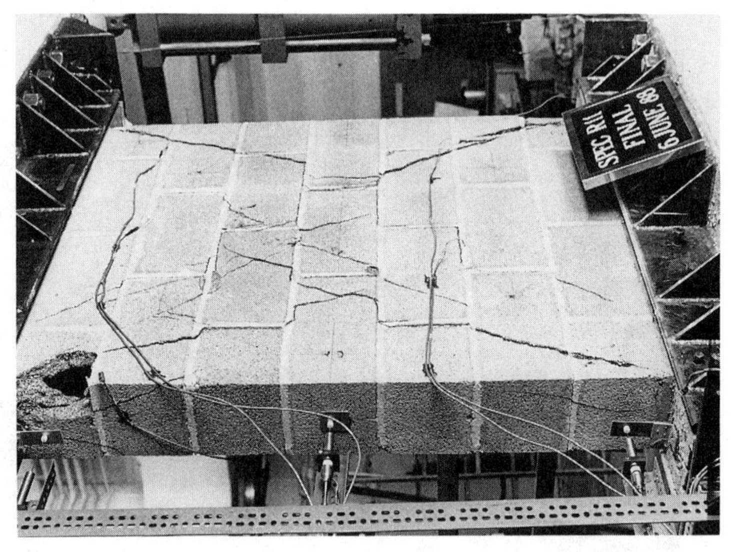

(a) North side

Figure 12--Pictures of Rwall 11

correctly represents the performance that can be generally expected from unreinforced walls. Four three-block-wide walls tested by Kyle Woodward [5,6] provide additional information on the behavior of unreinforced walls under similar conditions. Contrary to the current tests, those walls were eight courses high. Type S mortar was used in all four tests. Three used the same 2000 psi (13.8 MPa) stretcher blocks. The fourth used a weaker 1300 psi (8.97 MPa) block. Because Woodward's walls were monotonically loaded to failure, they provide only an upper boundary on the ultimate displacement expected under cyclic loading. His shear load-deflection plots show that a ductility of one can be frequently expected but also indicate that the ductility sometimes approaches two. If the effect of higher axial load on two of the walls is considered, all four walls have ultimate strengths comparable to unreinforced wall R1.

Axial Load Effect

For similarly designed walls, the relationship between vertical and lateral load at ultimate shear is linearly proportional. In this series of tests, ultimate shear was 0.425 times axial load at the time of failure (Figure 13). Previous tests in the series done by Woodward [7] resulted in a coefficient of 0.376. Tests done by Matsumara [8] resulted in coefficients of 0.233 to 0.241. In testing done on brick masonry by Epperson and Abrams [9] at the University of Illinois, the coefficient was 0.21. Tests at other facilities show a vastly different slope. Aspect ratio, grouting, loading history, and vertical reinforcement may be factors in this difference. The unsettling possibility is that the difference may be due to end constraints or loading method.

Ductility

These tests resulted in what is considered to be small ductilities. Ductilities ranged from 1.0 for the unreinforced wall to 3.0 for the two walls reinforced only in their bedjoints. There are numerous reasons for this. These tests were performed with the intent of investigating shear behavior and shear failures. These are less ductile than flexural failures. Secondly, the testing included many loading cycles, including "degradation cycles", which would reduce the integrity of the wall. Also, because these walls had no vertical reinforcement, the reinforcement percentage was relatively low.

Ductility was also calculated based on first yield of the reinforcement. This is determined from strain gage data which can be unreliable. Additionally, reinforcement yield can occur with negligible influence on the strain where the gage is located. Despite these possible inconsistencies, strain gage data is the most accurate indicator of the actual yield of the wall. Statistical data on reinforcement, displacements, and ductilities is given in Tables 1 and 2. Ductilities calculated on this basis correlated exactly with the FME based ductility for eight of the ten walls.

Further analysis of the test data indicated that ductility calculations made by any method were of limited usefulness in comparing the performance of the walls. There are numerous reasons for this. One factor is the large number of degradation cycles. This reduced ultimate deflection, and therefore ductility, to smaller values. The loading pattern is dependant on the occurrence of FME and therefore most walls had different loading patterns. These problems were caused by the method of testing used but the largest difficulties in using ductilities for comparison cannot be avoided. The occurrence of FME or yield of reinforcement appear to be highly uncertain and may be affected by weaknesses in the masonry which do not affect performance greatly after the masonry has developed

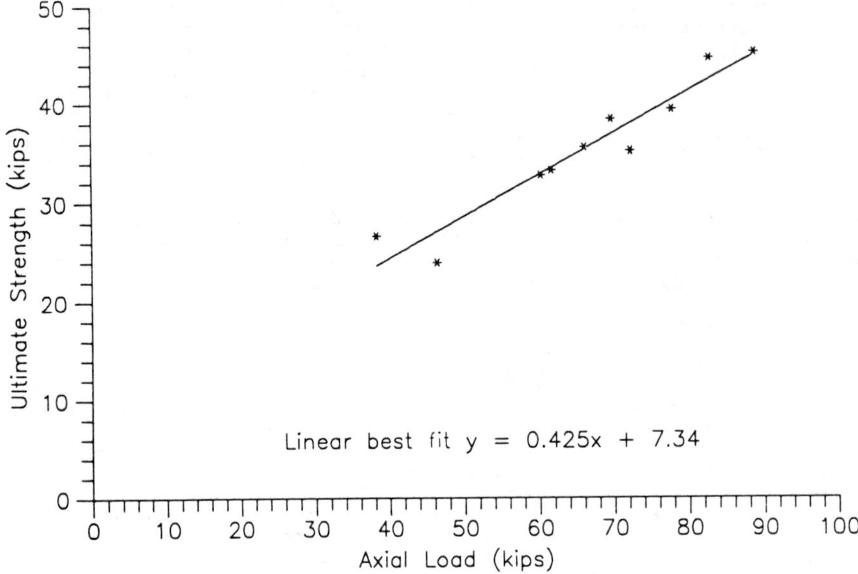

Figure 13--Effect of axial load on ultimate shear strength

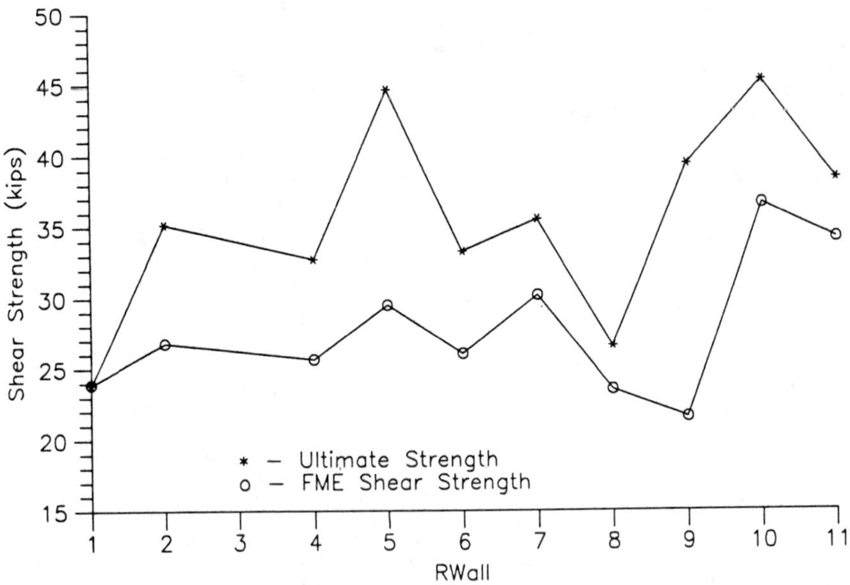

Figure 14--FME and ultimate shear strengths

TABLE 2--RWall test results

wall	FME (in.)	yield (in.)	DeflUlt (in.)	shear (kips)	axial (kips)	DuctTest	DuctYield
		1 inch = 2.54 cm		1 kip = 4.45 kN			
R1	0.10	...	0.10	23.9	46.4	1.0	...
R2	0.10	0.10	0.30	35.2	72.2	3.0	3.0
R4	0.12	...	0.36	32.8	60.3	3.0	...
R5	0.12	0.12	0.30	44.6	82.8	2.5	2.5
R6	0.12	...	0.24	33.3	61.7	2.0	...
R7	0.18	0.18	0.27	35.6	66.1	1.5	1.5
R8	0.12	0.12^1	0.30	26.6	38.2	2.5	2.5
R9	0.15	0.15	0.375	39.5	77.8	2.5	2.5
R10	0.24	0.30^2	0.60	45.3	88.7	2.5	2.0
R11	0.20	0.16^3	0.25	38.5	69.6	1.25	1.56

[1] Strain at zero deflection point drifted, first negatively and then positively, between cycles.

[2] Relationship between deflection and strain was unclear during first fifteen cycles. Yield at deflection of 0.30 inch is definite.

[3] Peak strain did not increase linearly with deflection increases on successive cycles starting at a deflection of .06 inch. Strain on individual cycles was nearly linear. It is likely due to slippage. Yield occurred on the fourth cycle at this deflection.

cracks. Secondly, although not consistent, it is expected that more heavily reinforced masonry will be able to undergo larger deflections before the behavior becomes non-linear. This "penalty" tends to reduce the differences in ductilities between walls. A good example of this problem is the ductility of R10. It is essentially average. But this is because the FME, yield, and ultimate strength deflections were all about double that of the other walls. Wall R10 performed much better but the ductility does not reveal this.

<u>Energy Dissipation</u>

Approximate energy related calculations can be made by calculating the area contained in the shear load - deflection curves. It was originally expected that this would provide information even more useful than the ductility. Comparison of cycles prior to FME led only to one expected conclusion. The more heavily walls are stiffer and therefore dissipate more energy. The post-elastic energy calculations are difficult to compare. In cycles where different walls were deflected the same distance, differences in prior cycle deflections are always obviously the main cause of the difference in energy dissipation.

<u>Strength of Walls</u>

A comparison of the FME and ultimate strength of the walls is made in Figure 14. It should be noted that the FME and yield displacement, and strengths, were the same for eight of the ten walls (See section <u>Reinforcement Strain Gages</u>).
Prediction of the shear strength of walls is limited by current knowledge of the behavior of masonry and the effect of design

variables such as block and mortar strength, aspect ratio, reinforcement quantity and placement, grouting, and loading history. Additionally, if and when these known factors are quantified, it is expected the performance of masonry will remain inconsistent and unpredictable. Figure 15 shows the lack of direct correlation between reinforcement percentage and ultimate shear strength.

Matsumara's equation--Matsumara [8] used the results from his monotonic tests to develop an empirical ultimate shear strength prediction method. It was the most comprehensive method found in review of previous research. The results of this series of tests were compared to strength predictions using his equation. See Figure 16. Although some walls were almost twice the strength of his prediction, his equation does appear to provide a lower boundary which the actual strength can be expected to exceed. The only wall which does not exceed the prediction is wall R8 which was tested under axial load control. His equation does not account for the apparent improvement in performance resulting from the use of bed joint reinforcement.

Comparison of Reinforcement Methods

One objective of this series of tests was to determine the relative effectiveness of horizontal reinforcement. This reinforcement was provided by rebar in bond beams and small gage bed joint reinforcement. The results of this comparison were highly conclusive. Under the test conditions used, bed joint reinforcement consistently resulted in ultimate displacements equal to or greater than larger percentages of reinforcement in bond beams. It remains to be determined how much more reinforcement could be put in the bed joint and still perform favorably.

The four walls with bed joint reinforcement were subjected to horizontal displacements (labeled DeflUlt) as large or larger than any of the walls without bed joint reinforcement. See Table 2. Despite this, visual inspection quite clearly shows these four walls to be essentially in one piece and the walls reinforced only with bond beams all had blocks which detached from the wall. Although the bond beams could have become unbonded to the masonry and helped push the blocks apart, visual observation does not show detachment which is consistent with that cause. Visual observation does support the confining behavior of the bed joint reinforcement. The apparent ability of the bed joint reinforcement to confine the masonry and reduce falling projectile hazard is another advantage to its use.

Figure 17 shows the shear load versus deflection envelopes for the three walls without bond beams. The bed joint reinforcement clearly improved the performance of an otherwise unreinforced wall. Figure 18 shows the envelopes for walls containing 0.076 to .145 percent reinforcement. Wall R9 with bed joint reinforcement has the smallest reinforcement percentage. Despite this, wall R9 withstood a larger displacement. Conversely, it also shows the wall to be much less stiff than the walls only having bond beam reinforcement. This can be at least partially accounted for by the smaller reinforcement percentage in wall R9. Also, the grout in the bond beams increases the stiffness of walls R5, R7, and R11. Figure 19 shows the envelopes for walls R6, R8, and R10 with approximately .215 percent reinforcement. Wall R10, containing bedjoint reinforcement, has a stiffness similar to walls R6 and R8. Further testing is needed to determine the effect of bed joint reinforcement on wall stiffness.

Reinforcement strain gages--All walls with reinforcement were instrumented with strain gages on some of the reinforcement. Strain gages were used on both rebar and bed joint reinforcement. Wall R1 had no reinforcement and for walls R4 and R6, no usable data was

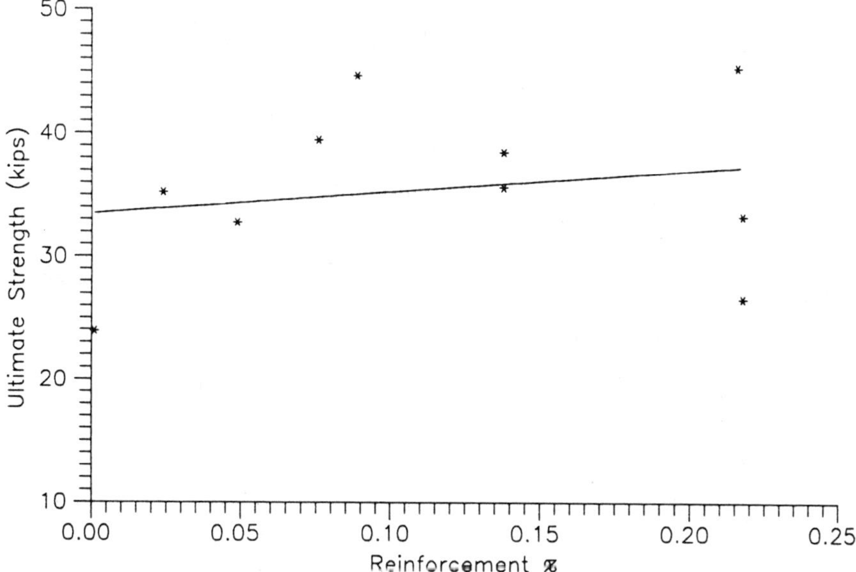

Figure 15—Ultimate strength versus reinforcement percentage

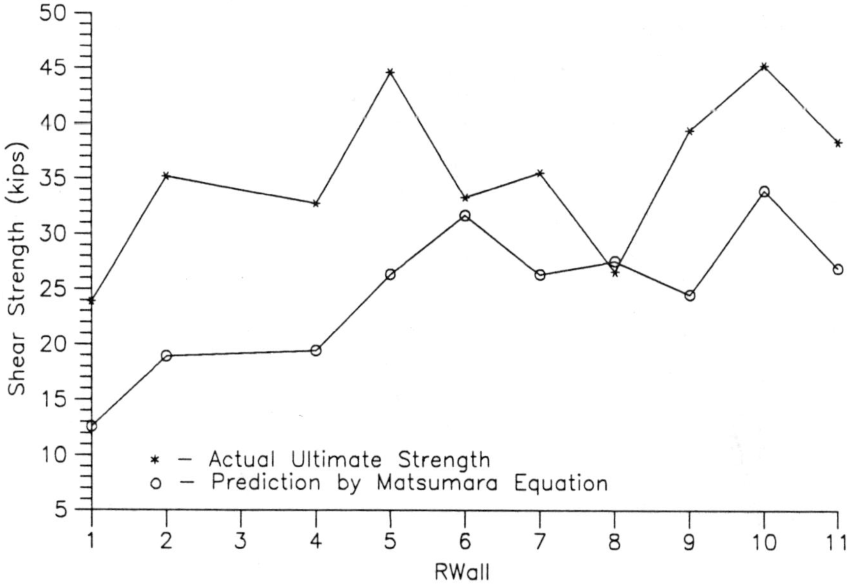

Figure 16—Matsumara's equation prediction

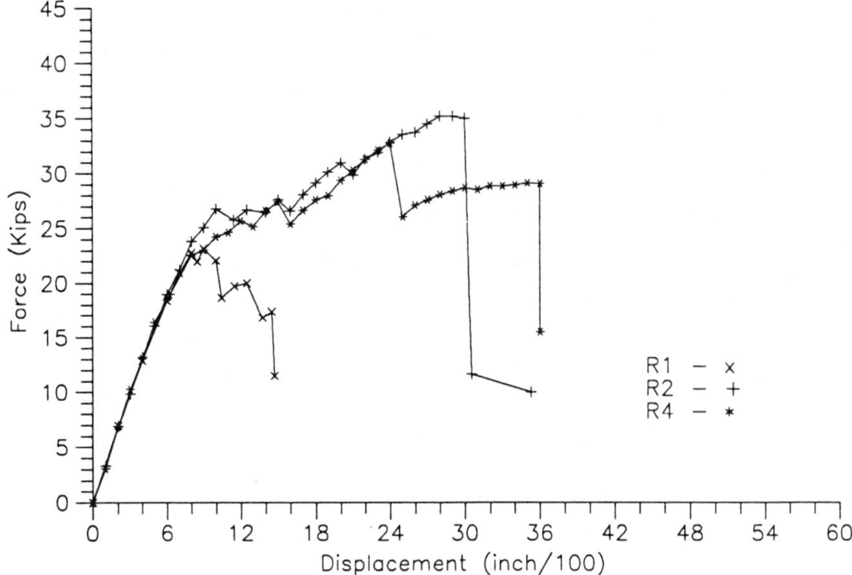

Figure 17--Shear load — deflection envelope comparison
of RWall 1, 2, and 4

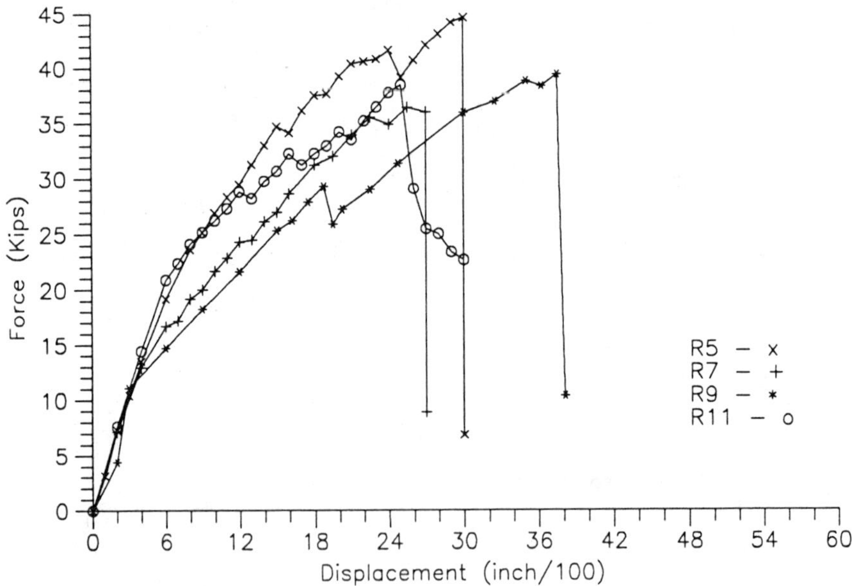

Figure 18--Shear load — deflection envelope comparison
of RWall 5, 7, 9, and 11

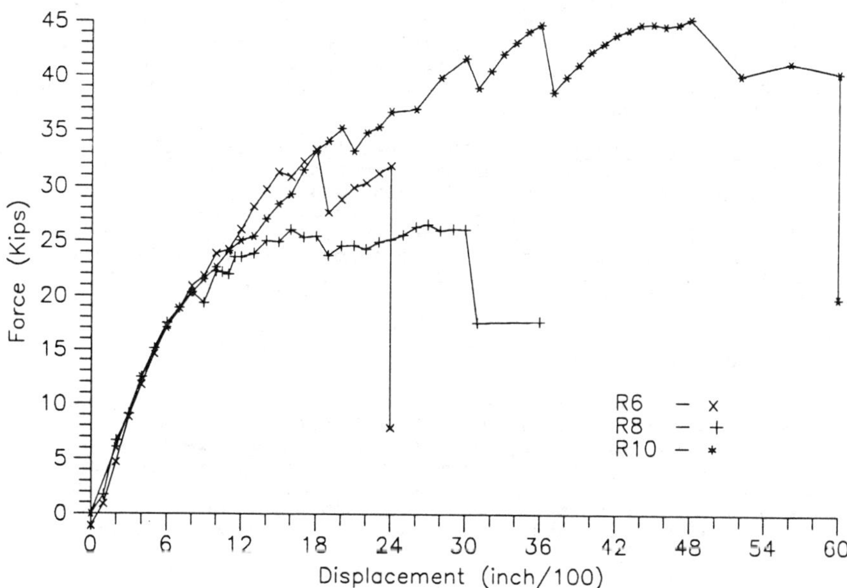

Figure 19--Shear load — deflection envelope comparison
of RWall 6, 8, and 10

obtained. For walls R2, R7, and R10, at the point of yield there was
a large increase in strain. For walls R5, R8 and R11, the yield was
marked by a drop in strain and corresponding loss in load. The
middle center gage of wall R9 failed and did not show yield but
reduced slope of the load-deflection curve after gage failure
indicates yield.

For walls with usable strain gage data, the yield points
indicated by strain gages corresponded very closely to the first
visual observation of cracking (FME). See Table 2. Because DuctTest
and DuctYield (ductility based on strain gage yield indication)
coincided so closely, only DuctTest was discussed in Chapter Three.
Reinforcement yield and FME occurred at the same deflection for walls
R2, R5, R7, R8, R9. Walls R10 and R11 had FME's and reinforcement
yield which differed by one increment in deflection. The concurrence
of these events strengthens confidence in the data. It also
indicates greater ability to monitor the progress of tests in the
future by visual inspection.

<u>Effect of load control</u>--The shear load - deflection envelope of
wall R8 differs from the other walls in Figures 17-19. This wall had
a much more gradual loss of load carrying ability after ultimate load
was attained. This was because load control was used for simulation
of the dead load instead of displacement control. Axial force was
not allowed to increase under load control. Under displacement
control, the axial load increased 50 to 100 percent and was a much
greater factor in the destruction of those walls.

CONCLUSIONS

These tests indicate that shear strength cannot be accurately
estimated for an individual wall. Unquantified factors have a large
influence. Displacement appears much more closely related to the
design of the wall. Further investigation of displacement at
ultimate shear and at yield load may help improve understanding of
masonry behavior. At this time, it is not expected that a usable
design procedure can be developed based on a displacement prediction
method.

The use of bed joint reinforcement greatly increased ultimate
displacement in these tests. All walls with bed joint reinforcement
had ultimate displacements of 0.30 inch (0.762 cm) or more. All
walls without bed joint reinforcement had ultimate displacements of
0.30 inch (0.762 cm) or less.

The effect of using stretcher blocks with the webs partially
removed instead of bond beam blocks is undetermined. The performance
of the bond beams was not good. This could be related to the use of
the stretcher blocks or because there was too much steel in too few
bond beams. This potential factor needs to be considered in the
conclusion that bed joint reinforcement performed better.

A linear relationship between axial load and ultimate shear
strength is substantiated by this series of tests. Testing done at
NIST, both present and past, has resulted in a substantially
different slope for this relationship versus tests done by others.
The reason for this is uncertain. Additionally, further
investigation is needed to determine how factors such as aspect
ratio, grouting, reinforcement, and loading history effect the
relationship and quantify that effect.

Section <u>Wall R6</u> raises the question of the effect of shear
strength and stiffness on flexural forces. This test series resulted
in far too little data on this subject to make any type of
evaluation. The planned test series on vertical reinforcement may
provide better information on this interaction.

The loading pattern was different for most walls. The

deflections after FME are based on the FME displacement. Analysis of the data revealed a close correlation between FME and yielding of the reinforcement. The disadvantage of basing subsequent deflections on the FME is that comparison of the performance of each wall is less direct. Deflection comparisons could be made but comparisons heavily dependant on the loading pattern, such as strength, stiffness, and energy dissipation, were more difficult.

The use of displacement control instead of load control on nine of the ten tests resulted in some extreme loading conditions. The axial load increased with displacement and varied greatly from the original dead load simulation. No information was found on the control or difficulties in controlling axial load in others' tests. Tests at University of California - Berkeley used springs to apply the axial load which should have held load constant. The test in the NBS series conducted under load control was slowed by the additional requirements placed on the computer control. The accuracy of these calculations was poor and resulted in erratic loading increments.

This paper reports the initial findings of the first part of a series of cyclic tests on in-plane shear behavior of concrete masonry under cyclic loading. Although this testing resulted in significant indications of the benefits of bed joint reinforcement, it was mainly a shakedown of the testing method. Future tests on the effect of vertical steel and combined horizontal and vertical reinforcement are planned. More testing will be useful in investigating questions raised in this series.

REFERENCES

[1] Porter, Max L., "Sequential Phased Displacement (SPD) Procedure for TCCMAR Testing," Third Meeting of the Joint Technical Coordinating Committee on Masonry Research, U.S.-Japan Coordinated Earthquake Research Program, Oct. 1987.

[2] Shing, P.B.; Noland, J.L.; Klamerus, E.; Spaeh, H., "Inelastic Behavior of Concrete Masonry Wall Panels Under In-Plane Cyclic Loads," 4th North American Masonry Conference, Aug. 1987.

[3] Shing, P.B.; Noland, J.L.; Klamerus, E.; Spaeh, H., "Inelastic Behavior of Concrete Masonry Shear Walls," ASCE Journal of Structural Engineering, Vol 115, No. 9, September, 1989.

[4] Shing, P.B.; Schuller, M.; Hoskere, V.S., "In-Plane Resistance of Reinforced Masonry Shear Walls," ASCE Journal of Structural Engineering, Vol 116, No. 3, March, 1990.

[5] Woodward, Kyle; Rankin, Frank; "Influence of Aspect Ratio on Shear Resistance of Concrete Block Masonry Walls," U.S. Dept. of Commerce, Jan. 1985.

[6] Woodward, Kyle; Rankin, Frank; "Influence of Block and Mortar Strength on Shear Resistance of Concrete Block Masonry Walls," U.S. Dept. of Commerce, April, 1985.

[7] Woodward, Kyle; Rankin, Frank; "Influence of Vertical Compressive Stress on Shear Resistance of Concrete Block Masonry Walls," U.S. Dept. of Commerce, Oct. 1984.

[8] Matsumura, A., "Shear Strength of Hollow Unit Masonry Walls," 4th North American Masonry Conference, Aug. 1987.

[9] Epperson, G.S., Abrams, D.P., "Non-Destructive Evaluation of Masonry Buildings," Advanced Construction Technology Center, Doc 89-26-03, University of Illinois at UC, Oct. 1989.

Raymond H. R. Tide[1], Norbert V. Krogstad[2]

ECONOMICAL DESIGN OF SHELF ANGLES

REFERENCE: Tide, Raymond H. R., Krogstad, Norbert V., "Economical Design of Shelf Angles," **Masonry: Design and Construction, Problems and Repair, ASTM STP 1180,** John M. Melander and Lynn R. Lauersdorf, Eds., American Society for Testing and Materials, Philadelphia, 1993.

ABSTRACT: General guidelines for the design and construction of shelf angles supporting masonry veneers in high-rise buildings can be found in several published documents. Examples of shelf angles in service have been found, however, where failures did not occur even though many of these guidelines were not followed. In one such case, no significant yielding or excessive deflection was observed even though the stresses in the steel predicted by traditional analysis greatly exceeded yield stresses and the computed deflection of the angle exceeded the expansion joint width.

This paper reviews the performance of shelf angles taking into account several boundary conditions and factors that are commonly overlooked in design but that have substantial effect on the actual performance. A procedure is outlined for the economical and rational design of shelf angles. Included are procedures for the sizing of shelf angles, for determining the spacing of connections, and for designing and detailing connections to the building structure. Examples are given to show how the procedure would be used in the design process.

KEYWORDS: Shelf angles, shims, masonry veneer, masonry arching, torsion, friction, inelastic deformation.

Masonry veneer in buildings taller than three stories are typically supported by shelf angles. These shelf angles are used to reduce loads in the masonry and to create horizontal expansion joints to accommodate differential movement between the masonry and structure. Shelf angles are generally designed using simplifying assumptions that produce conservative results. The masonry is assumed to load shelf angles uniformly at its centroid. Shelf angles are assumed to deform freely under this loading. In some buildings that have been encountered, shelf angles with very large bolt spacing have been used to support tall sections of masonry. Conventional structural analysis would predict failure of these shelf angles under loads significantly less than were actually supported. In order to explain the success of these examples, many factors were

---

[1]Senior Consultant, Wiss, Janney, Elstner Associates, Inc.

[2]Consultant, Wiss, Janney, Elstner Associates, Inc.

considered that are commonly overlooked in design but which have a significant effect on structural behavior.

CASE STUDY

The case study that will be reviewed throughout this paper is illustrated in Fig. 1. Twenty-three feet (seven meters) of masonry was supported by an L6x4x3/8 (L152mm x 102mm x 9.5mm) shelf angle, long leg vertical.   The 30 ft (9.1m) shelf angle was anchored to a concrete structure with bolts spaced at 6 ft 6 in. (1980mm) in the most severe case.   The wall was clay masonry with Type N masonry cement mortar. Because of tolerance problems encountered in the original construction, the masonry was placed near the edge of the shelf angle, having approximately 1 5/8 in. (41.3mm) of bearing. A continuous relief joint was provided beneath the shelf angle. Observations indicated that this joint was present and was free of obstructions.

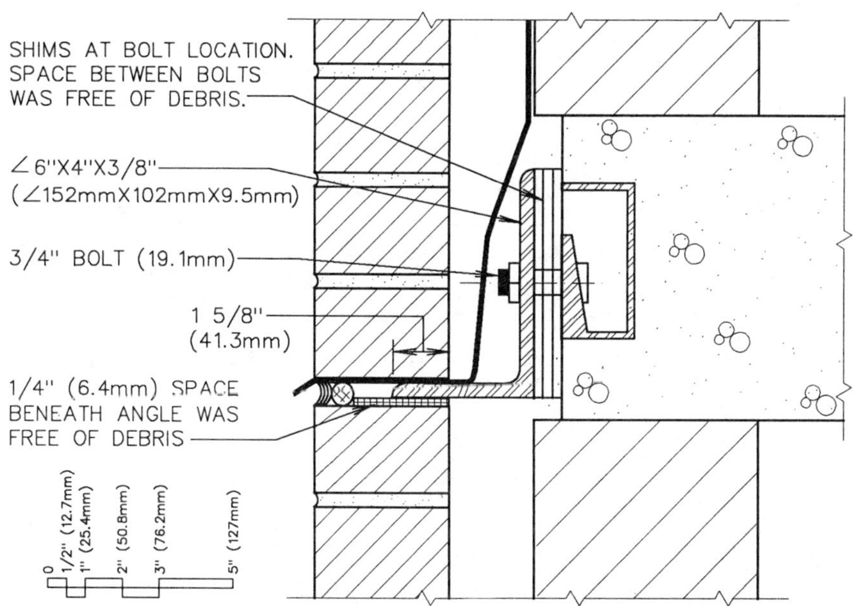

Fig. 1 - Details of construction for case study

The masonry in this location was free of cracking. When the masonry was removed, the shelf angles were found to be free of any obvious areas of failure including inelastic deformations. The shelf angle was held away from the concrete slab by shims placed at the anchor bolts. The resulting space behind the shelf angle was unobstructed. The yield strength of the shelf angle and bolts was not determined. The angle was specified to be ASTM A36 steel.

REVIEW OF TRADITIONAL METHODS OF ANALYSIS

Perhaps the most common method of analyzing shelf angles involves assuming that the uniform load acts along the total length. The load is generally assumed to act at the centroid of the masonry wall. This would occur only if the masonry remained plastic throughout the construction of the wall and followed the deflected shape of the angle. The shelf angle and bolts are then checked for stresses and deflections based on these assumptions. The limiting steel stresses would be those allowed in the American Institute of Steel Construction, Inc. (AISC) Specification [1]. Limiting deflections are listed in the Brick Institute of America's (BIA) Technical Note No. 28B and No. 31B [2,3]. These state that the total downward deflection of the shelf angle shall not exceed the smaller of L/600 or 0.3 in. (7.5mm). Both Technical Notes also limit the total rotation at the tip of the shelf angle to 1/16 in. (1.6mm).

Based on the previous assumptions and assuming continuity at the supports, the flexural stress at midspan (including unsymmetrical bending) and deflection were calculated at 10.9 ksi (75.2mPa) and 0.02 in. (0.5mm), respectively. In addition, the outstanding leg of the shelf angle would deflect approximately 0.01 (0.3mm) due to the load eccentricity. These stresses and deflections are within the limits established by AISC and BIA. However, the deflection due to shelf angle twist at midspan was computed to be 0.47 in. (11.9mm) which greatly exceeds the limit established by BIA. The total computed deflection at midspan was 0.50 in. (12.7mm). In the case study the joint beneath the angle was open and there were no indication of excessive deflection.

At the support, the flexural stress was calculated at 11.7 ksi (80.7mPa). The average cantilever bending stress assuming a 12 in. (305mm) effective length of shelf angle is 13.5 ksi (93.1mPa). The flexural shear stress and torsional shear stress are 2.31 ksi (15.9mPa) and 25.3 ksi (174mPa), respectively, for a total of 27.6 ksi (190mPa). This shear stress exceeds both the AISC allowable of 14.5 ksi (100mPa) and shear yield (approximately 22 ksi) (152mPa) for ASTM A36 steel.

The computed cantilever deflection of the outstanding leg of the shelf angle at the support is 0.01 in. (0.3mm). As indicated previously, the computed deflection at the centerline was 0.50 in. (12.7mm). Because 23 ft (7m) high masonry wall is very stiff, such a large difference in deflection between the midspan and the support can only occur if the masonry remained plastic during construction.

The eccentric loading on the shelf angle results in tension and shear forces of 6.0 kips (26.7kN), acting on the ASTM A307 3/4 in. (19mm) diameter bolts. These stresses exceed the allowable values established by AISC for shear and combined tension and shear, but not for tension alone. However, the predicted ultimate combined stress capacity of the bolt is just adequate to support the masonry wall. This is another factor that would explain why no distress was observed in the wall.

A second approach to the design of shelf angles has been presented by Grimm[4]. This approach uses a load distribution that concentrates loads at the support rather than being uniformly distributed. The downward deflection of the shelf angle is assumed to be uniform. The masonry load is assumed to act along a line located one-third of the bearing width from the interior face of the masonry. This would occur only if the angle was shored during construction. Once the masonry is set and the shores are removed, the very stiff masonry wall would force the angle to deflect in a uniform manner with the load applied near the back face of the masonry.

In this method, the effective length of shelf angle leg at the support for cantilever bending is set equal to the length of the outstanding width of the shelf angle.

By using the method described by Grimm[4], the computed cantilever bending stress in the outstanding leg of the case study was approximately twice the assumed yield stress. No obvious signs of yielding, however, were observed in the case study. It should be also noted that the paper by Grimm [4] states that the procedure may not be valid for bolt spacings greater than 3 ft (914mm) on center. For the shelf angle under consideration in the case study, the bolt spacing was 6 ft 6 in. (1980mm)

DISCUSSION

Traditional methods of analysis and the Grimm[4] procedure make opposite assumptions concerning the stiffness of the masonry at the time the masonry initially loads the angle. Traditional methods assume unshored conditions where the masonry remains plastic during construction. Grimm[4] assumes that the masonry is infinitely stiff at the time load is applied to the angle. In most applications within the United States, shelf angles are not shored during construction. Therefore, some deflection will take place during construction while the masonry is still plastic. The mortar, however, will begin to set long before the wall is completed. If it doesn't, the mortar will squeeze out of the completed joints below and the construction would have to stop. Actual loading conditions will be somewhere between that assumed in traditional methods and by Grimm[4]. This is likely to be the reason why there were no signs of excessive deflections or distress in the masonry wall or angle in our case study. Both methods predict failure at loads significantly less than the shelf angle was carrying.

Masonry walls will continue to gain strength throughout the construction. Several courses of masonry must be constructed before the wall can be considered to act as a beam spanning between anchor bolts. This beam action is dependent on the strength of the lower masonry courses and the effective depth of the beam. The greater the bolt spacing, the higher the constructed wall must be to resist the bending stresses in the masonry.

The reaction between the masonry and the shelf angle is likely to be located near the back edge of the masonry. This occurs because the masonry will not rotate to keep up with the twist of the shelf angle after the masonry has been constructed past the first wall tie. The eccentric load at the base of the masonry will create a moment in the masonry that is resolved by the wall ties.

As the shelf angle deflects and twists due to the applied load, the horizontal leg of the shelf angle will move downward and towards the interior of the building. The inward deflection, due to rotation, is restrained in part by a friction force that develops between the masonry and the shelf angle. Torsional stresses and rotations can be reduced as much as 20 percent by assuming the friction coefficient of 0.2 even when considering fabric flashings. Furthermore, this frictional force also acts as continuous lateral bracing for the shelf angle.

If some localized yielding occurs, the masonry load on the outstanding leg of the shelf angle at the support can be distributed to a greater length of shelf angle than assumed by Grimm[4]. In general, some redistribution of stresses after localized yielding is an accepted design philosophy for bolted and welded connections and is recognized by the AISC Specification[1].

CRITERIA FOR USING PROPOSED PROCEDURE

Before our procedure can be applied to actual structures, it is important that the following conditions are present in the structure.

1.   A working expansion joint must be installed beneath the shelf angle. This joint must be free of any obstructions which can affect the free movement of the shelf angle. Such obstructions can cause redistribution of loads within the masonry and result in localized cracking.

2.   The shelf angle must be properly anchored to a structural element that will rotate and deflect very little. The total downward deflection of the beam on which the shelf angle is attached plus the localized deflection of the angle should be less than L/600. This should include the long-term deflections of total dead load and immediate deflection from the masonry. If the shelf angle is attached to a flexible structure, cracking probably will occur in the masonry regardless of the design of the shelf angle.

3.   No vertical expansion joints can be located within the length of masonry that is assumed to span between bolts. If this condition occurs, the masonry should be designed as if it applied a uniform load to the shelf angle.

4.   Where needed, full height shims 2 to 3 in. (50 to 75mm) in width should be used at anchor bolt locations. If proper shimming practices are not followed, localized stresses and deflections at the support and midspan will be greatly increased and the underlying assumptions will not be valid.

5.   Joints in the shelf angle should be located 1/3 of the bolt spacing from the last bolt to balance the moments in the angle at the bolt. Otherwise, the assumptions based on continuity of the shelf angle will be invalid and the procedures will need to be modified accordingly. Joints in the shelf angle should coincide with expansion joints in the masonry.

6.   Corner joints should be installed in shelf angles and in the masonry to eliminate special conditions and reduce the potential for masonry cracking. If shelf angles are continuous around corners, procedures will need to be modified to account for variations in stresses and deflections.

7.   The shear and tension capacity of inserts in the concrete to connect bolts shall be equal to or greater than that of the bolt.

DESIGN METHODOLOGY

Functional Requirements

Assuming the conditions listed above are satisfied, the first step in designing shelf angles for new construction is to determine the necessary length of the outstanding leg of the shelf angle. In nearly all cases, the size of the shelf angle is determined by function rather than by structural requirements. In determining this length, the following conditions must be considered:

1.   The shelf angle should be designed so that at least two-thirds of the masonry is bearing on the shelf angle. If this is not the case, it will be very difficult to support the bottom course during construction.

2.   Three-eighths to 1/2 in. (10mm to 13mm) shim space should be provided behind the vertical leg of the shelf angle to accommodate construction tolerance. In actual construction the shim space may vary from zero to 1 in. (25mm). In general, it is not recommended to use more than 1 in. (25mm) of shims as significant bending stresses may be induced in the connection bolts if this limit is exceeded.

3.   A space must be provided between the front face of the bolt and the back face of the masonry. This is needed so that a compressible pad or protection can be provided between the edge of the bolt and the flashing to avoid flashing punctures during construction. This space should be a minimum of 1/2 in. (13mm) to accommodate construction tolerance.

4.   Most masonry veneers are designed with a 2 in. (50mm) cavity. The length of the outstanding leg of the shelf angle must accommodate this dimension.

5.   The front edge of the shelf angle should be held at least 1/2 in. (13mm) back from the face of the masonry to provide a space for sealant and backer rod.

Considering these factors, shelf angles to support nominal 4 in. (102mm) masonry (3 5/8 in. (92mm) actual depth) would be L5x5 (L127mm x 127mm) shelf angles. For 3 in. (76.2mm) thick masonry units, L4x4 (L102mm x 102mm) or L6x4 (L152mm x 102mm) shelf angles would normally be used. For most applications, the thickness of the shelf angle would either be 5/16 in. (7.9mm) or 3/8 in. (9.5mm). Because of the limited choices in shelf angles available, the structural design of shelf angles primarily involves determining the spacing of the support bolts. The spacing will be based on the strength and deflection characteristics of the angle and the strength of connections.

Structural Requirements

Once the size of the shelf angle is chosen, a trial bolt spacing should be selected. In most cases this spacing will be between 2 ft (610mm) and 4 ft (1220mm). The stresses and deflections are then calculated using the following procedure and compared with the allowable values.

1.   The section of shelf angle between bolts is assumed to be loaded by a uniformly distributed load, the height of which is equal to one-half the bolt spacing. The remainder of the masonry wall load, is distributed to the shelf angle at each bolt location over a distance calculated by extending a 45 degree angle from the top corner of the shims to the assumed reaction point of the supported masonry. The derivation of this distance will be presented in the finite element model (FEM) section. The line of reaction is assumed to be located 1/2 in. (13mm) in front of the back face of the masonry.

2.   The outstanding leg of the shelf angle at the connection bolt is checked for bending about the plane parallel to the vertical leg. The effective length of shelf angle on either side of the bolt is described above. Stresses at this plane should not exceed the allowable values given in the AISC Specification[1].

3.   Flexural stresses, torsional shear stresses, and flexural shear stresses are checked at the location of the connection bolts. Shear stresses are based on the total load applied to the shelf angle between the bolts. The cantilever deflection of the outstanding leg of the shelf angle at the support is computed and compared with the midspan deflection due to flexure, twist and cantilever action.

4.   Deflection of the toe of the shelf angle is verified at midspan. This deflection shall not exceed the lesser of the bolt spacing divided by 600 (L/600) or 0.3 in. (7.6mm) The total rotation deflection at midspan as measured at the toe of the shelf angle shall not exceed 1/16 in. (1.6mm)

5.   The section of shelf angle between the bolts is checked for bending stresses at the center line. Because bending is not about the principal axis, unsymmetrical bending provisions apply. These stresses may not exceed the allowable values given in the AISC Specification[1].

ANALYSIS OF DESIGN EXAMPLE

The shelf angle conditions observed in the case study were analyzed using the proposed procedure. As would be expected, the flexural stress at midspan, including biaxial bending, was calculated at 1.5 ksi (10.3mPa). When the restraining effect of friction was included, the flexural stress was reduced to 1.1 ksi (7.6mPa). These stress levels indicate that under normal conditions flexural stresses do not govern the design of shelf angles.

The shelf angle midspan deflection due to beam flexure, cantilever bending and torsion were found to be 0.0027 in. (0.069mm), 0.0011 in. (0.028mm) and 0.0986 in. (2.50mm), respectively. The three components result in a total outstanding leg deflection of 0.103 in. (2.62mm) which is less than the limits set by BIA (L/600=0.13 in. (3.3mm), 0.3 in. (7.5mm)). However, the 0.0986 in. (2.50mm) deflection due to rotation exceeds the 1/16 in. (1.6mm) limit but is unlikely to be noticed because of its magnitude.

At the support, the cantilever bending of the shelf angle occurs due to the 39 in. (991mm) high uniformly distributed loading and due to the concentrated load, equal to the remainder of the masonry load, that results from the masonry spanning between supports. The effective length over which the concentrated load acts was determined by the method described earlier, where a 45 degree line was projected from the top corner of the shim. For the case study, this gives an effective length of 10 in. (254mm). For computational purposes, this concentrated load is assumed to act at the center of this region. The cantilever bending and torsion deflection were calculated at 0.0283 in. (0.72mm) and 0.0649 in. (1.65mm), respectively. The combined deflection of 0.093 in. (2.36mm) at the support in slightly less than the midspan deflection of 0.103 in. (2.62mm).

The beam flexural stress was calculated at 2.3 ksi (15.9mPa) at the support. The flexural shear and torsional shear stresses were calculated at 2.31 ksi (15.9mPa) and 14.7 ksi (101mPa), respectively, for a total shear stress of 17.0 ksi (117mPa). This shear stress exceeds the AISC allowable of 14.5 ksi (100mPa) but does not exceed the shear yield stress (approximately 22 ksi) (152mPa). For the same 10 in. (254mm) length of shelf angle, the cantilever bending stress is 34.3 ksi (236mPa) in the vicinity of the fillet radius connecting the angles' two legs. This average stress is less than the nominal 36 ksi (248mPa) yield stress for ASTM A36 steel. In most cases, the actual yield stress of the steel in the shelf angles exceeds the nominal 36 ksi (248mPa). Although not necessarily recommended for design, this assumption is considered valid for analyzing an existing condition to explain why no distress was observed.

The calculated stress was based on an average condition over an assumed effective length of bending as determined from an evaluation of the FEM deflected shape. It is recognized that at the support the cantilever bending stresses would be somewhat higher, some localized yielding may have occurred with slightly increased deflection. Stresses in this location would be reduced if the increased section modulus at the fillet radius is included in the analysis. Because of the other assumptions used in the calculations, this refinement is not justified.

As previously discussed, the load on the ASTM A307 3/4 in. (19mm) diameter bolt exceeded the AISC allowable design values but not the ultimate strength of the bolt.

The case study computations are given in the Appendix. In addition, supplement computations are shown indicating the design procedure that should be followed to redesign the shelf angle which will provide proper support for the masonry and reduce the stresses in the shelf angle and supporting bolts.

VERIFICATION USING FINITE ELEMENT MODEL

A finite element method (FEM) analysis was performed to further examine the behavior of the shelf angle in the case study. The structural program used for this analysis was Supersap by Algor Interactive Systems, Inc.

The shelf angle was modeled using plate elements. Each element was 1 in. x 1 in. x 3/8 in. (25.4mm x 25.4mm x 9.5mm) thick. The bolt and shims were modeled using elastic springs. Continuity along the ends of the shelf angles was modeled by using rotational restraints about the x and y axes at each end.

The material properties assumed for the steel were a Young's modulus of 29 million (200000mPa), poisson's ratio equal to 0.30 and a unit weight of 490 pcf (77,000Ncm). The weight of the masonry was assumed to be 40 psf or 132 pcf (20,700Ncm).

The model was loaded with a uniformly distributed load equal to 39 in. (991mm) of masonry and the remainder of the load uniformly distributed in a 10 in. (254mm) region adjacent to the support. The 39 in. (991mm) of masonry represents one-half the bolt spacing. The deflected shape is shown in Fig. 2.

By examining the deflected shape, it is apparent that the angle begins twisting at the top edge of the shims. The downward deflection due to cantilever bending causes the shims above the bolt to be in compression. The torsional twist appears to cause the angle to bend about a line extending from the top of the shims downward at approximately a 45 degree angle.

The total deflection at the centerline was 0.123 in. (3.12mm). The deflection at the support was 0.074 in. (1.88mm). The difference in deflections was very similar to that when the shelf angle was loaded by only 39 in. (990mm) of uniformly distributed masonry. The localized stresses near the support exceed the nominal yield stress within a 2 in. (50mm) distance, as shown in Fig. 3. If the modeling reasonably represents actual conditions, some localized yielding and stress redistribution would occur. The average maximum principal stress over the 10 in. (254mm) region was approximately 30 ksi (207mPa). In conclusion, the deflections and average stresses of the FEM analysis, without considering the friction component, are in good agreement with the results obtained by the simplified analysis presented earlier.

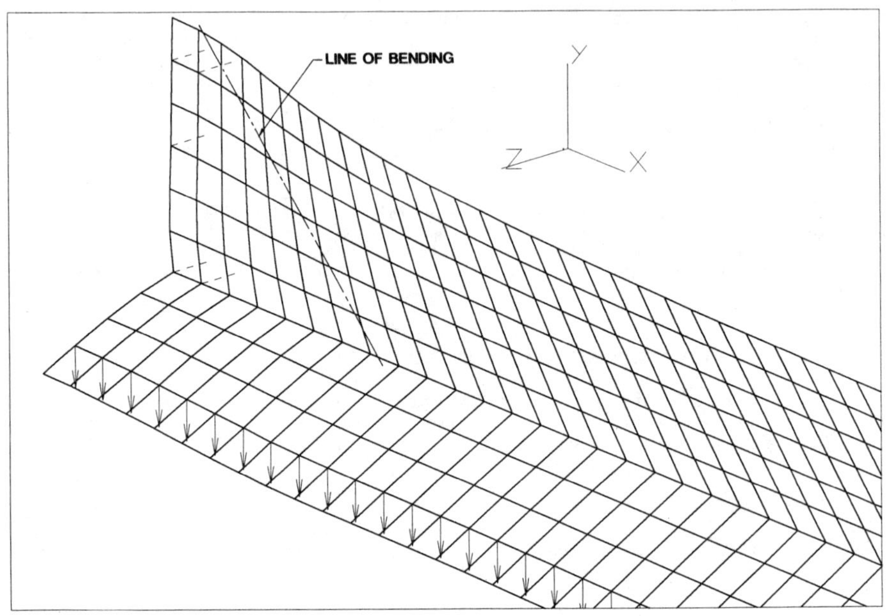

Fig. 2 - Deflection per assumed loading

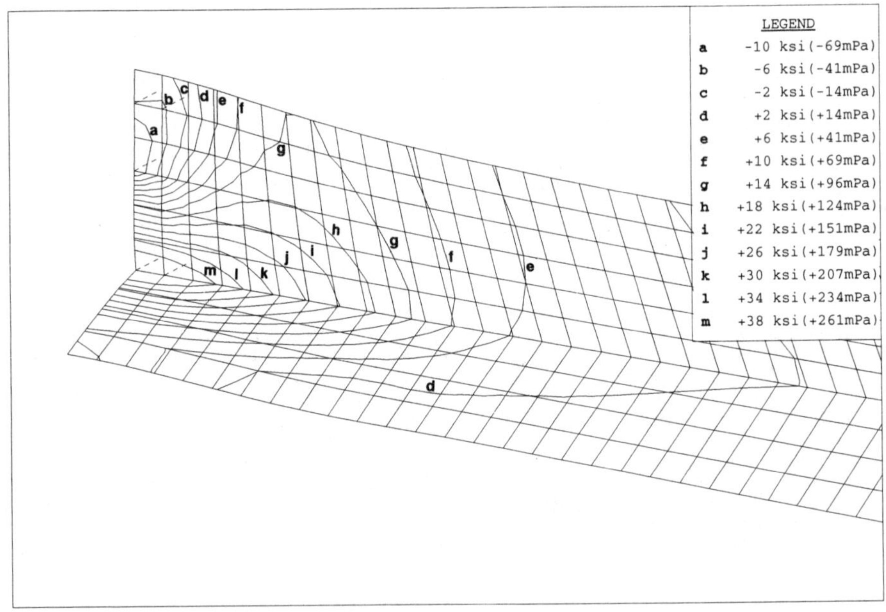

Fig. 3 - Stress contours per assumed loading

SUMMARY

The loading behavior of masonry walls on shelf angles are likely to fall somewhere between plastic (uniform load model) and very stiff (Grimm[4] model). Factors such as deformation of the lower courses of masonry during construction and the rigidity of masonry walls after the mortar begins to set cause shelf angles to be loaded by a combination of uniform and concentrated loads acting near the back edge of the masonry. Restraint provided by friction and the masonry ties reduce torsional forces and deflection and provide lateral support for the shelf angles. The number of connections and, in some cases, the thickness of angles may be reduced when procedures outlined in this paper are used. Generally, the size of a shelf angle will be determined by construction tolerances and functional requirements. Structural analysis is then used to determine stresses in the shelf angles and support bolts based on trial bolt spacing. The limiting stresses and deflections will likely occur in the area immediately around the connection bolts. In particular, bending stresses about a plane parallel to the vertical leg of the shelf angle or shear and tension in the attachment bolts will usually govern.

The proposed design recommendations are based on specific construction conditions that are commonly encountered, such as, shelf angle continuity, full height shims behind the shelf angle and other criteria listed in this paper. When these conditions are not encountered, the design procedure must be accordingly modified and other rational analysis methods adopted.

APPENDIX

CONVERSION TABLE FOR APPENDIX

UNIT	MULTIPLY BY	SI UNIT
ft	0.3048	m
in.	25.4	mm
lbs	4.448	N
ksi	6.695	mPa
pcf	157.1	Ncm
psf	47.88	Nsm

CASE STUDY

The case study shelf angle analysis is presented in a step-by-step format to facilitate performing future preliminary designs. Initially, all the basic parameters are established, then the support bolt capacity is compared to the supported load based on a chosen bolt spacing. The cantilever bending stress at the support is then computed and compared to the values allowed by the AISC Specifications[1]. Shear stresses due to flexure and torsion are computed and compared to the allowable values. Midspan deflections are calculated and compared to deflection limits. Generally, after these steps are completed, assuming all conditions are satisfied, is an indication that a satisfactory shelf angle and bolt spacing have been chosen. For completeness, the remaining stress conditions and deflections are computed for the case study.

Step 1 - Masonry wall 23 ft high supported by ASTM A36 L6x4x3/8, long leg vertical. ASTM A307 3/4 in. diameter anchor bolts spaced at 6 ft 6 in. (78 in.) and located 3 in. from heel of angle. Masonry weighs 40 lbs per sq ft (3 5/8 in. nominal thickness). Shelf angle properties I_x = 13.5 in.^{4}, S_x = 3.32 in.^{3}, J = 0.183 in.^{4}, assume coefficient of friction is 0.2.

Step 2 - Assume 3 in. finger shims behind anchors. Masonry located near edge of horizontal leg resulting in 3 in. eccentricity (the masonry reaction is assumed to act 1/2 in. in front of back face of masonry.) Effective tributary length of outstanding leg of shelf angle is one-half of the shim width plus height of vertical leg and eccentricity (45 degree angle from edge of finger shim.) Z = 1/2 (3) + 6 + 3 = 10.5 in. (Use 10 in.).

Height of uniformly distributed load (before mortar sets up) equals 1/2 bolt spacing, h = 1/2 (78) = 39 in. Uniformly distributed load, W_u = (40x39)/(12x12) = 10.83 lbs/in.   Shear at bolt for one span, V = (40x23x78)/(2x12) = 2990 lbs.   Concentrated load above the uniformly distributed load, P_c = 2990 - (40x39x78)/(2x12x12) = 2568 lbs.

Step 3 - Check capacity of ASTM A307 3/4 in. diameter bolt for combined tension and shear. Because eccentricity (3 in.) and gage (3 in.) are equal, the total shear and tension load from adjacent spans are of equal magnitude. T_b = V = 2x2990 = 5980 lbs,

$$f_a = f_v = 5980/0.442 = 13.5 \text{ ksi.}$$

$$\underline{\text{AISC ASD}} - F_t = 26 - 1.8f_v \leq 20 \text{ ksi}$$

$$F_t = 26 - 1.8 \times 13.5 = 1.7 \text{ ksi}$$

Because 1.7 ksi is less than 13.5 ksi, the bolt is inadequate.

$$\underline{\text{AISC LRFD}} - F_t = 59 - 1.9 \, f_v \leq 45 \text{ ksi.}$$
$$\emptyset = 0.75$$

$$F_t = 59 - 1.9 \times 13.5 = 33.35 \text{ ksi}$$
$$\emptyset \, F_t = 33.35 \times 0.75 = 25 \text{ ksi}$$

Because 25 ksi is greater than 13.5 ksi would explain why no distress was observed in the masonry even though service load stress in the bolt was exceeded.

Step 4 - Compute cantilever bending stress at the support. Section modulus for 10 in. of shelf angle leg, S = 10x(3/8)^2/6 = 0.234 in.^{3}. M = P_ce + ZW_ue = 2568x3 + 10x10.83x3 = 8030 lb-in.

$$f_{bc} = 8030/0.234 = 34.3 \text{ ksi}$$

This stress exceeds the allowable of 27 ksi (0.75 Fy) permitted by the AISC Specifications but is less than the yield value.

Step 5 - Compute shear stress due to flexure and torsion.

a)   Flexure shear due to total shear, f_v = 1.5V/Anet.

$$f_v = 1.5 \times 2990/[(6 - 13/16)3/8] = 2.31 \text{ ksi.}$$

b)   Torsion shear due to uniform distributed load,
T = W_ue, T_s = TL/2, f_{tu} = T_st/J.

$T = 10.83 \times 3 = 32.5$ lb-in./in. without friction, and because $e = g = 3$ in. $T = 10.83 \times 3 - 10.83 \times 0.2 \times 3 = 26$ lb-in./in. including friction. $T_s = 26 \times 78/2 = 1014$ lb-in.,

$$f_{tu} = 1014(3/8)/0.183 = 2.08 \text{ ksi.}$$

c)   Torsion shear due to concentrated load (arching of masonry) at support. Although distributed over 10 in., for computational purposes the load is assumed to be concentrated 5 in. from the support.

    $T = P_c e = 2568 \times 3 = 7704$ lb-in., without friction.
    $T = 2568 \times (1 - 0.2) \times 3 = 6163$ lb-in., including friction.

$$f_{tc} = Tt/J = 6163(3/8)/0.183 = 12.63 \text{ ksi}$$

Total shear stress = $f_v + f_{tu} + f_{tc} = 2.41 + 2.08 + 12.63 = 17.0$ ksi, which is greater than the AISC allowable of 14.5 ksi, but less than shear yield of approximately 22 ksi.

<u>Step 6</u> - Compute remaining stress conditions.

a)   Flexural stress at midspan $M = W_u L^2/24 = 10.83 \times (78)^2/24 = 2745$ lb-in.

$$f_b = M/S_x = 2745/3.32 = 0.83 \text{ ksi}$$

Because the load does not act through the shear center biaxial bending occurs. However, the bending stress is so small that the refinements are not justified. Biaxial bending stresses with and without friction were computed at 1.5 ksi and 1.1 ksi, respectively.

b)   Flexural stress at the support, $M = W_u L^2/12 = 10.83(78)^2/12 = 5492$ lb-in.,

$$f_b = M/S_x = 5492/3.32 = 1.65 \text{ ksi}$$

c)   Cantilever bending stress at midspan

$$M = ZW_u e = 10 \times 10.83 \times 3 = 325 \text{ lb-in. for 10 in. length}$$

$$f_{bc} = M/S = 325/0.234 = 1.39 \text{ ksi}$$

<u>Step 7</u> - Compute flexural deflection at midspan.

$\Delta_f = W_u L^4/384EI_x = 10.83(78)^4/(384 \times 29,000,000 \times 13.5) = 0.0027$ in.

<u>Step 8</u> - Compute cantilever bending deflection at tip of shelf angle due to 39 in. high uniformly distributed load.

a)   At midspan, deflection due to uniformly distributed load, $I = 10(3/8)^3/12 = 0.0439$ in.^{4}, for a 10 in. length.

$\Delta_d = We^2(3l-e)/6EI = 10 \times 10.83 \times 3^2(3 \times 4-3)/(6 \times 29,000,000 \times 0.0439) = 0.0011$ in.

b)   At the support, deflection due uniformly distributed load plus the concentrated load.

$\Delta_d = 0.0011 + [2568 \times 3^2(3 \times 4-3)]/(6 \times 29,000,000 \times 0.0439)$

$$\Delta_d = 0.0011 + 0.0272 = 0.0283 \text{ in.}$$

Step 9 - Compute midspan deflection at tip of shelf angle due to twisting caused by torsion.

a)   Uniformly distributed load, see Step 5b.
          T = 26 lb-in./in., including friction.

$$\theta_{cu} = TL^2/8GJ = (26 \times 78^2)/(8 \times 11,200,000 \times 0.183) = 0.0096 \text{ radians}$$

$$\Delta_{cu} = \theta_{cu} \times 1 = 0.0096 \times 4 = 0.0386 \text{ in.}$$

b)   Concentrated load acting 5 in. from support, see Step 5c., T = 6163 lb-in.

$$\theta_{cp} = TZ/2GJ = (6163 \times 10)/(2 \times 11,200,000 \times 0.183) = 0.0150 \text{ radians}$$

$$\Delta_{cp} = \theta_{cp} \times 1 = 0.0150 \times 4 = 0.0600 \text{ in.}$$

This deflection includes effect of concentrated load at both ends of span.

The total deflection due to twist of the shelf angle is equal to 0.0386 in. plus 0.600 in. or 0.0986 in.

Step 10 - Compute deflection at tip of shelf angle near support due to twisting caused by torsion.

a)   Uniformly distributed load, see Step 5b, T = 26 lb-in./in. and T_s = 1014 lb-in. Torsion at Z(10 in.) from support = 754 lb-in. Compute twist at Z/2 from support using average torsion.

$$\theta_{su} = (T_s + T_z) \, Z/4GJ = [(1014 + 754)10]/(4 \times 11,200,000 \times 0.183) = 0.00216 \text{ radians}$$

$$\Delta_{su} = \theta_{su} \times 1 = 0.00216 \times 4 = 0.0086 \text{ in.}$$

b)   Concentrated load, see Step 5c, T = 6163 lb-in.

$$\theta_{sp} = T(Z/2)(L-Z/2)/LGJ = (6163 \times 5 \times 73)/(78 \times 11,200,000 \times 0.183) = 0.0141 \text{ radians}$$

$$\Delta_{sp} = \theta_{sp} \times 1 = 0.0141 \times 4 = 0.0563 \text{ in.}$$

The total deflection due to twist of the shelf angle at the support is equal to 0.0086 in. plus 0.0563 in. or 0.0649 in.

SHELF ANGLE REDESIGN

The essential steps of the redesign of the shelf angle will be outlined in the following steps.

Step 1 - Same design conditions as for the Case Study except that an ASTM A36 L5x5x3/8 will be chosen. As a trial, ASTM A307 3/4 in. diameter anchor bolts at 4 ft (48 in.) spacing will be chosen. The bolt is located 3 in. from the heel of the angle. Shelf angle properties - I_x = 8.74 in.^{4}, S_x = 2.79 in.^{3}, J = 0.183 in.^{4}. Assume a coefficient of friction of 0.2.

Step 2 - Effective length of shelf angle supporting masonry. Assume 3 in. finger shims. Z = 1/2(3)+5+3 = 9.5 in. (Use 9 in.) Height of uniformly distributed masonry load h = 1/2(48) = 24 in. Uniformly distributed load, W_u = (40x24)/(12x12) = 6.67 lb-in. Shear load at bolt for one span, V = 40x23x48/2x12 = 1840 lbs. Concentrated load above the uniformly distributed load, P_c = 1840 - (40x24x48)/(2x12x12) = 1680 lbs.

$\underline{\text{Step 3}}$ - Verify adequacy of ASTM A307 3/4 in. diameter bolt. Because eccentricity (3 in.) and gage (3 in.) are equal, the total shear and tension load from adjacent spans are of equal magnitude. T_b = V = 2x1840 = 3680 lbs.

$$f_a = f_v = 3680/0.442 = 8.3 \text{ ksi}$$

$$\underline{\text{AISC ASD}} - F_t = 26 - 1.8f_v \leq 20 \text{ ksi.}$$

$$F_t = 26 - 1.8 \times 8.3 = 11 \text{ ksi.}$$

Because 11 ksi is greater than the applied tension of 8.3 ksi, the bolt is adequate. An adequate margin of stress remains to accommodate a prying force.

$\underline{\text{Step 4}}$ - Compute cantilever bending stress at the support. Section modulus for 9 in. of shelf angle leg, $S = 9(3/8)^2/6 = 0.211$ in.^{3}. $M = P_c e + 9W_u e = 1680 \times 3 + 9 \times 6.67 \times 3 = 4860$ lb-in.

$$f_{bc} = 4860/0.211 = 23.0 \text{ ksi}$$

which is less than the allowable of 27 ksi (0.75 Fy) permitted by the AISC Specifications.

$\underline{\text{Step 5}}$ - Compute shear stress due to flexure and torsion.

a)   Flexure shear,

$$f_v = 1.5V/Anet = 1.5 \times 1840/(5-13/16)3/8 = 1.76 \text{ ksi.}$$

b)   Torsion shear due to uniformly distributed load (including friction).

$$T = W_u e = 6.67 \times 3 \times 0.8 = 16 \text{ lb-in./in.}$$

$$T_s = TL/2 = 16 \times 48/2 = 384 \text{ lb-in.}$$

$$f_{tu} = T_s t/J = 384(x3/8)/0.183 = 0.79 \text{ ksi}$$

c)   Torsion shear due to the concentrated load at support (including friction).

$$T = P_c e = 1680 \times 3 \times 0.8 = 4032 \text{ lb-in.}$$

$$f_{tc} = Tt/J = 4032(3/8)/0.183 = 8.26 \text{ ksi}$$

Total shear stress = $f_v + f_{tu} + f_{tc}$ = 1.76 + 0.79 + 8.26 = 10.8 ksi, which is less then the AISC allowable of 14.7 ksi.

$\underline{\text{Step 6 through Step 10}}$ - The remaining steps are not repeated because they generally do not control the design. Reducing the span length from 78 in. to 48 in. will reduce all deflections to within acceptable levels.

REFERENCES

[1]   *Specification for Structural Steel Buildings, Allowable Stress Design and Plastic Design*, American Institute of Steel Construction, Inc. Chicago, IL., June 1, 1989.

[2]   "Brick Veneer Steel Stud Panel Walls" *Technical Notes on Brick Construction* No. 28B Revised II, Brick Institute of America, Reston, VA, February 1987.

[3]   "Structural Steel Lintels" *Technical Notes on Brick Construction*, No. 31B Revised, Brick Institute of America, Reston, VA, Reissued May 1987.

[4]   Grimm, C.T., and J. A. Yura, "Shelf Angles for Masonry Veneer," *Journal of Structural Engineering*, Vol 115, No. 3, American Society of Civil Engineers, New York, NY, March 1989.

Michael P. Schuller[1], Michael I. Hammons[2], Richard H. Atkinson[1]

INTERIM REPORT ON A STUDY TO DETERMINE LAP SPLICE REQUIREMENTS FOR REINFORCED MASONRY

REFERENCE: Schuller, M.P., Hammons, M.I., Atkinson, R.H., "Interim Report on a Study to Determine Lap Splice Requirements for Reinforced Masonry," Masonry: Design and Construction, Problems and Repair, ASTM STP 1180, John M. Melander and Lynn R. Lauersdorf, Eds., American Society for Testing and Materials, Philadelphia, 1993.

ABSTRACT: An investigation of parameters which affect the strength of lap splices within reinforced masonry is currently being conducted as part of the U.S. Army Corps of Engineers program for Construction Productivity Advancement Research (CPAR). Results from the first phase of the investigation complement existing data regarding lap splices in reinforced masonry and provide a comprehensive review of the effects of masonry unit width, masonry unit type, reinforcing bar diameter, and lap length on both the strength and monotonic behavior of lap splices in masonry.

Results of the experimental investigation have shown that the linear relationship used by current working stress masonry design standards does not accurately describe ultimate splice capacity in some cases. An alternate model, adopted for use in the proposed Masonry Limit States Design Standard, provides a more rational approach to the determination of lap splice lengths in masonry. This method considers reinforcing bar diameter and yield strength, grout tensile strength, and masonry wall thickness when determining splice length. Experimental results are in good agreement with values provided by this analysis and have been used to further verify the applicability of the model.

KEYWORDS: reinforced masonry, grouted hollow concrete masonry, grouted hollow clay masonry, lap splice strength, experimental program, limit states design.

In reinforced masonry construction, reinforcing bars are spliced where long reinforcement lengths are required, and vertical bars are often spliced at the foundation level and each successive story level for ease of construction. Economical and effective splices can be obtained

---

[1]Atkinson-Noland & Associates, Inc., 2619 Spruce Street, Boulder, CO, 80302, USA.

[2]U.S. Army Engineer Waterways Experiment Station, 3909 Halls Ferry Road, Vicksburg, MS, 39180-6199, USA.

simply by lapping the bars at the location of the splice. The lap
splice relies on bond between the reinforcing steel and the grout,
rather than actual mechanical connection, to transfer stress from one
bar to the next. It is essential that the length of lap provided be
able to fully develop the steel stress required at that point and that
the splice be sufficiently ductile to provide additional capacity for
cases of structural overloads.

An experimental program is being conducted as part of the U.S.
Army Corps of Engineers program for Construction Productivity Advancement Research (CPAR) in conjunction with the Technical Coordinating Committee for Masonry Research (TCCMAR) to investigate lap splice behavior
in reinforced masonry. The goal of this project is to expand upon current knowledge regarding lap splices in reinforced masonry and provide a
rational assessment of current design provisions governing this subject.
This information will provide the engineer with better overall knowledge
of lap splice behavior, resulting in a safer, more efficient, and cost-
effective design.

The initial phase of this project has been completed, in which a
series of tests was conducted on specimens constructed using different
bar sizes, masonry unit sizes, masonry unit types, and lap lengths.
Data from this program was compared to information from similar investigations to provide an assessment of current design provisions and provide a basis for lap splice requirements to be included in the limit
states design standard now being developed for reinforced masonry [1].
The range of unit and reinforcement sizes tested allows development of
recommendations regarding reinforcement size limitations and required
lap lengths. Additional parameters which may also have an effect on lap
splice behavior, such as the effect of confinement reinforcement at the
splice, off-center bar placement, and grout strength will be examined
during the next phase of the program.

BACKGROUND

A comprehensive review of literature pertaining to reinforcement
bond and anchorage has been compiled by Scrivener [2] as background material for the TCCMAR program. This effort indicated that research on
lap splices in reinforced masonry has been limited; the majority of past
investigations concentrated on reinforcement behavior in concrete. Reinforced masonry is sufficiently similar to reinforced concrete to allow
some comparisons to be made, but the tensile behavior of masonry is
known to be quite different. Reinforced masonry is a composite material, consisting of an assemblage of clay or concrete units, mortar,
grout, and reinforcement, and behavior is complicated by the interaction
of these materials during loading. In addition, masonry possesses regular planes of weakness at the mortar joints which tend to promote the
development of tension cracks at predetermined intervals. Several of
the more significant efforts investigating reinforcement anchorage and
lap splice behavior in reinforced masonry are described below.

Cheema [3] conducted a series of monotonic pullout tests on individual bars grouted in single-wythe concrete masonry wall specimens.
His tests showed that while anchorage failure mechanisms in masonry are
similar to that observed in concrete, there was an additional failure
mode of separation and uplift at the mortar joints. Cheema provided
suggestions for minimum bar spacing to prevent uplift failure and developed an analytical model to predict pullout strength of the bars.

Full-scale beam specimens were tested by Suter [4] in an investigation of the performance of lap splices in concrete masonry. The effect of grout type, misplaced bars, and multiple splices in a single
core were considered. All of Suter's tests showed considerably higher
bond strengths than could be predicted by current working stress and
limit states design theories, with all splices failing by yield of the
reinforcement. These results suggest that current provisions for lap
splice length may be over-conservative.

A comprehensive study of bond and slip in reinforced masonry has been conducted by Soric and Tulin [5] under the auspices of TCCMAR as part of the U.S.-Japan Coordinated Program for Masonry Building Research. Tests included specimens with both single bar anchorages and lap splices. Three analytical models were developed to describe bond stress distributions, considering both linearly elastic and cracking phases. A model based upon analysis by Cheema [3] was also developed to determine the appropriate lap length for spliced reinforcement in masonry structures [6]. This model provides the basis for the draft Masonry Limit States Design Standard described later.

Experimental results obtained by TCCMAR researchers Kubota [7], Watanabe [8] and Matsumura [9] provides additional information on lap splice behavior in reinforced hollow-unit masonry.

EXPERIMENTAL TEST PROGRAM

Test Specimens--Variables investigated during the CPAR program included the effects of reinforcing bar diameter, masonry unit width, masonry unit type, and lap splice length on lap splice behavior. A total of 70 specimens were tested for 35 different combinations of these parameters, with two replications of each type of specimen. Figure 1 shows the range of lap lengths and specimen sizes tested for both concrete and clay masonry specimens. A listing of all specimens tested is provided in Table 1.

Test specimens were fabricated using either hollow concrete masonry or hollow clay brick units with 3/8 in. (10 mm) fully bedded mortar joints and type S mortar. Hollow concrete masonry units with nominal widths of 4, 6, 8, 10, and 12 in. (102, 152, 203, 254, 305 mm) and hollow clay units with nominal widths of 4, 6, and 8 in. (102, 152, 203 mm) were used. Specimens were constructed in stack bond with a prism building jig [10] using half-units to provide a single vertical cell. Nominal specimen dimensions are provided in Figure 1. Lap splices were centered within the cell in the orientation shown in Figure 1. All specimens were fully grouted using a grout with volumetric proportions of 1:3:2 (cement:sand:gravel) with a water/cement ratio of approximately 0.7 to provide a slump of 9-1/2 to 10 in. (241 to 254 mm). Sika Grout-Aid expansive admixture was used in the grout to offset shrinkage due to migration of mix water from the grout to the surrounding masonry. The same grout mix design was used for both concrete and clay masonry specimens, however not all of the specimens were grouted from the same batch. A total of four separate grout batches were used. Grout and mortar material property values listed in Tables 2 and 3 are mean values from all batches combined.

Reinforcing steel used in this study included #4, #6, #8 and #11 (13, 19, 25, 35 mm) Grade 60 (413 MPa) bars, conforming to the requirements of ASTM A 615, Specification for Deformed and Plain Billet-Steel Bars for Concrete Reinforcement [Metric]. Deformations are in a diagonal pattern, at an angle of 70° to the longitudinal axis of the bar, with three longitudinal ribs.

Tests were conducted to obtain basic material properties and as a means of verifying quality control during construction of the test specimens. Average material property values are listed in Tables 2, 3, and 4 along with a listing of ASTM Specifications and Test Methods followed during material property determinations.

Test Apparatus--Tensile loads were applied monotonically in displacement control directly to the bars using hydraulically actuated tension grips. The test setup is shown in Figure 2 (a). Deformations were measured using electronic displacement transducers as shown in Figure 2 (b) to record specimen load-deformation behavior, bond slip, and relative slip between the spliced bars. Overall deformations between the tension grips were also measured using an electronic displacement transducer mounted between the tension grips.

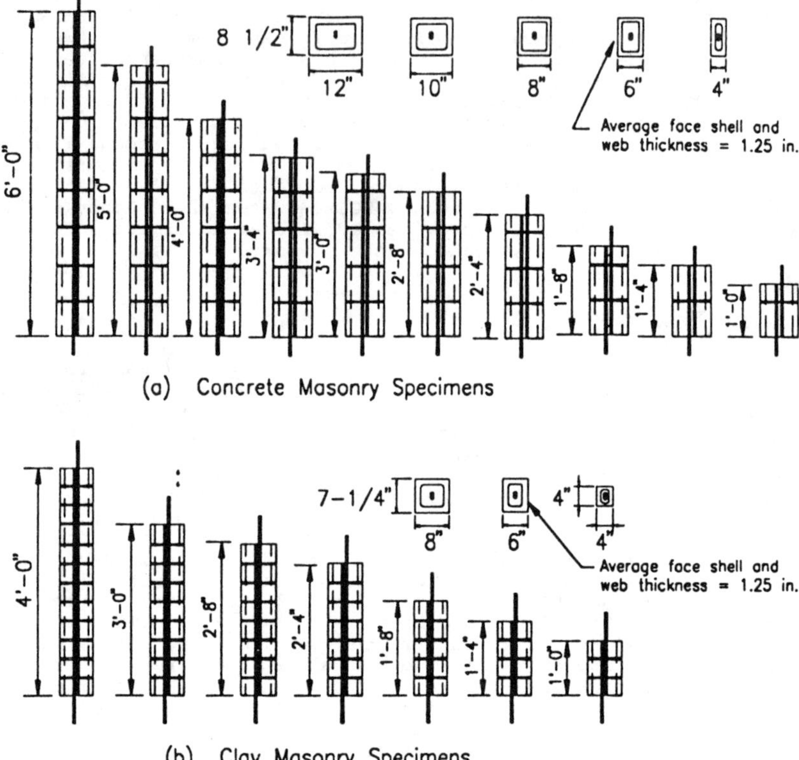

Figure 1--Typical lap splice specimens. Reinforcing bar sizes 4, 6, 8, and 11 were used for different specimens. (1 in. = 25.4 mm)

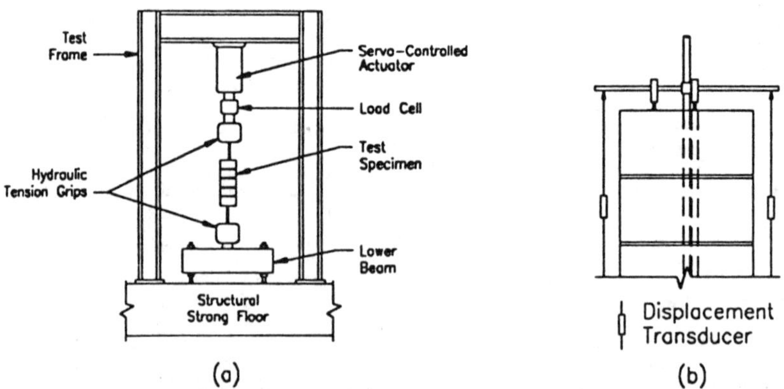

Figure 2--(a) Experimental test setup for tensile testing of lap splice specimens; (b) specimen instrumentation.

TABLE 1--Concrete and clay masonry test specimens. Lap lengths tested are shown as l/d_b=inches, where d_b is the reinforcing bar diameter. Clay masonry specimens designated as "CL".

Reinf. Size (No.)	4-inch Unit	6-inch Unit	8-inch Unit	10-inch Unit	12-inch Unit
4		CL - 24=12" CL - 32=16" CL - 40=20"	24=12" 32=16" 40=20"		
6	37=28" 48=36" CL - 37=28" CL - 48=36"	27=20" 37=28" 48=36" CL - 27=20" CL - 37=28" CL - 48=36"	27=20" 37=28" 48=36" CL - 27=20" CL - 37=28" CL - 48=36"	27=20" 37=28" 48=36"	27=20" 37=28" 48=36"
8		CL - 32=32" CL - 48=48"	32=32" 40=40" 48=48" 60=60"		
11			52=72"		

(1 inch = 25.4 mm)

TABLE 2--Average compressive strength of masonry materials.

Concrete Masonry[1] (Net Area Strength) Mean of 5 Tests	(psi)	Clay Masonry[2] (Net Area Strength) Mean of 5 Tests	(psi)
4"	3380	4"	15640
6"	3530	6"	14200
8"	3690	8"	10890
10"	2570	Mortar[3] (mean of 12 tests)	2270
12"	3480	Grout[4] (mean of 12 tests)	2910
8" Concrete Masonry Prism[5] (2 tests)	3120	6" Clay Masonry Prism[5] (2 tests)	4160

(1 inch = 25.4 mm; 145 psi = 1 MPa)
The following ASTM Test Methods were followed during material property testing:
[1] C 140-75, Method of Sampling and Testing Concrete Masonry Units
[2] C 67-87, Method of Sampling and Testing Brick and Structural Clay Tile
[3] C 109-87, Test Method for Compressive Strength of Hydraulic Cement Mortars
[4] C 1019-84, Method of Sampling and Testing Grout
[5] E 447-84, Test Methods for Compressive Strength of Masonry Prisms

TABLE 3--Average tensile yield strength of steel
reinforcing bars and tensile splitting strength of
masonry materials.

Reinforcing Bar Size[1] (3 tests each size)	psi	Material (2 tests ea. type)	psi
#4	72300	Grout[2] (in Concrete Units)	430
#6	66400	Grout[2] (in Clay Units)	500
#8	68700	Mortar[3]	250
#11	68000		

(1 inch = 25.4 mm; 145 psi = 1 MPa)
The following ASTM Test Methods were followed during material property
testing:
[1]A 615-84, Specification for Deformed and Plain Billet-Steel Bars for
    Concrete Reinforcement
[2]C 496-86, Test Method for Splitting Tensile Strength of Cylindrical
    Concrete Specimens (cylinders for testing obtained by coring 2 in.
    diameter by 4 in. cores from grouted hollow masonry units)
[3]C 496-86, Test Method for Splitting Tensile Strength of Cylindrical
    Concrete Specimens (cylinders for testing prepared in accordance
    with UBC 24-22, Field Test for Mortar)

TABLE 4--Absorption properties of concrete and clay units.

24 Hour Absorption	% by weight	Initial Rate of Absorption[2]	g/30 in[2]
Concrete Units[1] (mean of 15 tests)	12.0	Clay Units (mean of 9 tests)	17.5
Clay Units[2] (mean of 9 tests)	6.69		

(1 inch = 25.4 mm)
The following ASTM Test Methods were followed during material property
testing:
[1]C 140-75, Method for Sampling and Testing Concrete Masonry Units
[2]C 67-87, Method of Sampling and Testing Brick and Structural Clay Tile

EXPERIMENTAL TEST RESULTS

Typical Results and Failure Mechanisms

Complete load-displacement information was recorded for each of
the specimens. Selected results are presented here to provide a representation of the effect of lap length, reinforcing bar size, and masonry unit size and type on lap splice behavior.
Two replications each of 35 different lap splice specimens were
tested. In general, the tests showed good repeatability between identical specimens. The distribution of experimental precision is plotted in
Figure 3 for all 35 specimen pairs, where the precision is calculated as

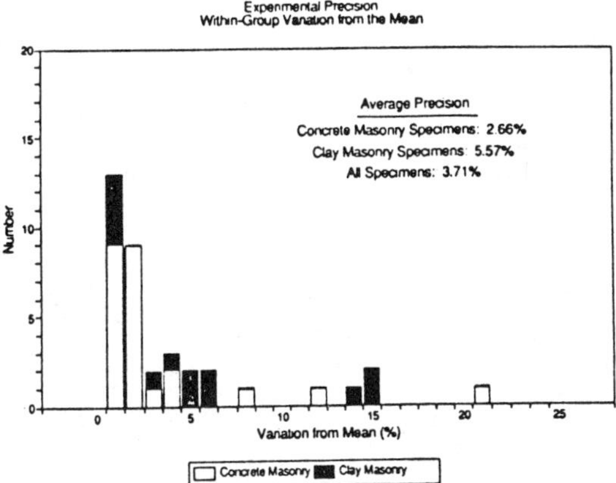

Figure 3--Distribution of experimental precision for lap splice speci-
    mens: Precision is measured as the percentage variation in
    strength values recorded for each test pair from the mean
    value for that pair.

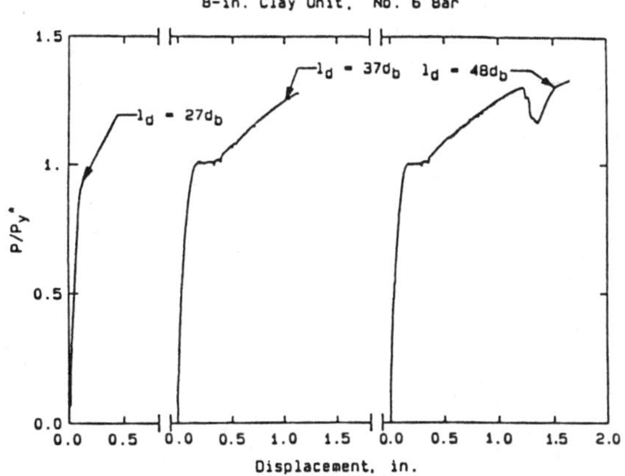

Figure 4--Typical lap splice load displacement behavior for number 6 bar
    in 8-in. clay masonry unit. (a) Short splice length, failure
    by longitudinal tensile splitting of the masonry; (b) medium
    splice length, reinforcement pullout and yield preceding ten-
    sile splitting failure; (c) long splice length, failure by
    yield and fracture of the reinforcement. (1 inch = 25.4 mm)

* P = applied load
  P_y = yield load of the reinforcing bar

the percentage variation of the recorded splice strength from the mean
value for each pair. Concrete masonry specimen pairs displayed excellent repeatability, with lap splice strengths for each pair generally in
agreement to within 4%, with a mean variation of 2.66%. Clay masonry
specimens displayed a somewhat greater variation, averaging 5.57% for
each pair, presumably due to the predominance of relatively brittle
failures for these specimens. Tested lap splice strengths for all specimen pairs varied by an average of 3.71%.

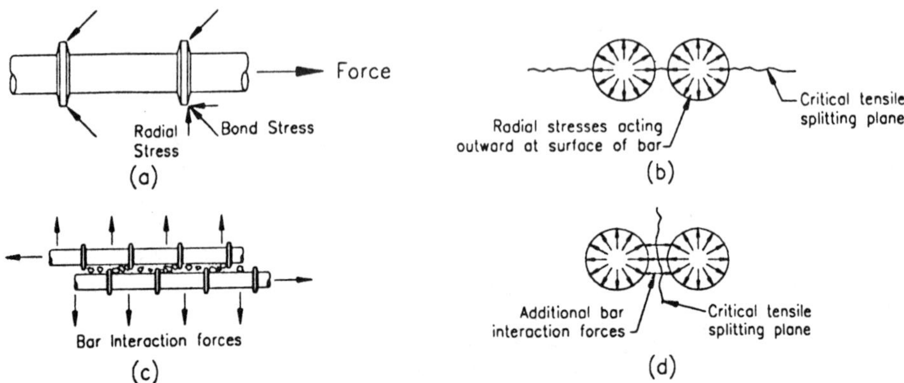

Figure 5--Mechanical bond stresses and bar interaction forces.

Failure Mechanisms--Several distinct failure modes have been identified for lap splices in reinforced masonry [3]. Observations from
this study and examination of load-deformation behavior identified three
failure modes for lap splices: (1) brittle tensile splitting of the
grout and unit; (2) yield and/or pullout of the bar preceding failure by
tensile splitting; (3) and yield of the bar with failure by pullout or
fracture of the reinforcement. The load-displacement curves for a #6
(19 mm) reinforcing bar with different lap splice lengths in 8-in. (203
mm) wide clay masonry specimens in Figure 4 illustrates the effect of
different failure modes on strength and ductility of the splice. It is
apparent that the mode of failure has a significant effect on both
strength and ductility of the splice.

As shown by curve (a) in Figure 4, a short lap length leads to
brittle failure at loads below the yield load of the reinforcement Py.
The mechanism leading to this type of failure results from mechanical
interaction between reinforcement deformations and grout. These bond
stresses develop a compressive grout stress inclined at an angle from
the longitudinal axis of the bar as shown in Figure 5 (a). The radial
component of this force induces circumferential tensile stresses in the
masonry. Splitting cracks occur when the circumferential tensile
stresses exceed the tensile strength of the grout, propagating along the
critical plane as shown in Figure 5 (b).

A stronger lap splice may be obtained by increasing the lap
length, as shown by the response in Figure 4 (b). Increasing the lap
length has the effect of reducing nominal bond stresses along the bar
such that yield and/or pullout of the reinforcement may occur without
exceeding grout tensile strength. However, relative movement between
two spliced reinforcing bars during yield or pullout develops additional
stresses within the masonry. Relative movement increases lateral tensile stresses within the surrounding masonry when one bar rides up on
the other, as shown in Figure 5 (c); these types of specimens were observed to occasionally fail along a different failure plane (Figure 5
(d)). Hence these types of specimen often experienced limited reinforcement yield, with failure ultimately resulting by longitudinal
splitting through the masonry.

The final failure mechanism is one where ultimate failure is by
yield and fracture of the reinforcing bars. Specimens failing in this
manner provided a strong and ductile lap splice, as shown in Figure 4
(c). This type of behavior can be obtained by providing a sufficiently
long lap length and also providing adequate cover to the splice. In-

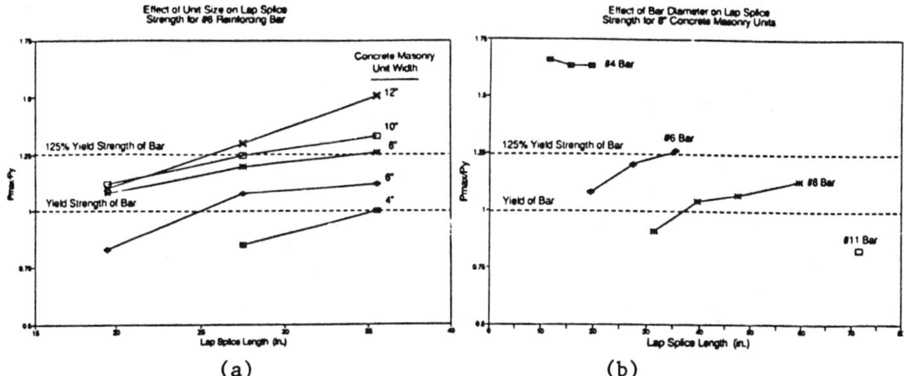

(a)   (b)

Figure 6--Typical lap splice test results: (a) effect of unit size on
lap splice strength for number 6 reinforcing bar; (b) effect
of reinforcing bar diameter on lap splice strength for 8-inch
concrete masonry specimens (1 inch = 25.4 mm).

creasing tensile strength of the grout may also serve to prevent splitting failure, however this effect was not investigated.

Effect of Masonry Unit Size on Lap Splice Strength--The effect of
varying masonry unit width on lap splice strength can be seen in Figure
6(a), which shows the ratio of the force resisted by the splice (Pmax)
to the yield force of the bar (Py) versus lap splice length for number 6
(19 mm) bars in different size concrete masonry units. Specimens constructed with a small unit width did not provide sufficient cover to the
lap splices to resist tensile splitting forces. For specimens constructed with 4 in. (102 mm) and 6 in. (152 mm) wide units, a large
increase in lap length resulted in only a small increase in splice
strength. Increasing the lap splice cover by using wider units
increased resistance to tensile splitting, resulting in a stronger
overall splice. Results for clay masonry specimens showed a similar
effect.

Effect of Reinforcing Bar Diameter on Lap Splice Strength--Results
from specimens constructed with different sizes of reinforcing bars in 8
in. (203 mm) wide concrete masonry units are plotted in Figure 6(b).
Increasing bar diameter increases the total force which must be resisted
by the masonry, and hence specimens with larger bar diameters failed at
a lesser fraction of their yield load. It appears that it may not be
possible to provide an effective lap splice for a number 11 bar: a lap
splice with a length of 6 feet (52 bar diameters) developed only 75 percent of the yield strength of the bar. Specimens constructed with small
diameter number 4 (12 mm) bars, on the other hand, were able to fully
develop the bar yield strength with a lap length of only 12 inches (305
mm) or 24 bar diameters.

Effect of Unit Type on Lap Splice Strength--Clay masonry units
used in this study had much greater compressive strengths and a greater
compressive modulus than the concrete masonry units. These differences
had a distinct effect on the measured lap splice strength. It appears
that, for short lap lengths where tensile splitting governs failure, the
stiffer and stronger clay masonry units had a confining effect on the
splice increased the overall resistance of the lap splice. This effect
is evident in the plot of Figure 7 even though the clay masonry speci-

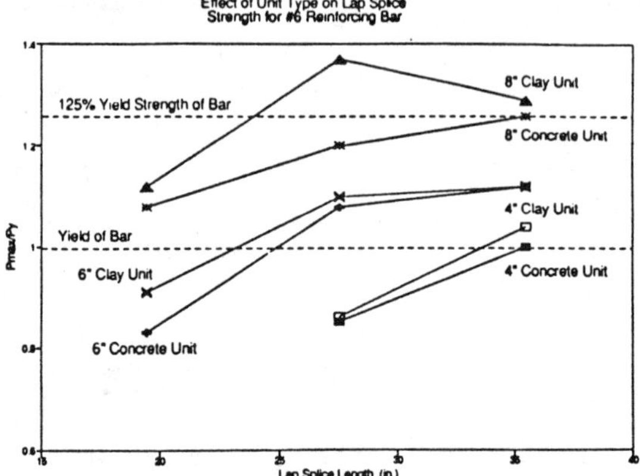

Figure 7--Lap splice test results for clay and concrete masonry specimens (1 inch = 25.4 mm).

mens had cross-sectional areas which were 16% to 49% less than their
concrete masonry counterparts (see dimensions provided in Figure 1).
This confining effect was much less pronounced as the lap length was increased and the failure shifted to one governed by pullout and yield of
the reinforcing bar, rather than longitudinal tensile splitting of the
masonry.

COMPARISON WITH DESIGN STANDARD REQUIREMENTS

Lap Splice Standard Requirements--The rationale behind design
standard requirements for lap splices in reinforced masonry is often unclear, but it appears that most requirements are derived from early
working stress criteria for reinforced concrete. Lap splices are required to develop a minimum tensile strength equivalent to 125% of the
yield strength of the reinforcing bars being spliced to provide adequate
splice ductility. A comparison of lap splice length requirements specified by various design standards is provided in Figure 8.
Currently the design of reinforced masonry is governed by two
working stress design codes: the Uniform Building Code (UBC) [11], and
ACI 530-88/ASCE 5-88 [12]. Criteria contained in these standards for
anchorage of reinforcement determine development lengths necessary to
develop working stress levels in the reinforcing bars based upon a limiting value for bond stress and consider the bond to be distributed
evenly along the length of the bar. Lap splices are generally designed
for a lower bond stress than single bar anchorages because large stress
concentrations at the cut-off end of the bar and the close proximity of
the bars both act to promote splitting failure.
Section 2409 of the UBC lists criteria for lap splices in reinforced masonry. A commentary to the UBC [13] states that the development length formula is based upon a maximum nominal bond stress of 125
psi (0.86 MPa) and a grout compressive strength of 2000 psi (13.8 MPa),
which is the minimum allowed for reinforced masonry. Lap splice length
for reinforcing bars in tension are calculated as the maximum of:

$$l_d = 30 \; d_b \tag{1}$$

or

$$l_d = 0.002 \; f_s \tag{2}$$

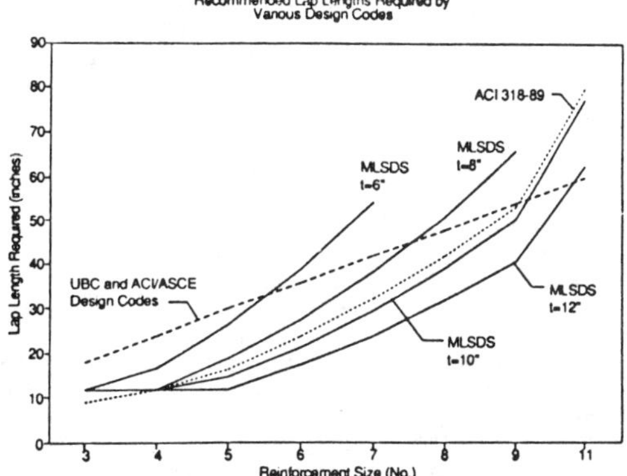

Figure 8--Lap splice lengths required by different design standards.
Note: MLSDS = draft, Masonry Limit States Design Standard,
shown are MLSDS lap splice requirements for different unit
thickness (1 inch = 25.4 mm).

where l_d is the lap splice length (inches), d_b is the diameter of the
reinforcing bar (inches), and f_s is the steel stress (psi) calculated at
the splice. For the case of Grade 60 (413 MPa) rebar, fully stressed to
its allowable limit of 24,000 psi (165 MPa), the lap splice length
specified by equation (2) becomes

$$l_d = 48\ d_b \tag{3}$$

Requirements listed by the ACI/ASCE masonry code are similar to
the UBC provisions. Lap length is determined by Equation 2 (above) with
the additional stipulation that the lap length be greater than 12
inches. Thus, for Grade 60 (413 MPa) reinforcement stressed to its
allowable limit of 24,000 psi (165 MPa), the required lap length is
again determined using equation (3).
It is useful to discuss requirements for lap splices in reinforced
concrete because these criteria appear to offer an improvement over the
simple linear relationships described above. ACI 318-89 [14] contains a
formula for lap splice length determination which considers the effect
of concrete tensile strength, represented as the square root of the compressive strength f'c, in addition to the area of the reinforcing bar
A_b:

$$l_d = 0.04 A_b f_y / \sqrt{f'_c} \tag{4}$$

If the reinforcement nominal yield stress is equal to 60,000 psi (413
MPa) and the concrete compressive strength is equal to 2000 psi (13.8
MPa) (for comparison with UBC and the ACI/ASCE masonry code above), this
formula reduces to

$$l_d = 42.1\ d_b{}^2 \tag{5}$$

Note that for this case the splice length is proportional to the square
of the bar diameter, which results in much greater splice lengths for
large diameter bars. This may account for the tendency of larger diame-

ter bars to fail at lower bond stresses by splitting rather than pullout or yield. An additional stipulation is that the splice length be greater than $0.0004d_b f_y$, or for Grade 60 (413 MPa) reinforcement,

$$l_d > 24\ d_b \qquad\qquad (6)$$

A Masonry Limit States Design Standard (MLSDS) for the design of reinforced masonry structures is currently being compiled by a committee consisting of members of The Masonry Society, American Concrete Institute, and American Society of Civil Engineers [1]. This standard appears to provide a more rational determination of lap splice length where lap length is based upon a formula which, in its original form, considers reinforcing bar diameter, the expected yield strength of the reinforcement, expected grout tensile strength, and masonry thickness at the splice location. The expected values used in the MLSDS are mean values of material properties determined by physical testing.

The relationship for splice length used by the MLSDS was originally developed by Soric [6] for the case of lap splices in grouted hollow concrete masonry, and utilizes a model which regards the radial stress due to bond action on the grout as an outward acting hydraulic pressure [3]. The surrounding masonry resists this pressure by acting as a thick-walled pressure vessel, and failure occurs when the circumferential tensile stress exceeds the tensile strength of the masonry. In its original form, the required lap length is:

$$l_d = \frac{C\ d_b{}^2 f_y}{(t-d_b) f_{gt}} \qquad\qquad (7)$$

where t is the masonry thickness, d_b is the reinforcing bar diameter, f_y the yield strength of the steel, f_{gt} is the grout tensile strength and C is a coefficient accounting for nonuniformity of bond stresses along the length of the bar. Note that this formula is nondimensional and may be used with either SI or English units. Soric conducted experimental tests with #4 and #7 (12 and 22 mm) bars in 6 in. (152 mm) hollow concrete units and calculated a mean value of 1.75 for the coefficient C. This determination is based upon the criteria that the lap splice develops a strength which is greater than 125% of the reinforcement yield strength. The MLSDS adopted this value for C and assumed a grout tensile strength of 400 psi (2.75 MPa):

$$\emptyset\ l_d = \frac{0.0045\ d_b{}^2\ f_{ye}}{(t-d_b)} \qquad\qquad (8)$$

where the capacity reduction factor $\emptyset = 0.8$, and f_{ye} is the expected yield strength of the reinforcement. The splice length must be 12 in. (305 mm) or greater. Splice length curves for different masonry unit thickness using this formula are plotted in Figure 8 along with lap lengths required by other design standards. The MLSDS requirements are significantly different than the UBC and the ACI/ASCE masonry code requirements and generally require shorter lap lengths for small bars in large units. In addition, the long lap lengths required for large bars in small units may act to prevent brittle splitting failure which is prevalent in this type of splice. Both the UBC and ACI/ASCE masonry codes are somewhat less conservative for lap splices than the MLSDS requirements in this case.

The formula adopted in the draft Masonry Limit States Design Standard (Equation 8) provides a rational approach to the determination of lap splice length and considers the important parameters of grout tensile strength, reinforcement yield strength, and the thickness of the

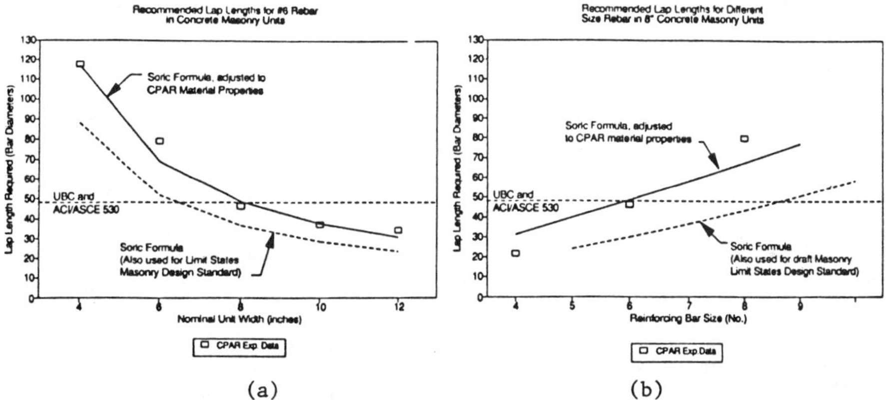

(a)   (b)

Figure 9--Comparison between lap splice lengths required by current design standards CPAR experimental results. Note that the lap
length required is expressed in terms of bar diameters. (a)
Recommended lap splice lengths for number 6 reinforcing bars in
concrete masonry units with different widths; (b) lap splice
lengths for different size reinforcing bars in 8-inch concrete
masonry units (1 inch = 25.4 mm).

grouted masonry. This approach provides for economical splices of small
bars and poses a severe penalty for large bars in small units, which
exhibit a tendency towards splitting failure. However, this model
relies on an accurate determination for the coefficient C, which must be
verified for additional combinations of bar diameter, unit size, and
grout strength. Results from the CPAR study are used to investigate the
value of this coefficient and the overall validity of the model.

  Lap Length Requirements--Current masonry design standards require
mechanical splice connections to develop a tensile strength greater than
125% of the reinforcement yield strength. This criteria has been
adopted here, as well, as performance criteria for ultimate strength of
lap splices. Experimental results can be used to determine lap lengths
required to develop 125% of the yield strength by either interpolation
or extrapolation of the experimental data in Figures 6 (a) and (b). The
points where the curves cross the line representing 125% of the reinforcement yield strength are then used to determine the appropriate lap.
These points are plotted in Figures 9 (a) and (b) along with the UBC,
ACI/ASCE 530-88, and Soric's requirements (in the form which has been
adopted by the MLSDS). Immediately obvious is the fact that the linear
relationship provided by the UBC and ACI/ASCE masonry design codes,
which are based upon working stress design principles, do not adequately
describe the ultimate capacity of lap splices.
  The experimental data obtained during the CPAR program fits the
general shape of the design formula developed by Soric, yet appears to
be shifted slightly upwards. This effect can be explained by the manner
in which Soric determined the coefficient C required in Equation (7)
above. Rearranging Equation (7), C is determined as a function of material properties and geometric dimensions:

$$C = \frac{l_d \, (t-d_b) \, f_{gt}}{d_b^2 f_y} \tag{9}$$

where l_d is the lap splice length required to develop 125% of the reinforcement yield strength as determined experimentally. The main discrepancy between Soric's original formulation and the CPAR results arises in the values used for grout tensile strength: Soric measured a direct grout tensile strength of 439 psi (3.0 MPa) (tested using the briquet specimen of ASTM C 190, Test Method for Tensile Strength of Hydraulic Cement Mortars) and grout cylindrical tensile splitting strength of 732 psi (5.0 MPa) (measured in accordance with ASTM C 496, Test Method for Splitting Tensile Strength of Cylindrical Concrete Specimens) corresponding to a grout compressive strength of 3760 psi (25.9 MPa). An average grout tensile splitting strength of 430 psi (2.9 MPa) for concrete masonry specimens and grout compressive strength of 2910 psi (20.0 MPa) were measured during the CPAR program. Soric determined a mean value of C=1.75 based upon the grout tensile strength using the briquet specimens, however only the grout tensile splitting strength was measured during the CPAR program. The relationship between grout compressive strength and tensile splitting strength is approximately equivalent for both sets of data, hence it is possible to recalculate Soric's values for the coefficient C to (a) account for the fact that the CPAR grout strengths were lower; and (b) base the model upon tensile splitting strength rather than direct tensile strength. This approach may in fact be more useful for design purposes because the direct tensile test is rarely conducted in practice.

Recalculation based upon grout tensile splitting strengths provides a value of 3.20 for the formula coefficient C, resulting in a relationship which closely models the experimental results shown in Figure 9 (a) for number 6 (19 mm) bars in concrete masonry units of different width. The model accurately describes the tendency of splices in areas with a large reinforcement ratio (i.e. large diameter bars in thin units) to fail by tensile splitting and requires lap lengths of 100 bar diameters and greater for these cases. The model also requires a shorter lap length in splices where the reinforcement ratio is small. The practical lower limit for splice length is approximately 30 bar diameters for this particular combination of material properties and reinforcement size.

The adjusted model's lap length requirements for different reinforcing bars in 8-in. (203 mm) concrete masonry construction is shown with experimental data points in Figure 9 (b). The model appears to be over-conservative for number 4 (13 mm) bars and slightly under-conservative for number 8 (25 mm) bars. It is unclear at this time if these variations are due to experimental techniques or deficiencies in the analytical model, however this effect will be investigated further with tests on additional specimens during the next phase of the CPAR program.

Reinforcement Limitations--The UBC limits the reinforcement ratio to 6% of the cell area for hollow unit construction, or 12% at lap splice locations. The MLSDS allows a maximum reinforcement ratio of 4% of the cell area, however does not explicitly address the reinforcement ratio at lap splice locations. For consistency, reinforcement ratio at lap splices is calculated here as the ratio of the area of one of the lapped bars to the net area of the grouted cell.

Preliminary observations based upon specimen load-displacement response and failure mechanisms indicate that the maximum reinforcement size is dependent upon reinforcement cover, as based upon unit width. Specimens which provided adequate cover to the reinforcement (i.e. low reinforcement ratio) required relatively small lap lengths to provide the necessary splice strength. Specimens with a large reinforcement ratio were more prevalent to fail by brittle tensile splitting and re-

quired excessively long lap lengths to develop the required strength.
Data collected to date indicates that specimens with number 6 (19 mm)
bars in 6-in. (152 mm) units and number 8 (25 mm) bars in 8-in. (203 mm)
units may be used to designate the practical upper limit for lap splice
reinforcement ratios. These specimens both require a lap length of
approximately 80 bar diameters and correspond to reinforcement ratios of
1.98% and 2.62%, respectively, which is somewhat less than the maximum
allowed by design standards as described above. Specimens to be tested
in the next phase of the CPAR program will be used to provide additional
information on maximum reinforcement ratios at lap splice locations.

CONCLUSIONS

Current criteria for design of lap splices within reinforced
masonry provided by the Uniform Building Code and ACI530-88/ASCE 5-88
are based upon working stress design principles and do not provide consistent margins of safety for ultimate lap splice capacity as compared
to the results reported in this paper. These formulae do not take into
account the effect of masonry unit thickness on splice capacity and are
over-conservative for small reinforcement ratios, and provide insufficient lap to prevent brittle splitting failures in areas where the reinforcement ratio is large.

The draft Masonry Limit States Design Standard provides a more
rational approach for lap splice design, and accurately predicted experimental results obtained in this study for lap splices in reinforced
masonry. However, the model adopted for use in the Standard is not
entirely consistent for all specimens and should be verified for cases
where the reinforcement ratio is either very small or very large. Testing of additional specimens is planned to further investigate the validity of the model for these conditions.

Tensile strength of the grout is thought to have a direct effect
on lap splice strength and this parameter should be retained in lap
splice design formulae. Currently the Masonry Limit States Design Standard generalizes this effect by assuming a value of 400 psi (2.76 MPa)
for grout tensile strength, which was the tensile strength obtained by
Soric when the corresponding grout compressive strength was 3760 psi
(25.9 MPa). If this parameter is not explicitly retained within lap
splice design formulae, the grout tensile strength should instead correspond to that expected for grout with a compressive strength of 2000 psi
(13.8 MPa), which is the minimum allowable by the Standard. The next
phase of study will include a limited number of specimens with varying
grout strengths to provide additional information on this subject.

Preliminary data indicates that the reinforcement ratio at a lap
splice location (calculated as area of a single lapped bar divided by
the area of the grouted cell) should be limited to approximately 2.0 to
3.0 percent. This stipulation would encourage the use of distributed
small reinforcing bars rather than a lesser number of large diameter
bars. The effect of reinforcement ratio on splice failure mechanism and
lap splice requirements will be investigated further during the next
phase of the program.

ACKNOWLEDGMENT

The tests and analysis described and the results presented herein
were obtained from research conducted under the Construction Productivity Advancement Research (CPAR) Program of the U. S. Army Corps of Engineers. Permission was granted by the Chief of Engineers to publish this
information.

REFERENCES

[1]   Masonry Standards Joint Committee, _Draft, Masonry Limit States Design Standard_, The Masonry Society, American Concrete Institute, American Society of Civil Engineers, April, 1992.

[2]   Scrivener, J.C., "Bond of Reinforcement in Grouted Hollow-Unit Masonry: A State-of-the-Art", _TCCMAR Report No. 6.2-1_, U.S.-Japan Coordinated Program for Masonry Building Research, July 1986.

[3]   Cheema, T.S., "Anchorage Behavior and Prism Strength of Grouted Concrete Masonry", PhD Dissertation, University of Texas at Austin, May 1981.

[4]   Suter, G.T., Fenton, G.A., "Splice Length Tests of Reinforced Concrete Masonry Walls", _Proceedings of the Third North American Masonry Conference_, University of Texas at Arlington, June 1985.

[5]   Soric, Z., Tulin, L.G., "Bond & Splices in Reinforced Masonry", _TCCMAR Report No. 6.2-2_, U.S.-Japan Coordinated Program for Masonry Building Research, August 1987.

[6]   Soric, Z., Tulin, L.G., "Length of Lap for Spliced Reinforcement in Masonry Structures", _Proceedings of the 8th International Brick/Block Masonry Conference_, Dublin, Ireland, 1988.

[7]   Kubota, T., and Kamogawa, N., "On the Lap Joint of Flexural Reinforcement in Concrete Masonry", _Proceedings of the First Joint Technical Coordinating Committee on Masonry Research_, U.S.-Japan Coordinated Program for Masonry Building Research, Tokyo, 1985.

[8]   Watanabe, M., "Bond and Anchorage of Reinforcement in Masonry", _Proceedings of the First Joint Technical Coordinating Committee on Masonry Research_, U.S.-Japan Coordinated Program for Masonry Building Research, Tokyo, 1985.

[9]   Matsumura, A., "Test on Lap Splices of Relatively Large Size Bars in Masonry", _Proceedings of the Sixth Meeting of the U.S.-Japan Joint Technical Coordinating Committee on Masonry Research_, Port Ludlow, Washington, 1990.

[10] Atkinson-Noland & Associates, _Preparation of Stack-Bond Masonry Prisms_, Boulder, Colorado, May 1985.

[11] International Conference of Building Officials, _Uniform Building Code_, Chapter 24, "Masonry", Whittier, California, 1988.

[12] ACI-ASCE Committee 530, _Building Code Requirements for Masonry Structures (ACI 530-88/ASCE 5-88)_, 1988.

[13] TMS Codes and Standards Committee, _Commentary to Chapter 24, Masonry, of the Uniform Building Code 1988 Edition_, The Masonry Society, Boulder, Colorado, 1990.

[14] ACI Committee 318, _ACI 318-89, Building Code Requirements for Reinforced Concrete_, American Concrete Institute, Detroit, Michigan, 1989.

Roger D. Flanagan[1], Richard M. Bennett[2], and James E. Beavers[3]

SEISMIC BEHAVIOR OF UNREINFORCED HOLLOW CLAY TILE INFILLED FRAMES

REFERENCE: Flanagan, R. D., Bennett, R. M., and Beavers, J. E., "Seismic Behavior of Unreinforced Hollow Clay Tile Infilled Frames," _Masonry: Design and Construction, Problems and Repair, ASTM STP 1180_, John M. Melander and Lynn R. Lauersdorf, Eds., American Society for Testing and Materials, Philadelphia, 1993.

ABSTRACT: The U.S. Department of Energy's Y-12 Plant[*], located in Oak Ridge, Tennessee, is conducting a comprehensive research program to evaluate the seismic capacity of unreinforced hollow clay tile infilled steel frames. This paper summarizes analyses and the results of in-situ and laboratory testing. Evaluations of unit tile, mortar, masonry assemblages, and large-scale building components are presented. Building components tested include infilled frames loaded in-plane and out-of-plane.

KEYWORDS: hollow clay tile, infilled frame, masonry infill, seismic

---

A typical building construction of older industrial facilities at the Y-12 plant is structural steel framing with infilled unreinforced hollow clay tile (HCT) walls. Figures 1 and 2 show 200 mm and 330 mm nominal thickness walls built with running bond and using full width tile units or a staggered combination of tile units. The HCT have been laid with the cores horizontal, approximately 13 mm full width bed joints, and only face shell mortar in the head joints. The 330 mm combination walls exhibit no vertical collar joint except occasional mortar that has fallen between the tile units.

Girders and columns that surround the HCT walls are generally connected using simple framing details, Figure 3. Although some rotational resistance is present, the steel frames by themselves are quite flexible and weak under lateral loads. Little or no cross-bracing exists in the buildings which results in the infilled HCT walls becoming the primary lateral load resisting mechanism. Figure 4 presents typical

---

[1]Senior Engineer, Martin Marietta Energy Systems, Inc., Oak Ridge, TN 37831.

[2]Associate Professor, The University of Tennessee, Knoxville, TN 37996-2010.

[3]Director, Center for Natural Phenomena Engineering, Martin Marietta Energy Systems, Inc., Oak Ridge, TN 37831.

[*]Managed by Martin Marietta Energy Systems, Inc. for the U.S. Department of Energy under contract DE-AC05-84OR21400.

column-infill interfaces, with the exact details hidden from view. Walls butting against the column flange are also found.

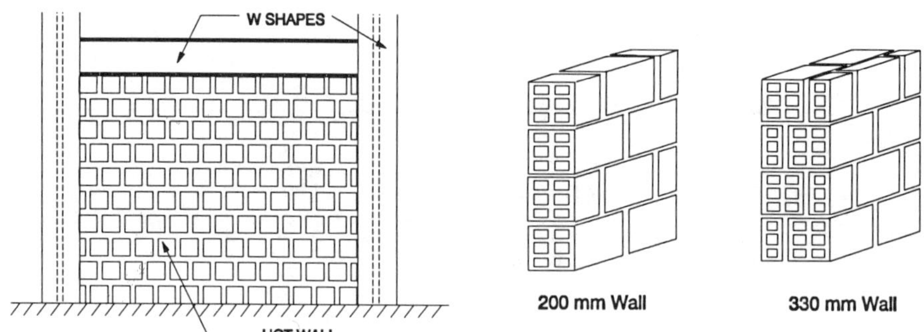

FIG. 1--Infilled Frame   FIG. 2--Wall Sections

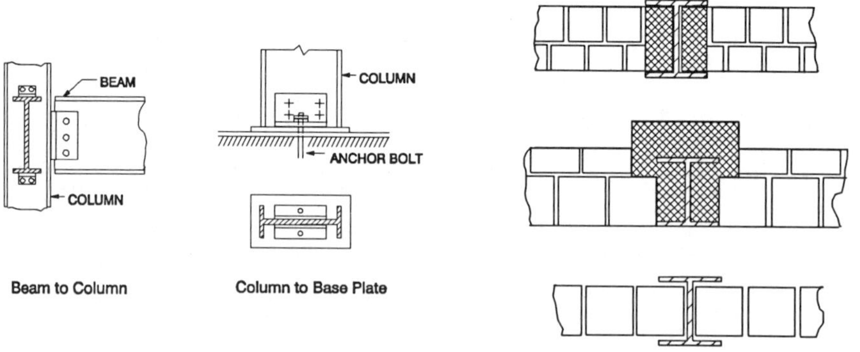

FIG. 3--Connection Details   FIG. 4--Column Interface

As part of the safety evaluation of these facilities, the resistance to various natural phenomena, particularly earthquake, must be determined. A review of past analyses indicates the steel framing may not be adequate to resist seismic loads. Thus, it is necessary to count on the HCT infills to resist lateral forces.

On-site investigations immediately after the 1985 Mexico City earthquake indicated that modern medium- and high-rise buildings with nonstructural masonry infills performed better than otherwise similar structures without infills [1]. The unreinforced infills were not tied well into surrounding frames, but appeared to have prevented structure collapse while sustaining significant damage. This was attributed to reduced inertial forces resulting from shorter periods of vibration of the stiffened frames and the particular soil conditions of the region.

A review of technical literature [2] indicates that while substantial research has been conducted on plain masonry and masonry infilled frames, little of this research involved unreinforced HCT. Of equal importance, a review of current masonry analytical methods presents the need for more advanced constitutive material modeling and a better understanding of the HCT structural system behavior.

The experimental part of the HCT research program consists of both in-situ and laboratory testing. Activities include the systematic characterization of existing construction (early 1940s) complemented with laboratory testing of similar conditions. This paper presents recent portions of both laboratory and in-situ testing along with preliminary analysis of the results.

MATERIALS

<u>Unit Tile</u>

Light red burned clay masonry units, manufactured in 1987, were sampled and tested in accordance with ASTM Method of Sampling and Testing Brick and Structural Tile (C 67) and ASTM Test Method for Splitting Tensile Strength of Masonry Units (C 1006). Two sizes of hollow clay tile units were tested, nominally 100 mm and 200 mm width. Forty individual tiles were sampled, twenty 100 mm and twenty 200 mm. Each specimen's size and void area was measured and initial rate of absorption (suction) determined. Of the 40 specimens, 20 were tested to failure to determine compressive strength and 20 were tested to failure to determine splitting tensile strength.

The results of the compressive strength tests with comparison to other published clay tiles values are given (Table 1). The orientation of the tiles are given as tested with a vertical compressive load. Tiles tested on their end (load parallel to cores) are denoted vertical and tiles tested on their edge (load perpendicular to cores) are denoted horizontal. Consistent with the findings of other researchers, the coefficients of variation of each set of compressive test results ranged from 5 to 25 percent.

A significant increase in compressive strength over time is indicated (Table 1). The oldest test data found in the literature is from 1918 [3]. The EDGe [4] results are from 1940s vintage tiles extracted from walls at Y-12. Additional results of the present unit tile testing are presented (Table 2).

While the fundamentals of HCT manufacture have not changed significantly, specific fabrication details, material additives and processing have changed dramatically. Use of finer graded clays, superior bonding agents, more uniformly controlled kilns, and other manufacturing enhancements have led to increased strength and economy of current clay tile products.

TABLE 1--<u>HCT Compressive Strength (kN)</u>

Specimen Orientation	Current Findings	EDGe [4]	Johnson & Matthys [5]	Hathcock & Skillman [3]
100 mm Horizontal	618	...	...	356
100 mm Vertical	1085	712[*]	...	592
200 mm Horizontal	845	...	463	440
200 mm Vertical	3037	970[*]	1081	1134

[*] Value is twice the average reported strength of half blocks tested.

<u>Mortar</u>

Representative mortar samples from existing facilities were evaluated by an external testing laboratory to determine their composition. Petrographic investigations were conducted using methods of ASTM Petrographic Examination of Hardened Concrete (C 856).

Soluble silica content of the mortar was determined by procedures in ASTM Portland Cement Content of Hardened Hydraulic-Cement Concrete (C 1084). Calcium oxide was determined by titration of filtrates from the soluble silica analyses. Hydrated lime was calculated after adjustments for lime in the portland cement. Magnesium oxide was determined by atomic absorption spectrophotometric methods and converted to magnesium hydroxide by molecular ratio calculations, corrected for magnesium hydroxide in the cement. Insoluble residue (essentially sand) content was determined by procedures of ASTM Standard Methods for Chemical Analysis of Hydraulic Cement (C 114). Portland cement content was calculated in accordance with ASTM C 1084.

The resulting mortar contents varied considerably. One set of samples conformed to the proportions of type N while the other three sets did not meet the requirements of any type, but were closest to type N as defined in ASTM Specification for Mortar for Unit Masonry (C 270). Thus, type N mortar was chosen for construction of laboratory specimens.

TABLE 2--<u>HCT Unit Test Results</u>

Parameter	HCT Size	
	100 mm	200 mm
Length	291.3 mm	293.1 mm
Width	93.0 mm	194.8 mm
Height	292.4 mm	294.4 mm
Net Area	14 135 mm^2	24 613 mm^2
Splitting Tensile Strength (Cores Horizontal)	55.8 kN	78.8 kN
Splitting Tensile Strength (Cores Vertical)	51.5 kN	71.8 kN
Initial Rate of Absorption (Cores Horizontal)	3.29 g/min/30in^2	2.70 g/min/30in^2
Initial Rate of Absorption (Cores Vertical)	5.22 g/min/30in^2	5.32 g/min/30in^2

ASSEMBLAGES

Little information describing the strength of clay tile prisms is available in published literature. Plummer [6] gives the results of tests on 200 mm width side constructed clay tile sections. Average gross prism compressive strengths were 2.3 MPa. Johnson & Matthys [5] have reported 4.3 MPa for 200 mm width side constructed prisms.

Four clay tile prisms were constructed and tested to determine basic compressive properties. Each prism was nominally 610 mm by 1220 mm and tested with the compressive force in the longer direction of the prism. Each specimen was constructed with the cores horizontal. Two prisms were built four courses high and two tiles wide and tested in that orientation under vertical load. The other two prisms were built two courses high and four tiles wide and rotated 90 degrees for vertical testing. Resulting gross compressive strengths normal and parallel to

the tile cores were 5.7 MPa and 2.8 MPa respectively. Gross modulii of
elasticity normal and parallel to the tile cores were 5210 MPa and 2730
MPa respectively.

     Prisms with their cores parallel to the load exhibited lower
compressive strengths than prisms with their cores normal to the load,
even though the unit tiles were stronger with the cores parallel to the
load. The lower strength for prisms with cores parallel is attributed
to the use of only faceshell mortar in the head joints. The small face
shell thickness (17 mm) made it difficult to obtain a good head joint.

BUILDING COMPONENTS

<u>Out-of-Plane Inertial Loads</u>

     For seismic loading, the out-of-plane response of masonry infilled
frames is comprised of the inertial effects resulting from the mass of
the panel and attachments (e.g. piping, etc.) and interstory drift
effects resulting from relative top-bottom displacements of the columns
and adjacent panels. Inertial loads are examined first.

     For determination of precracking response linear elastic analysis
is sufficient. Thurlimann and Guggisberg [7] have proposed an
elliptical interaction relationship of biaxial and twisting moments to
define cracking. Chua [8] found good correlation using this method with
the results of recent air bag tests of concrete masonry infills [9].

     The post cracking behavior of unreinforced masonry infill panels
subjected to inertial loads is membrane (arching) action. Seah [9]
proposed a modified fracture line method of calculating the ultimate
strength of infill panels subjected to uniform out-of-plane loading.
This procedure incorporates arching effects and the effect of frame
stiffness and deformation. Tensionless infill properties and a plastic
stress block for masonry in compression at regions of contact along
fracture lines and at panel boundaries are assumed. Various fracture
line patterns of the infill are proposed based on panel boundary
conditions. In a comparison with the results of nine air bag tests of
infilled frames, Seah found good correlation in predicting the ultimate
capacity.

     This method was used to predict the postcracking load-displacement
curve for a full-scale out-of-plane air bag test at the Y-12 plant. The
wall tested was 8535 mm long by 3658 mm high, consisting of 200 mm side
constructed single-wythe HCT units. The wall was enclosed by W360x211
columns, a W760x161 beam and a concrete floor slab. A comparison of
predicted and observed response at the wall center is given in Figure 5.

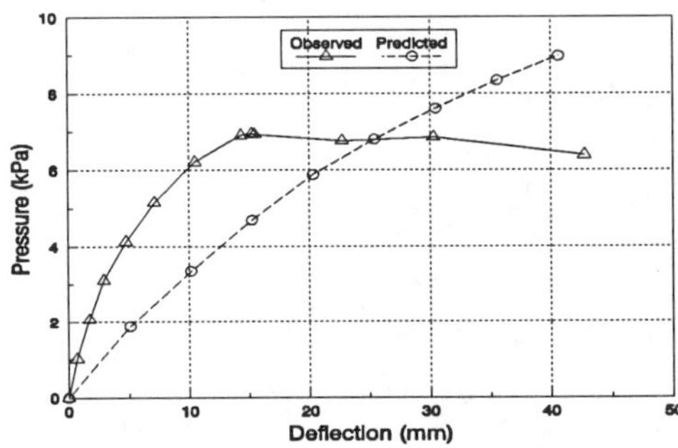

FIG. 5--Mid Panel Displacement of Full-Scale Air Bag Test

Parametric studies using this arching analysis method have indicated that variations in masonry compressive strength and confining frame restraint would significantly alter the ultimate capacity of the infill. The effects of modulus of elasticity, face shell thickness, and relative air bag size were found to be less significant.

In-Plane Capacity

The benefits of masonry infills in providing additional stiffness and strength of framing against lateral loads is well documented. Laboratory tests have been conducted separately by several researchers [10], [11], [12], and [13]. For the most part, this research has concentrated on the use of either brick and concrete masonry units, or micro-concrete for the infill.

A series of large-scale in-plane racking tests of HCT infilled frames was recently conducted. The effect of cyclic loading and varying frame stiffness was investigated. Three wall specimens nominally 200 mm wide and one specimen nominally 330 mm wide were tested to failure. Details of the test specimens are given (Table 3) and a typical configuration is shown in Figure 6.

TABLE 3--Summary of Test Specimens

Test No.	Wall		Frame		Relative Stiffness λh [14]
	Size (mm)	T (mm)	Beam	Column	
1	2240x2240	200	W310x52	W250x18	12.9
2	2240x2240	200	W310x52	W250x45	7.8
3	2240x2240	200	W310x52	W250x67	5.8
5	2240x2240	330	W460x68	W410x60	7.5

FIG. 6--In-Plane Racking Test Configuration

finite element parametric studies. They have suggested that the
equivalent strut width be reduced by 0.707 for pinned frames. The
results of this formulation are also provided (Table 6).

TABLE 5--Ultimate Capacity Displacement

Test	Compression		Tension	
	Load (kN)	Displacement (mm)	Load (kN)	Displacement (mm)
1	121.7	22.7	165.4	22.8
2	166.1	23.2	183.1	22.3
3	156.8	12.6	168.7	11.1
5	194.6	9.8	168.6	12.4

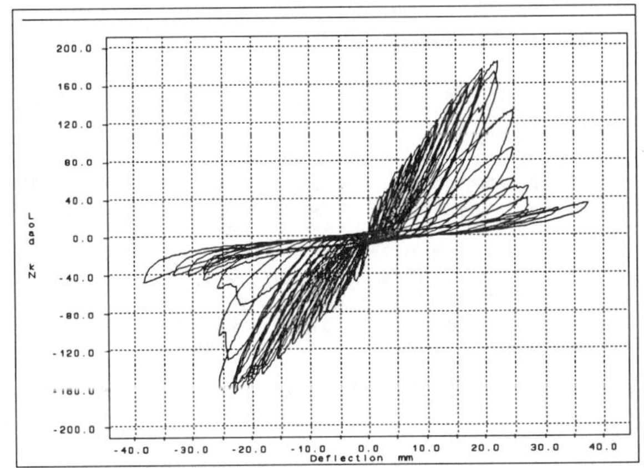

FIG. 7--Test 2 Hysteresis (Displacement at Top of Left Column)

TABLE 6--Initial Stiffness

Test	Observed Stiffness	Stafford-Smith and Carter [14] Equivalent Strut	Jamal et al. [15] Equivalent Strut
1	25 kN/mm	30 kN/mm	23 kN/mm
2	29 kN/mm	54 kN/mm	40 kN/mm
3	16 kN/mm[*]	73 kN/mm	56 kN/mm
5	45 kN/mm	89 kN/mm	66 kN/mm

[*] Inadvertent preload of specimen before data sampling.

Steel with a minimum specified yield strength of 248 MPa was used for the specimen frames. Actual mill test results indicated strengths ranging from 15 to 35 percent higher than the specified minimum. Framed beam connections consisting of clip angles were used as the beam-column connection. Racking tests of the bare frame of specimen 2 resulted in an observed load of 15 kN for a 25 mm displacement measured near the top of the column.

Hollow clay masonry units were used in the infill and type N mortar was used for construction. Average mortar compressive strengths are given (Table 4). Peak loads in tension and compression and their corresponding displacements are also presented (Table 5). These displacements were measured on the column near the top of the infill. Ultimate loads (larger of the two peaks) are underlined.

TABLE 4--<u>Average Mortar Compressive Strengths</u>

Test	Strength @ 30 Days	Strength @ Infill Test Date
1	12.5 MPa	12.4 MPa (75 days)
2	13.2 MPa	13.2 MPa (30 days)
3	10.2 MPa	10.5 MPa (66 days)
5	14.3 MPa	16.3 MPa (59 days)

Each of the three 200 mm wall panels responded in a similar manner. The behavior was characterized by mortar cracking and separation at the base and top of the panel followed by compression cracking in the mortar near the upper corners. Next, diagonal cracking throughout the mortar joints of the panels was observed and finally cracking of the clay tile as the faceshells of the upper row units split away from their webs. The 330 mm specimen exhibited a similar failure pattern. A significant separation of the wythes of the upper courses was observed as the tiles of each wythe failed independently in the latter portions of the load history. This follows logically as there was no appreciable collar joint or other special binding condition to prevent the wythes from responding separately. Figures 7 and 8 show a hysteretic curve of the response and the typical failure mode of the panels.

The load-deflection behavior of the infilled frames are characterized by fairly tight hysteretic loops of compression and tension. For specimen 2, the ultimate capacity of the infilled frame was approximately ten times the capacity of the bare frame tested at a similar displacement. Even though inelastic deformations were observed at low levels of load, the infill system behaved somewhat linearly until near ultimate capacity. The limited degrading strength and stiffness is most apparent in the load-deflection curves of the infill at or near ultimate capacity. Each infill exhibited significant strength (40-50% of ultimate) after several upper course tiles were destroyed and absent from the panel. However, the infill strength was significantly reduced during the next increasing displacement cycles. Finally, the results indicated little influence of varying frame stiffness on ultimate capacity of the combined system.

Initial stiffness of the infilled frames is presented (Table 6). The initial stiffness of each specimen softened notably after a small compression force. The reported stiffness is the secant stiffness of the first compression cycle of loading, approximately 1 mm displacement. A comparison is given of this observed stiffness to those calculated using the equivalent strut method proposed by Stafford-Smith and Carter [14]. Jamal et al. [15] have shown good correlation of the results of this equivalent strut method with both experimental results [9] and

FIG. 8--Ultimate Failure Mode of Racking Tests

Out-of-Plane Drift

     In addition to inertial loading, a wall must withstand the top-bottom relative displacements of interstory drift resulting from the transverse component of the earthquake. Little research has been performed on this condition. Of 83 references in a 1979 literature review of lateral loading on masonry infill panels [16], none dealt with top-bottom relative displacements. However, Benedetti and Benzoni [17] have indicated that interstory drift may be a more severe out-of-plane loading on typical masonry walls than inertial loading.

     As the steel framing and orthogonal shear walls deform an infill out-of-plane, the relative displacement capacity of the infill may be exceeded. Due to arching action, it is unlikely that the cracked infill will fall out of the enclosing frame. Another failure scenario is the loss of in-plane shear capacity of the infill due to out-of-plane displacements but before reaching the out-of-plane limit state.

     To investigate the effect of interstory drift, two large-scale infilled frame specimens, identical to in-plane test specimen 2, were constructed. The steel frames enclosing the infill panels were oriented with their weak axis in the plane of the wall. Each infill was bonded to the column web and to the beam lower flange of the specimen frame by snugly packing mortar between the steel and masonry. The panels were constructed with no offset between the wall and the frame centerline. No mortar was placed between the clay tile and the column flanges.

     One specimen (number 11) was cyclicly loaded out-of-plane with a hydraulic ram located near the top of the infill panel, see Figure 9. For this cantilever mode of deformation, a peak load of 57 kN was applied to the specimen with a corresponding displacement of 37 mm measured at the beam centerline. The specimen exhibited a pronounced horizontal crack along the base and hairline cracks along some of the lower course bed joints. The vertical interface along the panel and frame boundary remained intact as the wall moved with the columns. Observed out-of-plane displacements of the frame and infill (edge and midpanel) were nearly identical for the same height.

The other specimen (number 13) was cyclicly loaded out-of-plane to produce a beam type curvature, see Figure 10. The specimen was supported at the top and a hydraulic ram was placed at midheight. A peak load of 220 kN was applied to the specimen and a peak midheight displacement of 2.4 mm was measured. The specimen exhibited horizontal cracking in the bed joints near midheight and vertical hairline cracks in the head joints of the upper half of the panel. A horizontal crack along most of the panel and beam interface also developed. Again, similar out-of-plane movement of the frame and infill were observed with slightly higher displacements of the frame than the infill at midpanel.

FIG. 9--Cantilever Mode of Deformation

FIG. 10--Beam Mode of Deformation

<u>In-Plane Capacity Following Out-of-Plane Drift</u>

To measure the damage of drift loading on specimens 11 and 13, a low amplitude vibration (impact hammer) test was performed on both specimens before and after the out-of-plane loading. The use of this technique was employed since it is difficult to assess the damage to a masonry structure at low to moderate load levels. The percentage decrease in the first out-of-plane and the first in-plane frequencies of the panels due to the drift loadings are shown (Table 7).

TABLE 7--<u>Percentage Decrease in Frequency from Drift Loading</u>

Test	Out-of-Plane Frequency	In-plane Frequency
11 - Cantilever	16.8	28.7
13 - Beam	16.2	16.6

Specimens 11 and 13 were then loaded cyclicly in-plane to failure as specimen 2 had been tested. Observed failure modes were similar to those of the racking tests. Minimal degradation of in-plane strength and stiffness from the out-of-plane loading was observed (Table 8).

TABLE 8--<u>Ultimate Load and Initial Stiffness</u>

Test	Ultimate Load (kN)		Stiffness
	Compression	Tension	(kN/mm)
2 - Control	166.1	<u>183.1</u>	29
11 - Cantilever	<u>152.2</u>	149.3	27
13 - Beam	<u>186.0</u>	165.7	31

SUMMARY AND CONCLUSIONS

Testing has been performed of HCT units, mortar, assemblages, and large-scale building components to evaluate the capacity of typical Y-12 structures subjected to earthquakes. Arching action causes the panels of the infilled frame construction to have significant out-of-plane strength, thus the risk of panel failure is small. Wall damage due to out-of-plane inter-story drift effects had little influence on the in-plane stiffness and strength. Thus, traditional methods of treating orthogonal behavior of the infilled frames separately are adequate.

Depending on the characteristic frequencies of the ground motion, the presence of nonstructural unreinforced masonry infills may significantly improve the seismic behavior of structures. The HCT infills increase the stiffness, and therefore, the natural frequencies of the structures. The infills also increase the lateral strength of the otherwise unbraced frames. The experimental results showed an order of magnitude increase in lateral strength.

REFERENCES

[1]   Klingner, R.E., Beiner, R.J., and Amrhein, J.E., "Performance of Masonry Structures in the Mexican Earthquake of September 19, 1985," <u>Proceedings of the Fourth North American Masonry Conference</u>, 1987.

[2]   Bennett, R.M., and Flanagan, R.D., "Structural Behavior of Clay
       Tile Walls: Testing and Evaluation Program," Report Y/EN-4325,
       Martin Marietta Energy Systems, Inc., Oak Ridge, TN, 1992.

[3]   Hathcock, B.D., and Skillman, E., "Tests of Hollow Building
       Tiles," Technologic Papers of the Bureau of Standards No. 120,
       1918.

[4]   Engineering, Design & Geosciences Group, Inc., "Martin Marietta
       Prism Testing," Report Y/EN-4619, Martin Marietta Energy Systems,
       Inc., Oak Ridge, TN, 1989.

[5]   Johnson, F.B. and Matthys, J.H., "Structural Strength of Hollow
       Clay Tile Assemblages," Journal of the Structural Division,
       American Society of Civil Engineers, 99(ST2), 1973.

[6]   Plummer, H.C., Brick and Tile Engineering, Brick Institute of
       America, Reston, VA, 1962.

[7]   Thurlimann, B. and Guggisberg, R., "Failure Criterion for
       Laterally Loaded Masonry Walls: Experimental Investigations,"
       Proceedings of the Eighth International Brick and Block Masonry
       Conference, Dublin, Ireland, 1988.

[8]   Chua, L.S., "Out-of-Plane Behavior of Unreinforced Hollow Clay
       Tile Masonry Infilled Wall Panels," Master of Science Thesis,
       University of Tennessee, Knoxville, TN, 1991.

[9]   Seah, C.K., "Out-of-Plane Behavior of Masonry Infilled Panels,"
       Master of Science Thesis, University of New Brunswick, Canada,
       1988.

[10]   Dawe, J.L., and Seah, C.K., "Behaviour of Masonry Infilled Steel
       Frames," Canadian Journal of Civil Engineering, 16, 865-876, 1989.

[11]   Dhanasekar, M., Page, A.W., and Kleeman, P.W., "The Behavior of
       Brick Masonry under Biaxial Stress with Particular Reference to
       Infilled Frames," Proceedings of Seventh International Brick
       Masonry Conference, 814-824, 1985.

[12]   Riddington, J.R., "The Influence of Initial Gaps on Infilled Frame
       Behavior," Proceedings of the Institution of Civil Engineers, Part
       2, Vol 77, London, 295-310, 1984.

[13]   Stafford-Smith, B., "Model Tests of Vertical and Horizontal
       Loading of Infilled Frames," Journal of the American Concrete
       Institute, 65(8), 618-624, 1968.

[14]   Stafford-Smith, B., and Carter C., "A Method for the Analysis of
       Infilled Frames," Proceedings of the Institution of Civil
       Engineers, 44, 31-48, 1969.

[15]   Jamal, B.D., Bennett, R.M., and Flanagan, R.D., "Numerical
       Analysis for In-Plane Behavior of Infilled Frames," Proceedings of
       the Sixth Canadian Masonry Symposium, Saskatoon, Saskatchewan,
       Canada, 1992.

[16]   Lawrence, S.J., "Lateral Loading of Masonry Infill Panels: A
       Literature Review," Technical Record 454, Experimental Building
       Station, Sydney, Australia, 1979.

[17]   Benedetti, D. and Benzoni, G.M., "A Numerical Model for Seismic
       Analysis of Masonry Buildings: Experimental Correlations,"
       Earthquake Engineering and Structural Dynamics, 12(6), 1984.

Installation and Materials

John M. Melander[1] and John T. Conway[2]

COMPRESSIVE STRENGTHS AND BOND STRENGTHS OF PORTLAND CEMENT-LIME MORTARS

REFERENCE: Melander, J. M. and Conway, J. T., **"Compressive Strengths and Bond Strengths of Portland Cement-Lime Mortars,"** Masonry: Design and Construction, Problems and Repair, ASTM STP 1180, John M. Melander and Lynn R. Lauersdorf, Eds., American Society for Testing and Materials, Philadelphia, 1993.

ABSTRACT: ASTM Standard Specification for Mortar for Unit Masonry (ASTM C 270) requires mortar to be specified either by proportion specifications or property specifications. For portland cement-lime (PCL) mortars, it has been suggested that a significant difference exists between the properties of mortars mixed according to the proportion specifications and the requirements of the property specifications.

This paper presents data confirming that compressive strengths of PCL mortars according to current proportion limits do vary significantly from present property specification requirements. Alternative proportions that would bring the performance of cement-lime mortars more in line with the property specification requirements can be formulated based on the data.

The paper also investigates the relationship of portland cement content to flexural bond strength development of cement-lime mortars as obtained from tests conforming to UBC Standard 24-30. A mathematical model of the relationship is proposed and compared to additional experimental data.

KEYWORDS: mortar, portland cement, lime, hydrated lime, compressive strength, flexural bond strength

In the United States and Canada, mortar is specified by one of two alternate means - either by the proportion specifications or the property specifications. Under the proportion specifications of ASTM

---

[1]Masonry Specialist, Engineered Structures & Codes, Portland Cement Association, Skokie, IL 60077.

[2]Manager Quality Assurance, Research & Development, Holnam Inc., Dundee, MI 48131

Standard Specification for Mortar for Unit Masonry (C 270), mortar ingredients meeting the requirements of the materials section of ASTM C 270 are mixed according to the proportions indicated in Table 1 of that standard. Under the property requirements, a laboratory mixed and tested mortar prepared from the ingredients to be used in construction, at the proportions to be used on the job, must meet the property requirements of Table 2 of ASTM C 270.

While it has long been recognized that properties of mortars specified under the proportion specifications differ from those specified under the property specifications, an effort has recently been made to quantify those differences for portland cement-lime (PCL) mortars. Since compressive strength is the principal property of mortar used in ASTM C 270 to differentiate between the various types of mortar, this property has been the focus for comparison of the proportion and property specifications. The concern regarding this difference is limited to portland-lime mortars since masonry cement mortars have traditionally been qualified under the property specifications. ASTM C 270 property specifications requirements were the basis for the development of the physical requirements for masonry cements contained in the ASTM Standard for Masonry Cement (C 91).

Although compressive strength is used to differentiate between mortar types in ASTM C 270, the advent of engineered masonry and the design of some masonry using design procedures that consider the tensile resistance of the masonry poses an additional question concerning mortar properties. What difference in bond strength can be expected between mortars under the proportion specifications versus the property specifications? Investigation into the relationship of the proportions of cement-lime mortars to the relative compressive strength and bond strength of those mortars is thus the subject of this paper.

Data on these parameters were obtained from the following three sources:

1. ASTM Subcommittee C12.02 on Research (a subcommittee of ASTM Committee C12 on mortar) formed task group C12.02.10 to compile data on the compressive strengths of PCL mortars at various mix designs in the range of ASTM C 270 proportion specifications. These previously unpublished data are contained in Appendix A.

2. The Masonry Industry Code Committee (MICC) coordinated a test program to investigate the flexural bond strength of portland cement lime mortars yielding data published in The Masonry Society Journal in 1991 [1].

3. Additional flexural bond strengths of selected samples from the MICC test program were performed by Construction Technologies Laboratories CTL) under the sponsorship of the Portland Cement Association (PCA). These previously unpublished data are contained in Appendix B.

EXPERIMENTAL METHODS

<u>Compressive Strength Test Procedure</u>

Laboratories reporting data on the compressive strength of PCL mortars to task group C12.02.10 conducted the tests in accordance with the procedures of ASTM C 270 and ASTM C 91. A 50/50 blend of graded sand and 20 - 30 sand, each meeting the requirements of ASTM Standard Specification for Standard Sand (C 778), was used. Cement, lime, and dry sand were proportioned by weight corresponding to volume proportions in the range specified under the proportion specifications of C 270 using 80 lb (36.3 kg) of dry sand as equivalent to 1 cubic foot (0.028 cubic meters) of damp loose sand. The volumetric ratio of sand to cement plus lime was 3 to 1. Test mortars were mixed to a flow of between 110 ± 5 percent in a Hobart mixer.

In addition to the data on compressive strength of mortars using standard sand, ASTM C 270 test results of Type S and M PCL mortars using four different masonry sands were also reported to task group C12.02.10.

Laboratories participating in the MICC test program also reported compressive strengths for the mortars used in constructing the prism specimens for flexural strength testing. These tests were conducted using the same procedures noted above, with ASTM C 778 sand, except that mortar was mixed to a flow of 125 ± 5 in batch sizes of 1/2 cubic foot (0.014 cubic meters) and a dry sand unit weight of 100 lb/ft^3 (1602 kg/m^3) was used in proportioning mortar for this test series.

Flexural Bond Strength Tests

Tests of flexural bond strengths of concrete masonry prisms made using PCL mortars were conducted in research laboratories of the Construction Technology Laboratories, Inc. (CTL), the National Concrete Masonry Association (NCMA), and the University of Texas at Austin (UTA). Flexural bond strength tests were in accordance with the requirements of Uniform Building Code Standard No. 24-30 [2]. This procedure requires the use of standard concrete brick, controlled fabrication techniques, and controlled curing conditions in the preparation of specimens. Six prisms, each containing five joints, are constructed for each test series. The flexural bond strength is measured on the thirty joints using ASTM Standard Method for Measurement of Masonry Flexural Bond Strength (C 1072). The average of those thirty measurements is analyzed as a single bond test result in this paper. For further information on this test program, see reference [1].

The additional tests of flexural bond strength at higher lime contents performed at CTL utilized the same test method described above. The samples of cement and lime were retained portions of samples previously tested in the MICC test program.

Statistical Methodology

The methods used to analyze data in this paper are summarized in Appendix C [3], [4], and [5].

RESULTS

Compressive Strengths of Mortars

Seven laboratories supplied data on the compressive strength of PCL mortars made with three different brands of hydrated lime and six different brands of portland cement (Appendix A). All of the portland cement samples met ASTM Standard Specification for Portland Cement (C 150), Type I cement, except one that was an ASTM C 150 Type IA cement. The hydrated limes all met ASTM Standard Specification for Hydrated Lime (C 207), Type S lime. One lab added an air entraining agent to the

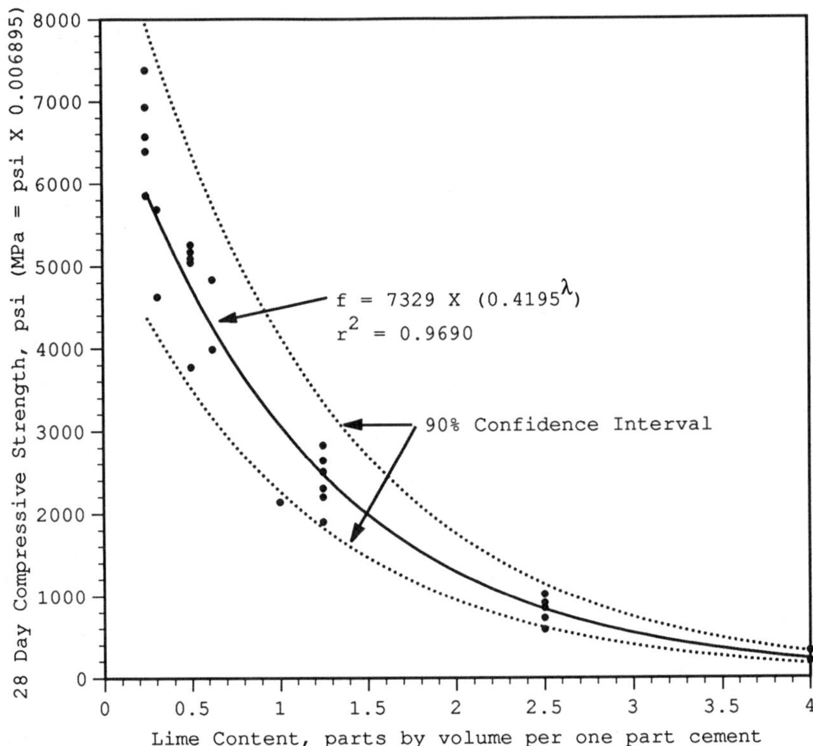

FIG. 1--Relationship of compressive strength to lime content
of PCL mortars using standard sand and mixed to an initial
flow of 110 ± 5 percent.

mortar mix of portland cement and lime to produce air contents ranging
from 9.1% to 12.0%. These air entrained cement-lime mortars are
included in the analysis of the compressive strength results since air
entrained cements and air entrained limes may be used under the
proportion specifications.

A plot of the compressive strength data as a function of the lime
content suggests a mathematical model for that relationship (Fig. 1).
Linear regression analysis indicates that the relationship is best
described by the following exponential function:

$$f = 7329 \; X \; (0.4195^{\lambda}), \tag{1}$$

where

λ = volume ratio of hydrated lime to cement, and

f = compressive strength of mortar.

The correlation coefficient (r^2) of 0.969 for the 29 data points
indicates that the model is significant and provides a good fit to the
observed data. Statistical analysis of variance confirms that the

regression is significant and that the lack of fit of the data to the
model is not significant.

Appendix C, Table C-1, contains the regression analysis statistics
for the data. These statistics can be used to calculate a confidence
interval for f(x) (compressive strength) values at given x (lime
content) values (see Appendix C for a more detailed discussion). A 90%
confidence interval has been calculated and plotted along with the model
regression equation for these data (Fig. 1). The lower confidence
interval represents a 95% confidence level for a one sided limit.

Similar analysis of compressive strength data from the MICC test
program indicates that an exponential equation is a good model for the
relationship of compressive strength to lime content. As one would
expect, the individual equation constants determined under these
alternate test conditions differ from those obtained using the C12.02.10
data. The mathematical equation best describing this data is:

$$f = 5825 \; X \; (0.2236^{\lambda}). \tag{2}$$

Flexural Bond Strengths

The MICC data provides information on the relationship of flexural
bond strength of masonry using PCL mortars to the proportions of

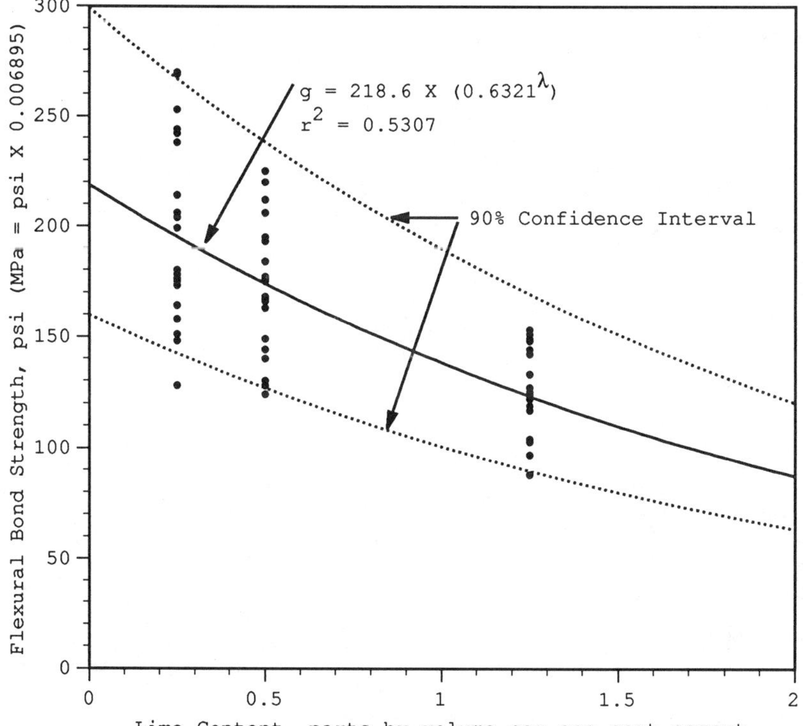

FIG. 2--Relationship of flexural bond strength to lime content
of mortars for MICC PCL data.

portland cement and hydrated lime in those mortars. However, it should be noted that the test conditions defined by UBC 24-30 are intended to eliminate as many variables as possible in determination of flexural bond strength. Therefore such variables as workmanship, curing, and unit properties are held constant in so far as possible. In actual practice these parameters may affect the flexural bond strengths as much or more than mortar material properties, including relative cement-lime proportions.

A plot of the bond strength data indicates that a relationship exists between the bond strength and lime content of PCL mortars (see Fig. 2). Once again, the best mathematical model for this relationship is an exponential equation as follows:

$$g = 218.6 \text{ X } (0.6321^{\lambda}), \qquad (3)$$

where

      g = flexural bond strength of mortar.

Although the coefficient of correlation is not very high ($r^2 = 0.5307$), analysis of variance indicates that the regression is significant and the exponential model is adequate.

There are two reasons that the coefficient of correlation is not

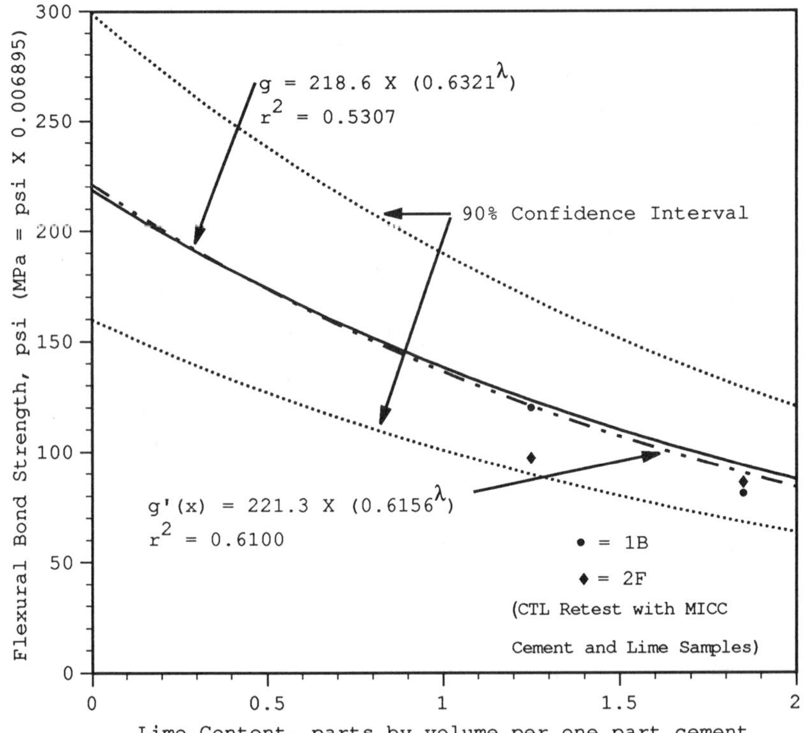

FIG. 3--Relationship of flexural bond strength to lime content of mortars for MICC PCL data and CTL data.

very high. First, the data are limited to three specific lime content values. Second, the scatter of test results at each value is significant. Analysis of variance indicates that most of the variability of the data from the model is "pure error" and not a lack of fit to the model.

As previously noted, while the MICC data include a fairly large number of data points, they are limited to three specific lime content values ranging from 0.25 to 1.25 parts by volume. Additional tests are needed to confirm that the model is meaningful at higher lime contents. To accomplish this, tests were run at CTL on two retained samples from the MICC study at 1.25 and 1.84 lime content levels by volume. The tests run at 1.25 lime content were to evaluate the condition of the materials while the 1.84 lime content was selected for comparison to the 1.875 lime content that had been proposed as a lower limit for Type N PCL mortar.

The results for flexural bond strength of the retest samples at 1.25 lime content were lower than those previously determined by CTL on those samples. However, the compressive strength values were in the range of previously determined values for one sample and higher for the other. This indicates that the samples had not physically deteriorated in storage and that variations noted are a result of testing variability.

Results of these tests are well within a 90% confidence interval of the model (Fig. 3). Regression analysis of the MICC data and the additional CTL data does not significantly affect the equation constants of the model but does increase the coefficient of correlation (r^2) from 0.5307 to 0.6100. We can conclude that the model is also meaningful at these higher lime content levels.

DISCUSSION

The compressive strength values for cement lime mortars mixed to the proportion specification limits are much higher than the minimum 28 day compressive strength requirements of ASTM C 270 Table 2. However, ASTM C 270 requirements are based on tests using masonry sand (ASTM C 144) rather than standard testing sand (ASTM C 778) The values obtained from tests using ASTM C 778 sand are expected to be somewhat higher due to the effect of the sand alone. Thus it is more proper to compare these results to the minimums required by ASTM C 91 for masonry cements. The ASTM C 91 twenty-eight day compressive strength minimums for Type N, S, and M masonry cements are 900, 2100, and 2900 psi (6.2, 14.5, and 20.0 MPa) respectively. A strength survey conducted by ASTM Subcommittee C01.27 on Strengths [6] also provides some additional information on what average strengths are required to meet ASTM C 91 and C 270 property specification minimums. The average 28 day compressive strength values for Type N, S, and M masonry cements reported in that survey were 1612, 2997, and 3451 psi (11.11, 20.66, and 23.79 MPa) respectively.

Inclusion of the ASTM C 91 values on the plot of compressive strength versus lime content (Fig. 4) provides an indication of what levels of lime content would more closely result in equalization of the proportion and property specification strength levels for PCL mortars. Comparing the lower 90% confidence interval values to the ASTM C 91 requirements suggests that Type M could be proportioned from 0.25 to 0.7 parts lime, Type S could be set in the range of 0.7 to 1.0, and Type N at 1.0 to about 1.9 parts hydrated lime by volume per part of portland cement.

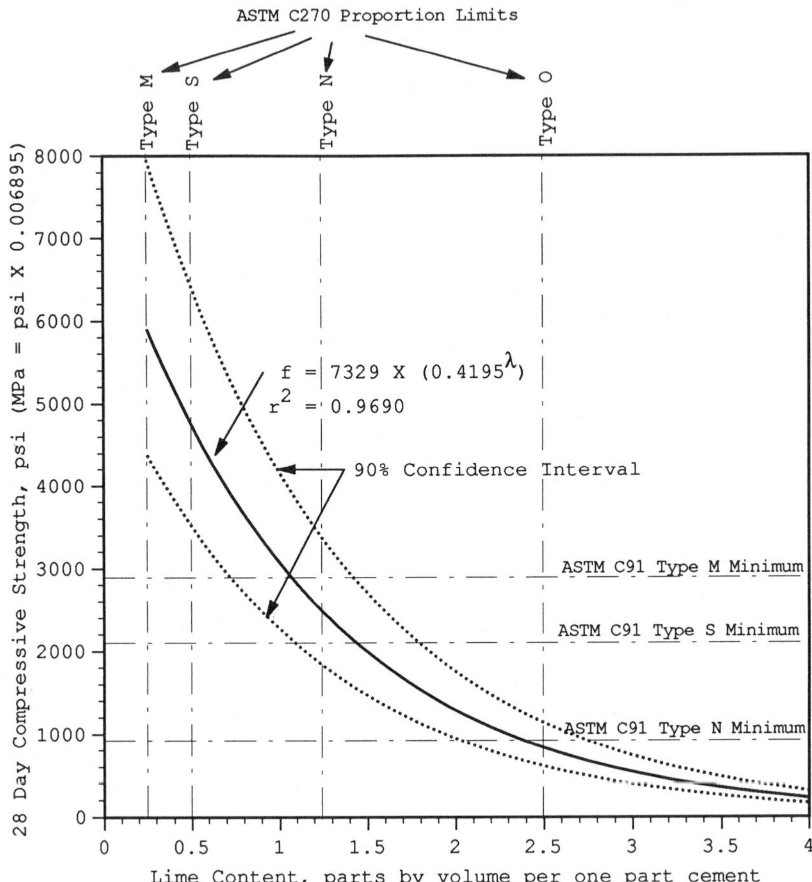

FIG. 4--Comparison of compressive strengths of PCL mortars to
ASTM C 91 requirements and current ASTM C 270 proportion
specification limits.

When discussing proportions of cement-lime mortars, the realities
of construction batching of materials and the dichotomy between ASTM C
270 specified unit weight for hydrated lime of 40 lb/ft^3 (640 kg/m^3)
compared to industry practice of packaging hydrated lime in 50 lb (22.7
kg) bags must be considered. Lime content ranges of 0.3125 to 0.625 for
Type M, 0.625 to 1.25 for Type S, and 1.25 to 1.875 for Type N mortars
would be more compatible with current bag weight convention and
construction batching procedures. Unfortunately, although 1.25 parts
lime corresponds to one bag of lime per bag of portland cement, it does
not coincide with the 1 part by volume limit suggested by the data. If
hydrated lime were packaged in 40 lb (18.1 kg) bags, the lime content
ranges of 0.25 to 0.5 for Type M, over 0.5 to 1 for Type S, over 1 to
1.5 for Type N and over 1.5 to 2 for Type O would comfortably exceed the
property specification requirements and be simple to proportion at the

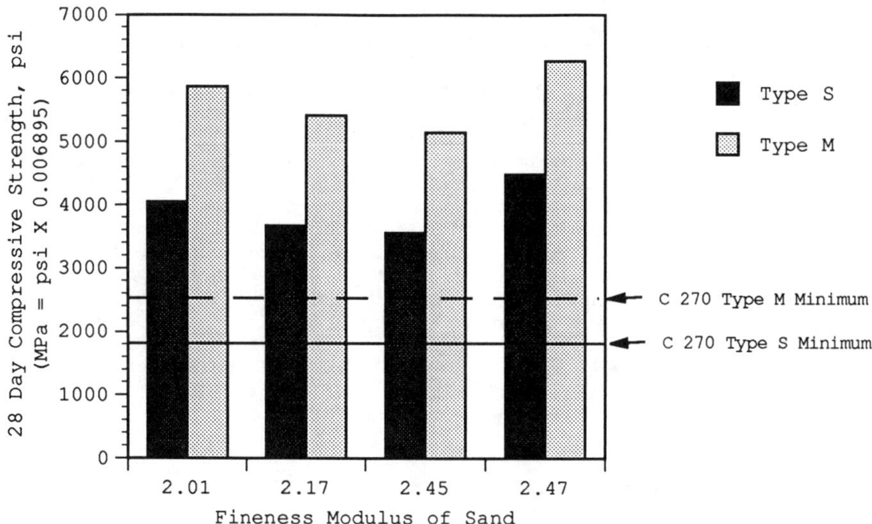

FIG. 5--ASTM C 270 test results.

construction site. Predicted compressive strength values and 90%
confidence interval limits at these proportions are listed in Table 1.

TABLE 1--<u>Predicted compressive strengths of PCL mortars
using standard sand.</u>

Lime Content[A]	Suggested Mortar Type	Compressive Strength, psi[B]		
		Average	Lower Limit	Upper Limit
0.25		5900	4360	7990
0.5	M	4750	3510	6420
1.0	S	3080	2280	4150
1.5	N	2000	1480	2690
2.0	O	1290	954	1744

[A]Lime content is expressed in parts per volume per one part
 portland cement.
[B]MPa = psi X 0.006895

The tests reported to task group C12.02.10 on ASTM C 270 tests of
Type S and M PCL mortars using four different masonry sands confirms
that compressive strengths for these mortars are much higher than needed
(see Fig. 5). The compressive strength results are over twice the
minimum values required by ASTM C 270 specifications. Such strengths
are not only not required, they may be achieved at the expense of other

desirable properties of the plastic and the hardened mortar, and thus be detrimental to the performance of the mortar during construction and in service.

The slope of the mathematical model for flexural bond strength over the range of lime contents from 0.25 to 2.0 is fairly flat (see Fig. 3). A change from 1.25 parts lime content to 1.84 at the 90% lower confidence limit corresponds to a change in flexural bond strength from about 90 psi (0.62 MPa) to about 70 psi (0.48 MPa) or approximately a 20 psi (0.14 MPa) reduction. Table 2 contains the expected average values as well as upper and lower 90% confidence interval limits for several selected levels of lime content. Clearly, specified changes in lime content from 0.25 to 0.3125 and from 0.5 to 0.625 are not significant to the level of flexural bond strength of PCL mortars. The range of the 90% confidence interval is rather broad, spanning 125 psi (0.86 MPa) at the 0.25 lime content level and approximately 80 psi (0.55 MPa) at the 1.25 lime content level. The high variability of flexural bond test results further diminishes the significance of minor changes in proportions to the bond strength performance of mortars.

TABLE 2--<u>Predicted flexural bond strengths for PCL mortars.</u>

Lime content[A]	Flexural Bond Stengths, psi[B]			
	Average	Lower Limit	Upper limit	Range
0.25	195	142	267	125
0.3125	190	138	259	121
0.5	174	127	238	111
0.625	164	120	225	105
1.0	138	101	190	89
1.25	123	90	169	79
1.5	110	80	151	71
1.84	94	68	129	61

[A]Lime content is expressed in parts per volume per one part portland cement.
[B]MPa = psi X 0.006895.

CONCLUSIONS

The following conclusions are made regarding the compressive strengths and bond strengths of PCL mortars:

1. Current proportion specification limits for PCL mortars have much higher cement to lime ratios than required, particularly at the Type M and Type S levels. Revisions to the proportion specification cement and lime content limits for Type M, S, and N mortars should be considered.

2. The laboratory compressive strength of PCL mortars as a function of its lime content follows an exponential relationship, decreasing with increasing lime content (see equations 1 and 2).

3. The flexural bond strengths of concrete masonry prisms made using PCL mortars also exhibits an exponential relationship to the lime content of mortar (see equation 3).

4. The scatter of flexural bond strength results of concrete masonry prisms made using PCL mortars is very broad even under carefully controlled test conditions.

5. A reasonable adjustment to the ASTM C 270 proportion limits would be to have lime contents for Type M at 1/4 to 1/2, Type S over 1/2 to 1, Type N over 1 to 1 1/2 and Type O over 1 1/2 to 2. These ranges would also be convenient for job site proportioning if hydrated lime were packaged in 40 lb (18.1 kg) bags.

ACKNOWLEDGEMENTS

The authors wish to thank Mr. Richard O. Hedstrom, chairman of task group C12.02.10, for initiating the investigation into the compressive strength of PCL mortars and compiling the data reported in Appendix A.

APPENDIX A: TASK GROUP C 12.02.10 COMPRESSIVE STRENGTH DATA

TABLE A1--<u>Test results for PCL mortars.</u>

Lab No.	Mortar Type	Notes	Lime Content[A]	Air Content, %	Compressive Strength, psi[B] 7 Day	28 Day
1	M		0.25	6.3	4660	6390
2	M	AE	0.25	9.1	5130	5850
3	M		0.25		5450	6931
4	M		0.25		5530	6570
5	M		0.25	5.5	6098	7376
6	(M)[C]		0.3125	5.4	4930	5685
6	(M)	IA	0.3125	11.6	4050	4625
1	S		0.5	5.2	3550	5040
2	S	AE	0.5	10.8	3110	3770
3	S		0.5		3963	5088
4	S		0.5		4440	5170
5	S		0.5	5.5	4396	5257
6	(S)		0.625	3.1	4600	4835
6	(S)	IA	0.625	8.3	3700	3985
2	N	AE	1	12.0	1810	2140
3	N		1.25		1641	2304
7	N		1.25		1620	1900
4	N		1.25		2380	2830
5	N		1.25	5.6	2001	2505
5	N		1.25		2503	2642
6	N		1.25	2.9	2015	2500
6	N	IA	1.25	6.3	1775	2200
3	O		2.5		533	731
7	O		2.5		520	590
4	O		2.5		790	1020
5	O		2.5	5.2	635	850
5	O		2.5		859	912
7	K		4		180	200
5	K		4		313	325

[A]Lime Content is expressed in parts per volume per one part portland cement.
[B]MPa = psi X 0.006895.
[C]Mortar types in parenthesis indicates proportions were based on unit weight of hydrated lime of 50 lb/ft^3 (80 kg/m^3) rather than 40 lb/ft^3 (64 kg/m^3).

TABLE A2--<u>Average Strength for each mortar Type.</u>

Mortar Type	Average Compressive Strengths, psi[A] Exclude AE & IA 7 Day	28 Day	N	All Data 7 Day	28 Day	N
K	247	263	2	247	263	2
O	667	821	5	667	821	5
N	2027	2447	6	1968	2378	8
S	4190	5078	5	3966	4735	7
M	5334	6590	5	5121	6204	7

[A]MPa = psi X 0.006895

TABLE A3--<u>ASTM C 270 test results for PCL mortars.</u>

Mortar Type	Sand No.	Fineness Modulus	Air Content, %	Water Retention, %	28 Day Compressive Strength, psi[A]
S	1	2.01	3.1	94.3	4050
S	2	2.17	4.6	92.7	3670
S	3	2.45	4.6	92.0	3560
S	4	2.47	3.5	92.9	4490
M	1	2.01	4.6	95.4	5870
M	2	2.17	4.5	94.7	5420
M	3	2.45	4.1	88.0	5150
M	4	2.47	5.5	93.8	6270

[A]MPa = psi X 0.006895

APPENDIX B: CTL PCL DATA

TABLE B1--<u>Mortar test results.</u>

Sample ID	Lime Content[A]	Initial Flow, %	Air Content, %	Water Retention, %	Compressive Strength, psi[B] 7 Day	28 Day
1BN	1.25	130	2.8	93.1	680	970
1BNN	1.84	130	2.1	94.2	310	630
2FN	1.25	130	2.5	89.2	610	940
2FNN	1.84	130	3.2	90.8	360	580

[A]Lime content is expressed in parts per volume per one part portland cement.
[B]MPa = psi X 0.006895

TABLE B2--<u>Flexural bond strength results, psi</u>[A].

SAMPLE ID = 1BN		SAMPLE ID = 1BNN		SAMPLE ID = 2FN		SAMPLE ID = 2FNN	
Joint No.	Bond Strength	Joint No.	Bond Strength	Joint No.	Bond Strength	Joint No.	Bond Strength
1.1	135	1.1	92	1.1	102	1.1	103
1.2	108	1.2	61	1.2	113	1.2	83
1.3	124	1.3	72	1.3	121	1.3	81
1.4	149	1.4	86	1.4	102	1.4	90
1.5	116	1.5	86	1.5	129	1.5	114
2.1	117	2.1	96	2.1	106	2.1	103
2.2	110	2.2	78	2.2	77	2.2	57
2.3	131	2.3	70	2.3	90	2.3	84
2.4	139	2.4	103	2.4	87	2.4	68
2.5	135	2.5	95	2.5	118	2.5	101
3.1	135	3.1	82	3.1	90	3.1	87
3.2	135	3.2	72	3.2	96	3.2	54
3.3	108	3.3	75	3.3	104	3.3	57
3.4	108	3.4	81	3.4	78	3.4	96
3.5	109	3.5	79	3.5	100	3.5	80
4.1	139	4.1	91	4.1	100	4.1	105
4.2	126	4.2	77	4.2	92	4.2	72
4.3	136	4.3	80	4.3	88	4.3	100
4.4	144	4.4	88	4.4	93	4.4	83
4.5	130	4.5	99	4.5	75	4.5	88
5.1	109	5.1	79	5.1	113	5.1	70
5.2	104	5.2	71	5.2	105	5.2	82
5.3	90	5.3	91	5.3	74	5.3	102
5.4	104	5.4	73	5.4	89	5.4	100
5.5	115	5.5	69	5.5	84	5.5	75
6.1	85	6.1	68	6.1	77	6.1	98
6.2	123	6.2	72	6.2	83	6.2	94
6.3	131	6.3	89	6.3	111	6.3	81
6.4	73	6.4	83	6.4	104	6.4	92
6.5	141	6.5	84	6.5	108	6.5	81
Mean	120		81		97		86
S.D.	18		10		15		15
COV %	15		13		15		18

[A]MPa = psi X 0.006895.

APPENDIX C: STATISTICAL METHODOLOGY

Each set of data was plotted on a scatter diagram to determine
what mathematical models might be appropriate. Regression analysis was
then performed on the data using linear, power, exponential and
logarithmic models. In each instance, the exponential model best
approximated the data, yielding the highest value for the coefficient of
correlation (r^2).

For the exponential equation of the form,

$$y = AB^x, \tag{C1}$$

the actual regression is based on the model,

$$\ln(y) = x\ln(B) + \ln(A). \tag{C2}$$

TABLE C1--<u>Regression analysis results.</u>

Statistic	Data Set		
	C12.02.10 Compressive Strength	MICC Compressive Strength	MICC Mean Bond Strength
N, number of data points	29	66	66
B, constant - slope	0.4195	0.2236	0.6320
A, constant - y intercept	7329	5825	218.6
$\ln(B)$	-0.8686	-1.4981	-0.4587
$\ln(A)$	8.8996	8.6699	5.3872
$Se_{\ln(B)}$, standard error of $\ln(B)$	0.02988	0.04046	0.05393
$Se_{\ln(A)}$, standard error of $\ln(A)$	0.04892	0.03199	0.04263
$Se_{\ln(y)}$, standard error of $\ln(y)$ estimate	0.1726	0.1397	0.1862
r^2, coefficient of correlation	0.9690	0.9554	0.5307
SS_{reg}, regression Sum of Squares	25.159	26.745	2.5078
SS_{res}, residual Sum of Squares	0.8039	1.2485	2.2178
df_{reg}, is the degrees of freedom of SS_{reg}	1	1	1
df_{res}, is the degrees of freedom of SS_{res}	27	64	64
F_1, F statistic used to test significance of the regression	845.02	1371.0	72.37
F_{tab}, tabular value of F statistic compared to F_1	7.68	7.06	7.06
α_1, level significance for F_1	0.01	0.01	0.01
F_2, calculated F statistic used to test the lack of fit of the regression,	1.90	2.18	0.19
df_{lf}, degrees of freedom of F_2 (1st parameter)	5	1	1
df_{pe}, degrees of freedom of F_2 (2nd parameter)	22	63	63
F'_{tab}, tabular value of F statistic compared to F2, and	2.66	4.00	4.00
α_2, level significance for F_2.	0.05	0.05	0.05

The regression statistics (Table C1) used to test the exponential model are therefore actually compared to the linear equation C2.

The confidence interval on the expected value of y at a given x' is calculated as,

$$\ln(y) = x'\ln(B)+\ln(A) \pm tSe_{\ln(y)}\sqrt{1+\frac{1}{N}+[x'-\bar{x}]^2\frac{[Se_{\ln(B)}]^2}{[Se_{\ln(y)}]^2}} \tag{C3}$$

for each x' value over the range of observations. The inverse natural logarithm of these values yields the confidence interval of y that was plotted along with the model exponential equation for each set of data.

The value of t is selected from a table of the t-distribution at the desired level of probability (two sided) and N-2 degrees of freedom. The t values at the 90% level of probability for 27 and 64 degrees of freedom are 1.703 and 1.67. This corresponds to a 95% level of probability for a one sided limit.

REFERENCES

[1]   Hedstrom, E.G., Tarhini, K.M., Thomas, R.D., Dubovoy, V.S., Klingner, R.E., Cook, R.A., "Flexural Bond Strength of Concrete Masonry Prism Using Portland Cement and Hydrated Lime Mortars," _The Masonry Society Journal_, Vol 9, No. 2, February, 1991, pp. 8-23.

[2]   _UBC Standard 24-30 (Standard Test Method for Flexural Bond Strength of Mortar Cement)_, International Conference of Building Officials, Whittier, CA, 1991.

[3]   Benjamin, J.R., and Cornell, C.A., _Probability Statistics, and Decision for Civil Engineers_, McGraw-Hill Book Company, New York, NY, 1970.

[4]   Ramberg, J.S, "Regression Analysis", _Quality Control Handbook - Third Edition_, J. M. Juran, Editor-in-Chief, McGraw-Hill Book Company, New York, NY, 1979, pp.26-1 to 26-29.

[5]   Juran, J. M., "Appendix II - Tables and Charts", _Quality Control Handbook - Third Edition_, J. M. Juran, Editor-in-Chief, McGraw-Hill Book Company, New York, NY, 1979, pp.1-48.

[6]   "Unpublished Compressive Strength Survey of Cement Produced in the U.S. Conducted by C01.27", American Society for Testing and Materials, Philadelphia, PA, 1991.

Andrew T. Krauklis[1]

A STUDY OF THE COMPATIBILITY OF BRICK AND MORTAR FOR MAXIMIZING MASONRY BOND STRENGTH

REFERENCE: Krauklis, A. T., "A Study of the Compatibility of Brick and Mortar for Maximizing Masonry Bond Strength", <u>Masonry: Design and Construction, Problems & Repair, STP 1180</u>, John M. Melander, and Lynn R. Lauersdorf, Eds., American Society for Testing and Materials, Philadelphia, 1993.

ABSTRACT: With the increasing prevalence of steel-stud brick-veneer curtain wall construction, the requirement for reliable bond strength has become critical for sound masonry performance. This paper presents a study, including test data and the result interpretation, used in the selection of a compatible brick and mortar mix for a particular project and a comparison with current design practice. Eight (8) brick types and 24 mortar mixes were examined. Of the 24 mortar mixes, 20 were Type "S." The study was conducted in three parts. The first part commenced in May of 1987, and the third part continues today, with prisms over 1,600 days old. The results of 110 flexural tests conducted in accordance with ASTM Test Method for Flexural Bond Strength of Masonry (E 518), and over 1,700 bond-wrench tests conducted in accordance with ASTM Method for Measurement of Masonry Flexural Bond Strength (C 1072) are presented. This study reinforces the need for research and compatibility testing in the selection of the masonry components to be used on a project where reliable bond strength is required.

KEYWORDS: masonry, design, specifications, construction, walls, cladding, bond, mortar, cement

---

In 1985, this author was given the opportunity to study and analyze a failed steel stud brick veneer (SS/BV) wall system as part of a repair and restoration program. The wall system of this 15-story residential building was designed and constructed per the relatively new guidelines of the metal stud industry in 1975. This study allowed for the assessment of the structural mechanism and the advantages and disadvantages of the system.

---

[1]Consultant/Associate, Raths, Raths & Johnson, Inc., 835 Midway Drive, Willowbrook, IL 60521.

The structural mechanism of the brick and backup wall system has three parts that affect the distribution of load. These include the brick masonry, the backup wall, and the ties between them. The stiffness of each part affects the distribution of the stresses between them.

The principal advantage of the SS/BV is cost. The SS/BV is a cost-effective system for quickly enclosing the envelope of a building. Enabling the interior trades to start earlier has the potential for significant savings by compressing the construction schedule. However, there are some downsides to the SS/BV system. Replacing the traditional concrete block backup with a more flexible system, which is susceptible to water damage and corrosion, is a disadvantage that must be addressed. This can be overcome with a complete and thorough waterproofing system. In order to ensure the performance of the waterproofing, the contractor must include a comprehensive quality control program. In addition, the owner needs to institute a quality assurance program to verify the contractor QC program.

The major difference between the concrete block backup and the SS/BV system is the change in the structural system. The traditional stiff block backup is the structural support member of the wall, and the brick wythe serves as an architectural feature and the initial barrier against water infiltration, with little strength requirements. The stiff block backup results in relatively low flexural stresses in the brick wythe seldom approaching the code allowables.

The flexible steel backup in the SS/BV system results in higher brick stress often approaching or exceeding the code allowables. The guidelines used in the design of a SS/BV system are based on a combination of stud height and deflection. Although these requirements have gotten more stringent since 1975, they do not include the stiffness of the masonry veneer, and the effects of the various available wall ties. Both will affect the flexural stress levels of the masonry.

Based on the assessment of the structural mechanism of the SS/BV wall system, a rational design approach to the analysis is required. The brick masonry in the SS/BV wall is a structural flexural member that requires similar consideration as other structural members. One (1) part of a rational approach to the design of a SS/BV wall system is the determination of bond strength capabilities of the selected materials and their reliability. This paper presents observations made during a compatibility study for bond strength.

ACCEPTANCE CRITERIA

Unpublished flexural bond strength tests, conducted by SCPI and presented in the *Recommended Practice for Engineered Brick Masonry* [1], indicate an average Modulus of Rupture (MOR) in the range of 144 psi (993 kPa) for a Type "S" mortar. This range of MOR has a factor of safety on the average test value of 4 for the 36 psi (248 kPa) allowable for flexural tension stresses normal to the bed joints. Designers often increase this allowable to 48 psi (331 kPa) for wind loadings, thereby decreasing the factor of safety to 3.

When considering a member as a structural element, the reliability of the member strength is critical. When determining the member strength by testing, the test results must be examined for both the average capacities and the variation from the average. Widely scattered test results must account for this variation in defining the member's strength. Standard deviation of the data is a measurement of the actual amount of variation present in the set of test data.

Selection of a performance level is a function of the consequences of failure. When the failure of a single member is catastrophic, the confidence level must be higher than if the structural system contains multiple members, providing redundancy.   The level of confidence allows for a variation in the acceptance levels based on test results. The selection of the confidence level for masonry walls is a function of the length of the wall. The strength of a long wall approaches the average test strength of a series of the single brick prism. The strength of short walls approaches the minimum test values.

The acceptance criteria for bond strength on this project was based on a normal distribution. A high performance level was selected because a large number of building details contained only two brick bonds. The building wall system repair design was predicated on the allowable flexural bond strength of 36 psi (248 kPa) increased to 48 psi (331 kPa) for wind loading.

A 96 psi (622 kPa) lower strength limit, for the 99th percentile of the test sample, was selected for the project based on a factor of safety of 2 above the required design strength of 48 psi (331 kPa). Using a normal distribution, the 99th percentile lower strength limit of the test sample is defined as:

$$F_b = \overline{X} - 2.33\,S \tag{1}$$

Where:

$$F_b \quad = \text{the minimum test value}$$

$$\overline{X} \quad = \text{the test mean}$$
$$S \quad = \text{the standard deviation}$$
$$2.33 \quad = \text{standard normal distribution value for the 99th percentile}$$

The above statistical treatment does not define the lower tolerance limit of the population at a 99th percentile with a 99% confidence, but only defines the 99th percentile of the test sample. Referring to "Experimental Statistics" [2], the lower limit is defined as:

$$X_\ell = \overline{X} - K\,s \tag{2}$$

Where:

$$X_\ell \quad = \quad \text{the lower tolerance limit}$$
$$\overline{X} \quad = \quad \text{the test mean}$$
$$s \quad = \quad \text{the test standard deviation}$$
$$K \quad = \quad \text{table value [2] as a function of sample size (n), confidence level } (\gamma), \text{ and percentile (P).}$$

In addition, in order to establish the lower limit mean value of the population based on the sample test results, the following equation is utilized:

$$X_\ell = X - \frac{ts}{\sqrt{n}} \tag{3}$$

Where:

X_ℓ = the minimum mean value of the population

X = the test mean value

s = the test standard deviation

n = the sample size

t = the table value [2] as a function of confidence level (t), and degrees of freedom (n-1)

Equations (2) and (3) are affected by the sample size. The larger the sample size, the less influence the statistical mathematics have on defining the lower limit values of the population. Future studies should consider larger sample sizes, similar to Pilot Test #1.

TEST PROGRAM

The purpose of the test program was to pre-qualify the masonry for use in the replacement of the exterior building walls of a high-rise condominium project. The test objectives included maximizing the Modulus of Rupture (MOR) of the brick masonry to an average strength of not less than 144 psi (993 kPa), and a strength limit of the 99th percentile of the test sample greater than or equal to 96 psi (622 kPa).

Three (3) pilot test programs were conducted. The results of the first tests were used in defining the second test, and the results of the second test were used in defining the third test. In all cases, the mortars tested were comprised of Portland Cement, hydrated lime, and sand. Masonry cement was not considered.

Modulus of Rupture Tests

The Modulus of Rupture (MOR) was determined by two test methods. Prisms were tested using ASTM Standard Test Method for Flexural Bond Strength (E 518). Those portions of the prisms that survived the ASTM E 518 flexural tests were then tested using the ASTM Standard Method for Measurement of Masonry Bond Strength (C 1072).

For the flexural tests (ASTM E 518), the prism samples were placed on the reaction frame shown in Figure 1. Thin bearing pads and triangular steel blocks support the prism. The load is applied by a hydraulic ram through a load cell centered on a spreader beam supported at third points on the prism, creating a uniform moment between the two load points. The hydraulic pressure is uniformly applied by a hand

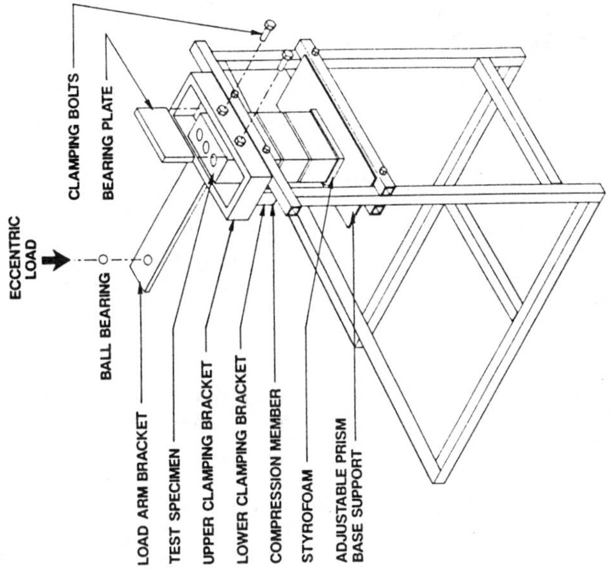

FIG. 2 -- Bond Wrench set-up ASTM C 1072.

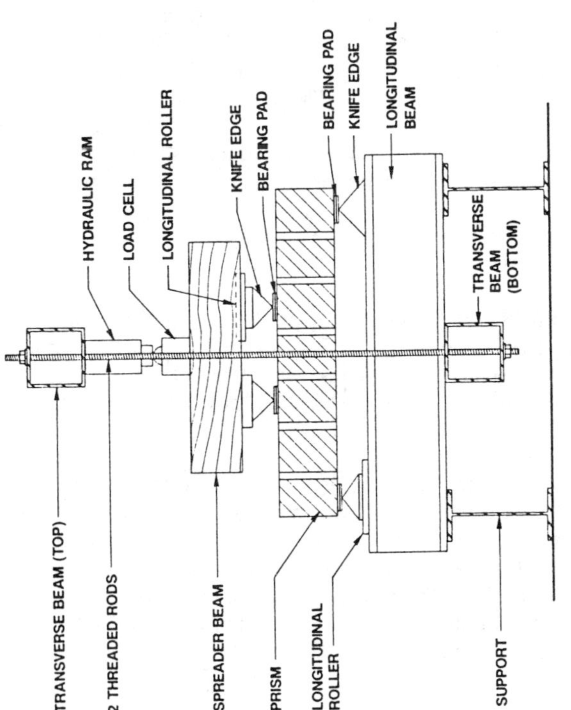

FIG. 1 -- Flexural test set-up ASTM E 518.

pump and the maximum load at failure is recorded. Measurements of the failed cross-section of the masonry are used to determine the tensile flexural stress at failure.

For the bond wrench tests (ASTM C 1072), the surviving portions of the flexural tests are then placed in the bond wrench frame shown in Figure 2. A pressure plate clamps the prism in place just below the mortar joint leaving one course of masonry projecting above. The loading arm bracket is then clamped to the projected brick leaving the mortar joint exposed. A load is applied through a ball bearing at the extended end of the loading arm with a screw jack. The magnitude of the failure load is recorded by a trace needle on a calibrated force gage. Measurements of the failed cross-section of masonry are used to determine the flexural tensile stress at failure.

Materials

The project repair specifications included requirements for Grade SW and Type FBX brick. The dimensions of the existing masonry were to be matched. The BIA Guide Specification [1] indicates that brick with an initial rate of absorption (IRA) in excess of $20g/min \cdot 30in^2$ ($20g/min \cdot 194cm^2$) shall be pre-wetted so that, when laid, the rate of absorption is less than this amount. ASTM Standard Specification for facing brick (Solid Masonry Units Made from Clay or Shale) (C 216) explanatory Note #2 requires brick and an IRA exceeding $30g/min \cdot 30 in^2$ ($30g/min \cdot 194cm^2$) should be well wetted prior to laying.

Common practice indicates that a Type "S" mortar should be used where maximum flexural bond strength is required. In order to verify this, Types "S", "N" and a hybrid between the "S" and the "N" were included in the first pilot test program. All mixes contain Portland cement and hydrated lime, per ASTM Standard Specification for Portland Cement (C 150) and ASTM Standard Specification for Hydrated Lime for Masonry Purposes (C 207) respectively. Air-entraining additives were not permitted. Local sands were tested, and local sands modified to meet the gradation requirements of ASTM Standard Specification for Aggregate for Masonry Mortar (C 144) were also tested.

PILOT TEST #1

The same bricks from two different production runs were included in the first test. These were labelled Bricks P and S. The Initial Rate of Absorption (IRA) was found to be $15.2g/min \cdot 30in^2$ ($15.2g/min \cdot 194cm^2$) and $11.9g/min \cdot 30in^2$ ($11.9/min \cdot 194cm^2$) for Bricks P and S respectively. Both bricks were laid without prewetting. Table 1 presents a breakdown of the materials.

Eight (8) mortar types were tested. Four (4) were considered to be Type S, two were considered to be Type N, and two were considered to be between Types N and S. Two (2) cements were tested. Seven (7) of the mortars contained colored cement matching the existing mortar, and one contained gray cement. Seven (7) of the mortars contained locally available sand and one contained the local sand modified to meet the gradation requirements of ASTM C 144. The local sand was finer than the gradation requirements of ASTM C 144.

TABLE 1 -- PART 1 MATERIAL SUMMARY

| TEST I.D. | BRICK | | | MORTAR | | | | VOLUME ft^3(dm^3) | | | WEIGHT lbs(kg) | | SAND TO CEMENT +LIME |
	IRA	PRESOAK	PATTERN	TYPE CEMENT	LIME	SAND		CEMEN	LIME	SAND	CEMENT	LIME	
P1	15.2	NO	STACKED	S	GRAY	TYPE S	LOCAL	1 BAG	1/2 BAG	4.00(113)	94(43)	26.50(12	2.41
P2	15.2	NO	STACKED	S	COLORED	TYPE S	LOCAL	1 BAG	1/2 BAG	4.00(113)	94(43)	26.50(12	2.41
P3	15.2	NO	STACKED	N	COLORED	TYPE S	LOCAL	1 BAG	1 BAG	5.00(142)	94(43)	53.00(24	2.15(U)
P4	15.2	NO	STACKED	N/S	COLORED	TYPE S	LOCAL	1 BAG	3/4 BAG	4.50(127)	94(43)	39.75(18	2.26
P5	15.2	NO	STACKED	S	COLORED	TYPE S	MOD	1 BAG	1/2 BAG	4.00(113)	94(43)	26.50(12	2.41
S2	11.9	NO	STACKED	S	COLORED	TYPE S	LOCAL	1-1/2(42)	3/4(21)	6.00(170)	118(54)	27.50(12	3.09(O)
S3	11.9	NO	STACKED	N	COLORED	TYPE S	LOCAL	1-1/2(42)	1-1/2(42)	7.50(212)	118(54)	54.90(25	2.85
S4	11.9	NO	STACKED	N/S	COLORED	TYPE S	LOCAL	1-1/2(42)	1-1/8(32)	6.75(191)	118(54)	41.20(19	2.95
CP1	15.2	NO	RUNNING	S	GRAY	TYPE S	LOCAL	1 BAG	1/2 BAG	4.00(113)	94(43)	26.50(12	2.41

NOTES:

MOD = LOCAL SAND MODIFIED TO MEET ASTM C 144 GRADIATION

VOLUME METHOD = LOOSELY PACKED IN MEASURING BOX

RATIO BASED ON CEMENT AT 94 LBS/FT^3(1505 kg/m^3), AND LIME AT 40 LBS/FT^3(640 kg/M^3)

(U) = UNDERSANDED

(O) = OVERSANDED

Three (3) of the eight mortars were mixed using a volume box and the remaining five were mixed using the bag volume. The prisms and wallette containing Brick P were constructed using the mortar mixed by the bag volume, and the prisms and wallettes containing Brick S were constructed using the mortar mixed by volume box.

In both cases, the materials were weighed. The bags of cement contained 94 lbs (43 kg) of material. This weight indicates less than a cubic foot of Portland cement for those cements containing color additives. The actual amount of cement was not determined. It was also found that 1-1/2 ft^3 (42 dM^3) of cement weighs 118 lbs (54 kg) when loosely placed in the measuring box.

The bags of hydrated lime were labeled 50 lbs but contained 53 lbs (24 kg) of material. It was also found that 1.0 ft^3 (28 dM) of lime weighs 36.6 lbs (17 kg) when loosely placed in the measuring box. ASTM Standard Specification for Mortar for Unit Masonry (C 270) indicates that hydrated lime weighs 40 lbs/ft^3 (640 kg/M^3). Therefore, both methods of proportioning result in mixes with excessive lime. ASTM C 270 requires the volume of sand to be between 2.25 and 3.0 times the sum of the volume of cement and lime. Two of the mixes fell outside the sand parameters, when back calculated using the ASTM C 270 material densities.

All of the test specimens were constructed by professional masons local to the project. Fifty-seven (57) wallettes were constructed in stack bond. Four (4) wallettes were cut from the top of the P1 permeability test wall. These panels were constructed in running bond. The prisms were tested in flexure according to ASTM E 518, and the remaining parts were tested using the bond wrench method of ASTM C 1072.

Results

The results of the 61 flexural tests, and 338 bond wrench tests are presented in Table 2. The project requirements for a minimum average flexural bond test strength of 144 psi (993 kPa) were not obtained in any of the test cases. The best average strength for those prisms tested in stack bond, was 72 psi (496 kPa) with a maximum and minimum single test value of 134 psi (924 kPa) and 4 psi (28 kPa), respectively. The average test strength of those specimens cut from the permeability test walls was 74 psi (510 kPa) with individual high and low test values of 94 (648) and 60 psi (414 kPa).

The required minimum average test strength was also not obtained by the bond wrench tests. The best average test strength was 92 psi (634 kPa) with individual high and low test values of 195 psi (1,344 kPa) and 9 psi (62 kPa). The average test strength of those specimens cut from the permeability test walls was 119 psi (820 kPa) with individual high and low test values of 182 psi (1,255 kPa) and 58 psi (400 kPa).

These results were less than satisfactory. The wide variation in the test results indicates the traditional methods of masonry specification and construction could not adequately produce the masonry bond strength required for SS/BV construction. A more reliable bond strength was needed. More research and testing was required in order to consider SS/BV as a design solution.

TABLE 2 -- PART 1 TEST RESULTS

TEST I.D.	AGE DAYS	TEST DATA BOND WRENCH psi(kPa)					LOWER LIMIT 99 %-TILE	FLEXURAL psi(kPa)				
		N	MEAN	STD DEV	MAX	MIN			MEAN	STD DEV	MAX	MIN
P1	33	40	90(621)	31.12(215)	152(1048)	9(62)	18(124)	8	39(269)	23.92(165)	79(545)	11(76)
P2	32	40	67(462)	23.60(163)	118(814)	17(117)	13(90)	7	57(393)	13.78(95)	84(579)	45(310)
P3	32	40	62(427)	61.60(425)	102(703)	33(228)	23(159)	8	38(262)	17.31(119)	55(379)	4(28)
P4	34	35	80(552)	21.78(150)	130(896)	36(248)	30(207)	8	70(483)	31.54(217)	121(834)	28(193)
P5	33	39	92(634)	31.87(220)	165(1138)	27(186)	18(124)	7	38(262)	12.94(89)	55(379)	14(97)
S2	34	40	90(621)	31.81(219)	195(1344)	29(200)	17(117)	8	72(496)	45.90(316)	134(924)	13(90)
S3	35	28	46(317)	22.55(155)	85(586)	10(69)	0(0)	4	47(324)	17.70(122)	73(503)	35(241)
S4	35	35	83(572)	23.23(160)	149(1027)	36(248)	30(207)	7	69(476)	21.54(149)	107(738)	42(290)
CP1	34	41	119(820	31.21(215)	182(1255)	58(400)	47(324)	4	74(510)	14.18(98)	94(648)	60(414)

PILOT TEST #2

The parameters of this test were expanded to include more types of brick, cement and sand, while limiting the mixing proportion to one type. Five (5) bricks, two sands, and two cements were studied. The combination of materials including brick and mortar types are presented in Table 3.

Five (5) brick types were supplied by three manufacturers. Thirteen (13) prisms were constructed for each of the Bricks A, B, C, and D. The quantity of Brick E was limited and, therefore, only two prisms were constructed for testing.

Both cements were Portland Cement. One (1) contained color additives matching the existing color. The other was gray cement. Thirty of the mortar mixes were gray and 24 were colored. The sand locally available to the project was compared to Illinois #2 sand modified to meet ASTM C 144 gradation requirements. Thirty (30) of the mortar mixes contained the modified Illinois #2 sand and 24 contained the locally available sand.

The mortar mixing ratio was held to 1 bag of cement, 1/2 bag of Type "S" lime, and 4-1/2 ft^3 (127 dM^3) of sand, which corresponds to 94 lbs (43 kg) of cement, and 26.5 lbs (12 kg) of lime. All prisms were constructed with 1/2 joints in stacked bond, and tested at approximately 14 and 28 days.

In addition, the effects of presoaking the bricks prior to lay-up were studied. Saturation testing of each brick type indicated that 5 minutes of soaking resulted in a saturation level of approximately 50%, while 30 minutes of soaking resulted in approximately 80%. Presoaking was limited to between 5 and 30 minutes for each brick.

Five (5) prisms were laid up using bricks that were presoaked in a saturated lime solution. The bricks were presoaked in the solution for 1 minute and then allowed to dry for 24 hours. Prior to lay-up, the bricks were then presoaked for 1 minute in water.

For each mortar and brick group type, one prism constructed with the presoaked bricks, one prism constructed with the non-presoaked bricks, and one prism constructed with bricks soaked in the lime solution were tested at approximately 14 days. The remaining prism, constructed with the presoaked bricks for each group, was tested at approximately 28 days.

Results

The test result summaries are presented in Table 4. Fifteen (15) out of 16 prisms constructed with presoaked bricks and tested at 28 days had greater bond strengths than those tested at 14 days. This follows the traditional notion of strength gain for mortars constructed with Portland cement.

It was also observed that higher average bond strengths were achieved at 14 days in 14 out of the 16 prisms constructed with presoaked bricks over the prisms constructed

TABLE 3 -- PART 2 MATERIAL SUMMARY

| TEST | BRICK | | | MORTAR | | | | MIXING RATIO | | |
| | | | | | | | | CEMENT | LIME | SAND |
I.D.	IRA	PRESOAK	PATTERN	TYPE	CEMENT	LIME	SAND	lb(kg)	lb(kg)	ft^3(dm^3)
A-1A,A-1B	5.9	YES	STACKED	S	GREY	TYPE S	ILL#2	94(43)	26.5(12)	4.50(127)
A-1C	5.9	NO	STACKED	S	GREY	TYPE S	ILL#2	94(43)	26.5(12)	4.50(127)
A-1D	5.9	LIME	STACKED	S	GREY	TYPE S	ILL#2	94(43)	26.5(12)	4.50(127)
A-2A,A-2B	5.9	YES	STACKED	S	GREY	TYPE S	LOCAL	94(43)	26.5(12)	4.50(127)
A-2C	5.9	NO	STACKED	S	GREY	TYPE S	LOCAL	94(43)	26.5(12)	4.50(127)
A-3A,A-3B	5.9	YES	STACKED	S	COLOR	TYPE S	LOCAL	94(43)	26.5(12)	4.50(127)
A-3C	5.9	NO	STACKED	S	COLOR	TYPE S	LOCAL	94(43)	26.5(12)	4.50(127)
A-4A,A-4B	5.9	YES	STACKED	S	COLOR	TYPE S	ILL#2	94(43)	26.5(12)	4.50(127)
A-4C	5.9	NO	STACKED	S	COLOR	TYPE S	ILL#2	94(43)	26.5(12)	4.50(127)
B-1A,B-1B	27.3	YES	STACKED	S	GREY	TYPE S	ILL#2	94(43)	26.5(12)	4.50(127)
B-1C	27.3	NO	STACKED	S	GREY	TYPE S	ILL#2	94(43)	26.5(12)	4.50(127)
B-1D	27.3	LIME	STACKED	S	GREY	TYPE S	ILL#2	94(43)	26.5(12)	4.50(127)
B-2A,B-2B	27.3	YES	STACKED	S	GREY	TYPE S	LOCAL	94(43)	26.5(12)	4.50(127)
B-2C	27.3	NO	STACKED	S	GREY	TYPE S	LOCAL	94(43)	26.5(12)	4.50(127)
B-3A,B-3B	27.3	YES	STACKED	S	COLOR	TYPE S	LOCAL	94(43)	26.5(12)	4.50(127)
B-3C	27.3	NO	STACKED	S	COLOR	TYPE S	LOCAL	94(43)	26.5(12)	4.50(127)
B-4A,B-4B	27.3	YES	STACKED	S	COLOR	TYPE S	ILL#2	94(43)	26.5(12)	4.50(127)
B-4C	27.3	NO	STACKED	S	COLOR	TYPE S	ILL#2	94(43)	26.5(12)	4.50(127)

TABLE 3 -- PART 2 MATERIAL SUMMARY (CONTINUED)

TEST	BRICK			MORTAR				MIXING RATIO		
I.D.	IRA	PRESOAK	PATTERN	TYPE	CEMENT	LIME	SAND	CEMENT lb(kg)	LIME lb(kg)	SAND ft^3(dm^3)
C-1A,C-1B	15.2	YES	STACKED	S	GREY	TYPE S	ILL#2	94(43)	26.5(12)	4.50(127)
C-1C	15.2	NO	STACKED	S	GREY	TYPE S	ILL#2	94(43)	26.5(12)	4.50(127)
C-1D	15.2	LIME	STACKED	S	GREY	TYPE S	ILL#2	94(43)	26.5(12)	4.50(127)
C-2A,C-2B	15.2	YES	STACKED	S	GREY	TYPE S	LOCAL	94(43)	26.5(12)	4.50(127)
C-2C	15.2	NO	STACKED	S	GREY	TYPE S	LOCAL	94(43)	26.5(12)	4.50(127)
C-3A,CA-3	15.2	YES	STACKED	S	COLOR	TYPE S	LOCAL	94(43)	26.5(12)	4.50(127)
C-3C	15.2	NO	STACKED	S	COLOR	TYPE S	LOCAL	94(43)	26.5(12)	4.50(127)
C-4A,C-4B	15.2	YES	STACKED	S	COLOR	TYPE S	ILL#2	94(43)	26.5(12)	4.50(127)
CA-4C	15.2	NO	STACKED	S	COLOR	TYPE S	ILL#2	94(43)	26.5(12)	4.50(127)
D-1A,D-1B	11.9	YES	STACKED	S	GREY	TYPE S	ILL#2	94(43)	26.5(12)	4.50(127)
D-1C	11.9	NO	STACKED	S	GREY	TYPE S	ILL#2	94(43)	26.5(12)	4.50(127)
D-1D	11.9	LIME	STACKED	S	GREY	TYPE S	ILL#2	94(43)	26.5(12)	4.50(127)
D-2A,D-2B	11.9	YES	STACKED	S	GREY	TYPE S	LOCAL	94(43)	26.5(12)	4.50(127)
D-2C	11.9	NO	STACKED	S	GREY	TYPE S	LOCAL	94(43)	26.5(12)	4.50(127)
D-3A,D-3B	11.9	YES	STACKED	S	COLOR	TYPE S	LOCAL	94(43)	26.5(12)	4.50(127)
D-3C	11.9	NO	STACKED	S	COLOR	TYPE S	LOCAL	94(43)	26.5(12)	4.50(127)
D-4A,D-4B	11.9	YES	STACKED	S	COLOR	TYPE S	ILL#2	94(43)	26.5(12)	4.50(127)
D-4C	11.9	NO	STACKED	S	COLOR	TYPE S	ILL#2	94(43)	26.5(12)	4.50(127)
E-1A,E-1B	12.1	YES	STACKED	S	GREY	TYPE S	ILL#2	94(43)	26.5(12)	4.50(127)
E-1C	12.1	LIME	STACKED	S	GREY	TYPE S	ILL#2	94(43)	26.5(12)	4.50(127)

TABLE 4 -- PART 2 TEST RESULTS

TEST I.D.	AGE DAYS		TEST DATA BOND-WRENCH PSI(kPa) MEAN	SDEV	MAX	MIN	LOWER LIMIT 99 %-TILE	FLEXURAL PSI(kPa) N	MEAN
A-1A	17	5	92(637)	21.01(145)	121(834)	66(455)	44(304)	1	72(499)
A-1B	31	5	129(887)	5.90(41)	135(933)	121(833)	115(793)	1	107(740)
A-1C	17	5	46(315)	11.55(80)	61(418)	29(202)	19(132)	1	27(187)
A-1D	17	5	105(722)	24.51(169)	130(900)	70(485)	48(333)	1	89(616)
A-2A	17	5	79(545)	18.18(125)	98(675)	57(392)	37(256)	1	68(468)
A-2B	31	5	69(474)	21.26(147)	96(662)	47(326)	20(137)	1	68(467)
A-2C	15	5	71(489)	13.65(94)	94(645)	62(425)	40(273)	1	56(389)
A-3A	17	5	66(453)	19.42(134)	94(645)	47(324)	21(145)	1	60(411)
A-3B	31	5	90(622)	22.11(152)	123(846)	68(469)	39(272)	1	64(443)
A-3C	17	5	42(292)	10.26(71)	53(366)	31(212)	19(130)	1	57(393)
A-4A	17	5	70(481)	28.53(197)	94(645)	24(166)	4(29)	1	32(219)
A-4B	31	5	71(488)	16.15(111)	97(670)	56(385)	34(232)	1	66(452)
A-4C	17	5	59(404)	20.37(140)	74(512)	28(192)	12(81)	1	29(203)
B-1A	15	5	106(730)	22.61(156)	130(895)	85(584)	54(372)	1	128(886)
B-1B	31	5	140(964)	57.95(400)	201(1387)	77(529)	7(45)	1	82(567)
B-1C	17	5	87(601)	10.00(69)	102(706)	78(536)	64(442)	1	88(608)
B-1D	17	5	68(469)	8.74(60)	78(536)	59(405)	48(330)		N.T.
B-2A	20	5	115(791)	25.90(179)	148(1017)	86(594)	55(381)	1	80(554)
B-2B	26	5	140(968)	27.81(192)	169(1163)	107(736)	76(527)	1	71(493)
B-2C	15	4	92(636)	32.75(226)	134(924)	54(372)	17(116)	1	67(459)
B-3A	17	5	90(619)	59.82(412)	194(1339)	50(346)	0(0)	1	51(349)
B-3B	31	5	128(880)	62.15(429)	208(1433)	72(499)	0(0)	1	102(705)
B-3C	17	5	52(356)	32.07(221)	103(712)	20(137)	0(0)	1	47(323)
B-4A	17	5	87(597)	15.70(108)	111(762)	69(477)	50(347)	1	69(478)
B-4B	31	5	173(1190)	47.26(326)	248(1708)	126(871)	64(441)	1	62(427)
B-4C	17	5	68(466)	42.83(295)	117(806)	25(174)	0(0)		N.T.

TABLE 4 -- PART 2 TEST RESULTS (CONTINUED)

TEST I.D.	AGE DAYS		TEST DATA BOND-WRENCH PSI(kPa)				LOWER LIMIT 99 %-TILE	FLEXURAL PSI(kPa)	
			MEAN	SDEV	MAX	MIN		N	MEAN
C-1A	17	5	91(630)	47.01(324)	162(1118)	41(285)	0(0)	1	90(622)
C-1B	31	5	114(787)	18.84(130)	138(954)	94(651)	71(488)	1	152(1050)
C-1C	17	5	67(462)	11.29(78)	82(564)	53(368)	41(283)	1	77(529)
C-1D	15	5	98(673)	34.23(236)	141(972)	66(456)	19(131)	1	109(750)
C-2A	17	5	65(445)	12.54(86)	77(530)	44(302)	36(246)	1	104(715)
C-2B	31	5	117(808)	15.90(110)	142(982)	99(680)	81(556)	1	144(789)
C-2C	15	4	80(550)	9.07(63)	89(614)	70(484)	59(406)	1	48(328)
C-3A	17	5	94(647)	20.99(145)	124(852)	74(509)	46(315)	1	85(584)
C-3B	31	5	151(1041)	23.40(161)	191(1318)	133(916)	97(670)	1	134(924)
C-3C	17	5	66(453)	16.77(116)	88(608)	41(285)	27(187)	1	61(420)
C-4A	17	5	66(455)	37.89(261)	108(742)	18(126)	0(0)	1	49(339)
C-4B	31	5	118(812)	44.70(308)	168(1162)	72(493)	15(103)	1	84(581)
C-4C	17	4	55(381)	22.66(156)	78(540)	29(201)	3(22)	1	85(586)
D-1A	15	5	77(534)	35.07(242)	132(912)	42(287)	0(0)	1	62(425)
D-1B	31	5	105(726)	8.60(59)	116(799)	95(652)	85(589)	1	117(807)
D-1C	17	5	35(240)	12.91(89)	49(341)	17(115)	5(35)	1	57(395)
D-1D	15	5	77(529)	22.59(156)	102(705)	42(287)	25(171)	1	48(330)
D-2A	17	4	39(266)	26.89(185)	73(501)	10(67)	0(0)	1	52(356)
D-2B	31	5	84(582)	21.79(150)	116(799)	59(407)	34(236)	1	64(438)
D-2C	15	5	44(306)	13.42(93)	58(399)	27(187)	14(93)	1	24(165)
D-3A	17	5	73(502)	23.11(159)	90(619)	35(243)	20(135)	1	27(187)
D-3B	31	5	100(692)	15.86(109)	128(880)	90(619)	64(440)	1	97(666)
D-3C	17	5	31(214)	20.32(140)	56(386)	9(62)	0(0)	1	13(89)
D-4A	17	5	75(520)	33.19(229)	123(850)	34(233)	0(0)	1	49(336)
D-4B	31	5	117(805)	26.65(184)	164(1130)	102(701)	55(382)	1	122(841)
D-4C	17	5	25(170)	21.17(146)	58(339)	5(34)	0(0)		N.T.
E-1B	31	5	103(712)	33.34(230)	142(981)	70(484)	27(183)	1	75(519)
E-1D	17	5	72(494)	12.72(88)	91(629)	59(406)	42(292)	1	107(738)

N.T.= NOT TESTED

with bricks laid dry. In some cases, the presoaked bricks resulted in more than double the 14-day strength of similar prisms laid dry.

The results of the presoaking of the bricks in the lime solution were found to be inconclusive. The average bond strengths of two prism groups were higher than the corresponding prisms constructed with brick presoaked in water, and two were lower. The average bond strength of the remaining prism was equal to its companion sample.

When comparing the bond strength gains of the two cements tested, six out of eight mortar mixes containing gray cement had higher average strengths at 14 days. However, at 28 days, the colored cements had higher average bond strengths in six out of the eight mortars tested. These results tend to indicate a slower, but ultimately higher, strength gain for this particular colored cement.

Although the tendencies discussed above were found to be interesting, the test results were that out of the 54 prisms tested, only two test mean values were greater than the project requirement, and two others met the 99th percentile lower strength limit of 96 psi (622 kPa). The test results also indicated a 99th percentile lower strength limit of 0 psi (0 kPa) for 11 prisms, raising a concern as to whether or not reliable bond strength could be achieved at the stress levels of SS/BV construction.

Mortar Flow Rate

During the first two pilot tests, additional research included the review of certain published and unpublished documents regarding bond strength. The effect of mortar flow was discussed in *"Brick and Tile Engineering"* by Harry C. Plummer [3]. The general consensus was that increasing the fluidity of the mortar will increase the bond strength. Tests were described that correlated the flow rate to flexural bond strength. ASTM Standard Test Method for Compressive Strength of Hydraulic Cement Mortars (using 2-inch, or 50-mm cube specimens) (C 109) presents the method of measuring the mortar flow using the flow table described in ASTM Standard Specification for Flow Table for Use in Tests of Hydraulic Cement (C 230). This method of monitoring the flow rate works well in the laboratory, but does not adapt well to field construction.

Sample mortar flow tests were conducted in order to become familiar with the correlation between flow rates and workability. The best flow rate obtained was one that remained workable for the masons. The optimum flow rate was a mix that just barely clings to the mason's trowel. This method of determining the maximum or optimum flow rate for the mix was used during Pilot Test #3.

PILOT TEST #3

The material summary of Pilot Test #3 is presented in Table 5.

This test program included three brick types produced by three different manufacturers and 13 variations on Type S mortar, resulting from two cements, two sands, and six mixing ratios. The two cements included one gray and one with color additives.

TABLE 5 -- PART 3 MATERIAL SUMMARY

| TEST | BRICK | | | MORTAR | | | | MIXING RATIO | | |
I.D.	IRA	PRESOAK	PATTERN	TYPE	CEMENT	LIME	SAND	CEMENT lb(kg)	LIME lb(kg)	SAND ft^3(dm^3)
F-1	20.47	YES	RUNNING	S	COLOR	TYPE S	LOCAL	94(43)	10(5)	2.81(80)
F-2	20.47	YES	RUNNING	S	COLOR	TYPE S	LOCAL	94(43)	10(5)	3.75(106)
F-3	20.47	YES	RUNNING	S	COLOR	TYPE S	LOCAL	94(43)	15(7)	3.09(87)
F-4	20.47	YES	RUNNING	S	COLOR	TYPE S	LOCAL	94(43)	20(9)	4.13(117)
F-5	20.47	YES	RUNNING	S	COLOR	TYPE S	LOCAL	94(43)	20(9)	3.38(96)
F-6	20.47	YES	RUNNING	S	COLOR	TYPE S	LOCAL	94(43)	20(9)	4.50(127)
Fa-3	20.47	YES	RUNNING	S	GREY	TYPE S	ILL#2	94(43)	15(7)	3.09(87)
Fa-4	20.47	YES	RUNNING	S	GREY	TYPE S	ILL#2	94(43)	20(9)	4.13(117)
Fa-5	20.47	YES	RUNNING	S	GREY	TYPE S	ILL#2	94(43)	20(9)	3.38(96)
Fa-6	20.47	YES	RUNNING	S	GREY	TYPE S	ILL#2	94(43)	20(9)	4.50(127)
G-1	23.18	YES	RUNNING	S	COLOR	TYPE S	LOCAL	94(43)	10(5)	2.81(80)
G-2	23.18	YES	RUNNING	S	COLOR	TYPE S	LOCAL	94(43)	10(5)	3.75(106)
G-3	23.18	YES	RUNNING	S	COLOR	TYPE S	LOCAL	94(43)	15(7)	3.09(87)
G-4	23.18	YES	RUNNING	S	COLOR	TYPE S	LOCAL	94(43)	20(9)	4.13(117)
G-5	23.18	YES	RUNNING	S	COLOR	TYPE S	LOCAL	94(43)	20(9)	3.38(96)
G-6	23.18	YES	RUNNING	S	COLOR	TYPE S	LOCAL	94(43)	20(9)	4.50(127)
H-1	23.34	YES	RUNNING	S	GREY	TYPE S	ILL#2	94(43)	10(5)	2.81(80)
H-2	23.34	YES	RUNNING	S	GREY	TYPE S	ILL#2	94(43)	10(5)	3.75(106)
H-3	23.34	YES	RUNNING	S	GREY	TYPE S	ILL#2	94(43)	15(7)	3.09(87)
H-4	23.34	YES	RUNNING	S	GREY	TYPE S	ILL#2	94(43)	20(9)	4.13(117)
H-5	23.34	YES	RUNNING	S	GREY	TYPE S	ILL#2	94(43)	20(9)	3.38(96)
H-6	23.34	YES	RUNNING	S	GREY	TYPE S	ILL#2	94(43)	20(9)	4.50(127)

The two sands included one local to the project (LOCAL) and one local to our laboratory (Ill #2). The bricks were designated F, G, and H. Nine (9) wallettes were constructed using Bricks F. Six (6) included colored cement, and three included gray cement and are designated Fa. Six (6) wallettes were constructed for both Bricks G and H. At 14 days, the wallettes were saw cut into test prisms labelled A, B and C. In addition, six permeability test walls were constructed using the Bricks F and three permeability test walls were constructed for each of the Bricks G and H. Four (4) of the six walls constructed with Bricks F included color mortar. The other two were constructed using gray mortar and are designated Fa. The walls constructed with the Brick G included colored mortar, while the walls constructed with the Brick H included gray mortar. These walls were saw cut into test prisms after the permeance testing and labeled 1 through 18. At 1,600 days, the individual labels faded to beyond recognition requiring re-labeling to 1, 2 and 3.

All of the mortar mixes were Type S. ASTM C 270 defines a Type S mortar as:

Portland Cement	=	$1ft^3$ ($28dM^3$)	= 94 lbs (42 kg)
Hydrated Lime	=	1/4 to 1/2 ft^3 (7 to 14 dM^3)	= 10 to 20 lbs (5 to 9 kg)
Sand	=	2-1/4 to 3 times cementitious material (cement + lime)	

All of the mortar mixes tested contained 94 lbs of cement. Mortar Mixes #1 and #2 contained 10 lbs of lime. Mix #3 contained 15 lbs of lime, and Mixes #4, #5, and #6 contained 20 lbs of lime. Mixes #1, #3, and #5 contained the minimum amount of sand or 2-1/4 times the cementitious material. Mixes #2 and #6 contained the maximum amount of sand or 3 times the cementitious material. The sand content in Mortar Mix #4 was 2.75 times the cementitious material. Mortars were re-tempered often, keeping the mix as fluid as possible.

Modulus of Rupture tests using both the bond wrench method and the flexural prism method were conducted on the prisms cut from the wallettes at 14 and 28 days. At 6 months, the walls constructed for the permeability tests were cut into 12 prisms each. Bond wrench tests were conducted on these prisms at approximately 250 days, 650 days, 850 days, and 1,600 days.

Results

The results of these tests are presented in Table 6. The bond wrench test results at 28 days indicated an average strength increase over results at 14 days in 10 out of the 12 prism groups constructed with colored cement, and 1 out of 9 constructed with gray mortar. The 28-day average bond strength of prism groups constructed with colored mortar varied from 101 (696) to 220 psi (1,517 kPa), while the prism groups constructed with gray mortar varied from 139 (958) to 228 psi (1,572 kPa). While virtually the same average test strength was reached for both colored and gray mortar, the gray mortar appears to have a more accelerated strength gain.

The flexural prism test results at 28 days indicated an average test strength increase over the 14-day results in 3 out of 12 prism groups constructed with colored cement, and 1 out of 9 prism groups constructed with gray cement. The flexural test strength results were generally lower than the average bond wrench test results, which follows

TABLE 6 -- PART 3 TEST RESULTS

| TEST | AGE | | TEST DATA | | | | LOWER | FLEXURAL | |
| | | | BOND WRENCH psi(kPa) | | | | LIMIT 99 | PSI (kPa) | |
I.D.	DAYS	N	MEAN	SDEV	MAX	MIN	%-TILE	N	MEAN
F-1-A	14	5	180(1241)	28.98(200)	210(1447)	147(1012)	113(781)	1	166(1145)
F-1-B&C	28	10	183(1259)	24.73(170)	215(1482)	146(1009)	126(867)	2	147(1014)
F-1-9	281	8	272(1875)	74.84(516)	396(2732)	196(1352)	100(689)		
F-1-7&8	655	16	189(1304)	66.62(459)	313(2159)	50(346)	36(247)		
F-1-2&?	860	16	269(1857)	76.96(531)	378(2603)	103(713)	92(637)		
F-1-1,2,3	1616	24	352(2427)	53.04(336)	417(2875)	213(1469)	230(1586)		
F-2-A	14	5	227(1564)	37.66(260)	283(1948)	189(1305)	140(967)	1	228(1572)
F-2-B&C	28	10	220(1518)	45.53(314)	315(2169)	165(1137)	116(797)	2	161(1110)
F-2-9	279	7	279(1925)	66.63(459)	339(2341)	166(1143)	126(869)		
F-2-2&8	656	16	250(1720)	66.03(455)	344(2374)	139(957)	98(673)		
F-2-1&6	861	16	265(1827)	67.39(465)	381(2627)	155(1069)	84(580)		
F-2-1,2,3	1618	24	319(2199)	55.24(381)	388(2675)	203(1400)	192(1324)		
F-3-A	14	5	183(1263)	20.26(140)	213(1467)	157(1084)	137(942)	1	154(1062)
F-3-B&C	28	10	184(1270)	32.54(224)	228(1570)	123(848)	109(754)	2	175(1207)
F-4-A	14	5	117(807)	16.09(111)	137(947)	92(636)	80(552)	1	106(731)
F-4-B&C	28	10	157(1079)	22.45(155)	192(1322)	126(868)	105(723)	2	179(1234)
F-4-5&11	223	16	208(1434)	55.93(386)	284(1958)	111(764)	79(547)		
F-4-8&9	638	16	173(1190)	46.43(320)	294(2027)	103(707)	66(453)		
F-4-2&7	861	16	219(1511)	55.47(382)	314(2164)	122(839)	92(631)		
F-4-1,2,3	1619	24	242(1669)	68.61(473)	354(2441)	55(379)	84(579)		
F-5-A	14	5	201(1384)	51.07(352)	238(1638)	112(774)	83(574)	1	193(1331)
F-5-B&C	28	10	206(1424)	35.48(245)	253(1747)	139(960)	125(861)	2	178(1227)

TABLE 6 -- PART 3 TEST RESULTS (CONTINUED)

TEST I.D.	AGE DAYS	TEST DATA BOND WRENCH psi(kPa)					LOWER LIMIT 99 %-TILE	FLEXURAL PSI (kPa)	
		N	MEAN	SDEV	MAX	MIN		N	MEAN
F-6-A	14	5	119(818)	7.96(55)	130(894)	110(757)	100(691)	1	106(731)
F-6-B&C	28	10	126(866)	32.55(224)	191(1317)	93(639)	51(350)	2	87(600)
F-6-5&8	224	16	232(1598)	60.58(418)	374(2579)	146(1008)	92(638)		
F-6-2&9	664	16	197(1359)	53.48(369)	292(2013)	109(751)	74(511)		
F-6-7&?	863	15	216(1490)	46.96(324)	269(1856)	122(842)	108(746)		
F-6-1,2,3	1623	24	286(1972)	58.39(403)	373(2572)	150(1034)	151(1041)		
G-1-A	14	5	163(1125)	36.28(250)	201(1384)	109(755)	80(550)	1	147(1014)
G-1-B&C	28	5	189(1305)	20.73(143)	211(1458)	167(1151)	142(977)	1	145(1000)
G-2-A	14	5	161(1112)	10.48(72)	177(1219)	151(1042)	137(945)	1	110(758)
G-2-B&C	28	10	169(1168)	20.44(141)	205(1412)	140(964)	122(844)	2	137(945)
G-2-2&11	217	18	233(1605)	42.02(290)	327(2255)	133(920)	136(939)		
G-2-9	278	8	281(1937)	68.48(472)	367(2532)	179(1238)	123(851)		
G-2-1&8	662	18	265(1824)	59.50(410)	367(2532)	136(935)	128(881)		
G-2-4&7	859	18	267(1840)	48.27(333)	365(2517)	184(1272)	156(1074)		
G-2-1,2,3	1622	27	342(2358)	39.50(272)	395(2723)	259(1786)	251(1731)		
G-3-A	14	5	178(1227)	31.42(217)	215(1479)	148(1022)	106(729)	1	110(758)
G-3-B&C	28	5	161(1113)	16.92(117)	188(1298)	147(1010)	122(844)	1	71(490)
G-4-A	14	5	108(748)	6.94(48)	114(786)	98(676)	93(638)	1	108(745)
G-4-B&C	28	10	113(780)	20.56(142)	141(976)	78(539)	66(454)	2	55(379)
G-4-2&10	222	18	190(1310)	42.41(292)	269(1852)	111(764)	92(637)		
G-4-8&9	665	18	209(1442)	36.85(254)	286(1969)	147(1013)	124(857)		
G-4-1&7	862	18	206(1422)	41.48(286)	294(2027)	139(955)	111(764)		
G-4-1,2,3	1637	26	297(2048)	45.03(310)	382(2634)	219(1510)	194(1338)		

TABLE 6 -- PART 3 TEST RESULTS (CONTINUED)

TEST I.D.	AGE DAYS	N	TEST DATA BOND WRENCH psi(kPa) MEAN	SDEV	MAX	MIN	LOWER LIMIT 99 %-TILE	FLEXURAL PSI (kPa) N	MEAN
G-5-A	14	5	163(1125)	24.33(168)	190(1311)	138(953)	107(739)	1	149(1027)
G-5-B&C	28	5	185(1277)	37.97(262)	218(1502)	125(864)	98(674)	1	108(745)
G-6-A	14	5	99(680)	13.12(90)	109(755)	76(526)	68(472)	1	86(593)
G-6-B&C	28	10	101(696)	17.18(118)	128(881)	61(421)	61(424)	2	83(572)
G-6-5&7	223	18	175(1204)	60.37(416)	280(1934)	75(515)	36(246)		
G-6-8&9	664	18	132(909)	42.16(291)	207(1424)	39(270)	35(240)		
G-6-2&4	863	18	229(1579)	31.64(218)	285(1964)	174(1203)	156(1078)		
G-6-1,2,3	1638	27	260(1793)	60.05(414)	377(2599)	141(972)	122(841)		
Fa-3-A	14	5	175(1204)	36.99(255)	207(1425)	115(790)	90(618)	1	157(1082)
Fa-3-B&C	28	5	160(1105)	43.53(300)	196(1354)	88(608)	60(415)	1	148(1020)
Fa-3-1&10	186	16	216(1489)	64.18(442)	337(2323)	104(719)	68(471)		
Fa-3-2&8	654	16	251(1730)	37.40(258)	312(2148)	191(1317)	165(1137)		
Fa-3-?&3	857	16	324(2234)	66.91(461)	389(2682)	169(1162)	170(1172)		
Fa-3-1,2,3	1616	24	296(2041)	71.87(496)	421(2903)	181(1248)	131(903)		
Fa-4-A	14	5	155(1068)	15.92(110)	177(1219)	137(946)	118(815)	1	137(945)
Fa-4-B&C	28	10	139(958)	28.68(198)	188(1299)	106(733)	73(503)	1	154(1062)
Fa-5-4&8	215	16	250(1722)	56.84(392)	341(2353)	153(1052)	119(821)		
Fa-5-2&9	651	16	314(2165)	52.12(359)	387(2666)	207(1429)	194(1339)		
Fa-5-3&6	859	16	332(2286)	60.29(416)	419(2888)	199(1374)	193(1330)		
Fa-5-1,2,3	1617	24	347(2392)	40.85(282)	391(2696)	262(1806)	253(1744)		
Fa-6-A	14	5	182(1256)	29.06(200)	226(1556)	158(1090)	115(795)	1	130(896)
Fa-6-B&C	28	10	161(1112)	29.69(205)	201(1384)	104(716)	93(641)	2	124(855)

TABLE 6 -- PART 3 TEST RESULTS (CONTINUED)

TEST I.D.	AGE DAYS	N	TEST DATA BOND WRENCH psi(kPa) MEAN	SDEV	MAX	MIN	LOWER LIMIT 99 %-TILE	FLEXURAL PSI (kPa) N	MEAN
H-1-A	14	5	283(1951)	71.15(491)	364(2510)	213(1466)	119(822)	1	188(1296)
H-1-B&C	20	5	228(1575)	11.16(77)	246(1693)	215(1485)	203(1398)	1	151(1041)
H-2-A	14	5	185(1278)	57.31(395)	261(1798)	129(891)	54(370)	1	149(1027)
H-2-B&C	28	5	195(1341)	35.01(241)	243(1674)	151(1043)	114(786)	1	124(855)
H-2-9	272	7	306(2108)	87.79(605)	406(2802)	184(1266)	104(716)		
H-2-2&8	644	19	281(1936)	76.33(526)	416(2871)	145(1002)	105(726)		
H-2-6&12	851	18	334(2303)	45.23(312)	394(2716)	251(1731)	230(1586)		
H-2-1,2,3	1606	27	338(2330)	53.87(371)	411(2834)	209(1441)	214(1475)		
H-3-A	14	5	210(1445)	47.53(328)	267(1839)	148(1023)	100(691)	1	210(1448)
H-3-B&C	28	5	188(1297)	21.80(150)	218(1500)	164(1132)	138(952)	1	119(820)
H-4-A	14	5	171(1179)	23.83(164)	213(1467)	153(1054)	116(801)	1	143(986)
H-4-B&C	28	10	154(1062)	17.65(122)	177(1221)	129(893)	113(782)	2	120(827)
H-4-9	273	9	335(2307)	55.46(382)	406(2802)	260(1792)	207(1428)		
H-4-2&8	649	18	310(2138)	83.75(577)	391(2694)	132(908)	117(810)		
H-4-3&12	851	18	353(2433)	57.19(394)	413(2847)	216(1486)	221(1526)		
H-4-1,2,3	1608	27	297(2048)	91.85(633)	417(2875)	70(483)	86(593)		
H-5-A	14	5	209(1444)	37.06(256)	261(1798)	161(1107)	124(856)	1	168(1158)
H-5-B&C	28	5	202(1392)	36.93(255)	246(1698)	146(1009)	117(806)	1	128(883)
H-6-A	14	5	182(1256)	29.06(200)	226(1556)	158(1090)	115(795)	1	151(1041)
H-6-B&C	28	10	157(1082)	26.52(183)	193(1329)	118(816)	96(661)	2	97(669)
H-6-7&10	210	18	240(1654)	56.59(390)	366(2520)	151(1040)	110(757)		
H-6-8&9	650	18	276(1904)	67.97(469)	383(2640)	176(1216)	120(826)		
H-6-1&4	852	18	255(1761)	75.52(521)	396(2733)	102(700)	82(563)		
H-6-1,2,3	1608	27	265(1827)	52.78(364)	399(2751)	162(1117)	143(986)		

the logic that the flexural test tends to find the weakest bond in a prism. However, the flexural test results were generally greater than the 99th percentile of lower strength limit requirement.

The project specifications required the average test strength to be equal to or greater than 144 psi (993 kPa). At 14 days, 9 out of the 9 prism groups constructed with gray mortar, and 8 out 12 prism groups constructed with colored mortar met this requirement. At 28 days, 8 out of the 9 prism groups constructed with gray mortar, and 9 out of the 12 prism groups constructed with colored mortar met this requirement. Two (2) of the prism groups that did not meet this requirement were constructed with mortar Type 6 (F-6 and G-6) and 2 were constructed with mortar Type 4 (Fa-4 and G-4). The prism constructed with gray mortar (Fa-4) was close, with an average strength of 139 psi (958 kPa).

The project specifications also required the 99th percentile of the lower strength limit to be greater than or equal to 96 psi (662 kPa). At 14 days, 7 out of the 9 prism groups constructed with gray cement, and 7 out of the 12 prism groups constructed with colored cement met this requirement. At 28 days, 6 out of the 9 prism groups constructed with gray cement, while 9 out of 12 prism groups constructed with colored cement met this requirement. Three (3) of the prism groups that did not meet this requirement were constructed with mortar Type 6 (F-6, Fa-6, and G-6), 2 were constructed with mortar Type 4 (Fa-4 and G-4), and 1 was constructed with mortar Type 3 (Fa-3).

At 28 days, the prism Groups F-6, G-6, and G-4 constructed with colored cement and prism Group Fa-4 constructed with gray cement failed to meet both of the specified strength requirements. These four prism groups were constructed with mortar containing the maximum amount of lime permitted under ASTM C 270 for a Type S mortar, and three prism groups contained the maximum amount of sand permitted.

Long-Term Test Results

At four times, between approximately 250 days and 1,600 days, a total of 885 bond wrench tests were conducted on 105 prisms in 12 groups. Ninety-two (92) of these tests were to a load of 500 lbs (227 kg), the limits of the test machine. The 500-lb (227 kg) load corresponds to a stress range between 367 (2,530 kPa) and 411 psi (2,834 kPa). The data presented includes the corresponding stress level at 500 lbs (227 kg). Including these results reduces both the average strength and the standard deviations.

All of the prism groups tested met the project specification for 144 psi (993 kPa) average bond test strength, while 34 out of 48 met the 96 psi (662 kPa) requirement. Eleven (11) prism groups constructed with colored mortars and three prism groups constructed with gray mortar failed to meet the 96 psi (662 kPa) requirement. Eight (8) of the 11 groups constructed with colored mortar included Type F brick. Two (2) of three constructed with gray mortar included Type H brick. The average bond strengths of the 14 prism groups that failed to meet this requirement included six at less than 200 psi (1,379 kPa), three between 200 (1,379) and 230 psi (1,586 kPa), and five greater than 230 psi (1,586 kPa). Six (6) out of the 14 prism groups that failed to meet this requirement had individual minimum test scores of less than 96 psi (662 kPa), and eight had minimum scores greater than 96 psi (662 kPa).

In all of the long-term tests, Prism Groups F-2, G-2, H-2, Fa-5 met both requirements for an average bond test strength greater than 144 psi (993 kPa), and the 99th percentile lower strength limit test value greater than 96 psi (662 kPa).

CONCLUSIONS

A wide variation in average bond strengths and standard deviations was observed in all three pilot tests. At 28 days, the Type S mortar average bond strengths varied from 67 (462) to 119 psi (820 kPa) for the first pilot program, from 69 (476) to 173 (1,193 kPa) psi for the second program, and from 101 (696) to 228 psi (1,572 kPa) for the third program. The standard deviation varied from 23.60 (163) to 31.21 psi (215 kPa) for the first pilot program, from 5.90 (41) to 62.15 psi (429 kPa) for the second program, and 11.16 (77) to 45.53 psi (314 kPa) for the third program. These wide variations continued through the long-term testing. At 1,600 days, the average bond strengths varied from 242 (1,669) to 352 psi (2,427 kPa) and the standard deviations varied from 39.50 (272) to 91.85 psi (633 kPa).

From these results, it is easy to conclude that all Type S mortars are not the same when it comes to reliable and predictable bond strengths. The allowable range of material content permitted under ASTM C 270 results in a significant variation in strengths as demonstrated by these test observations.

Items that tend to increase the bond test strength include:

- Presoaking of the brick prior to lay-up.
- Reducing lime to minimum allowable.
- Reducing the sand content.
- Maximizing the fluidity of the mortar mix.
- Laying the brick in running bond wallettes and saw-cutting them into test prisms.

RECOMMENDATIONS

Based on the observations and test data, the following recommendations are made. Specifications for construction using steel stud brick veneer technologies must include:

- Consideration of the brick portion of the SS/BV system as a structural element having a flexural bond strength requirement.
- Determination of the actual flexural bond strengths of the materials selected based on compatibility testing.
- The selection of an appropriate safety factor between the design requirements and the test results.
- A complete inspection and testing program confirming compliance with the specifications.

Requiring the brick portion of the SS/BV system to have a minimum flexural bond strength necessitates the designer to select a member size and strength based on a confidence level similar to other structural elements. A confidence level between 90% and 99% is common for members having a structural strength requirement.

Bond wrench testing results in the ultimate strength of the masonry and therefore an appropriate factor of safety must be included in determining the allowable design strength. A factor of safety of 4 on the mean is common. However, the large variation in test results (large standard deviation) observed during this study, indicates that this may not be adequate. After completing a confidence level study, a minimum test value and test sample size can be determined. A factor of safety of 2 on the minimum test value is recommended.

Once the design performance criteria, including allowable stress levels and factors of safety, are selected, compatibility tests must be conducted to determine the adequacy of the materials selected. The project specifications for materials and mixing ratios must be based on the results of these tests, and will most likely be more specific than just Type S mortar. This "tighter specification" will require a complete inspection and testing program during construction in order to verify continued compliance with the specifications.

REFERENCES

[1]   Gross, J. G., Dikkers, R. D., Crogan, J. C., <u>Recommended Practice for Engineered Brick Masonry</u>, Brick Institute of America, McLean, Virginia, 1975

[2]   Natrella, M. G., <u>Experimental Statistic</u>, NBS Handbook 91, U.S. Government Printing Office, Washington, D.C., 1966

[3]   Plummer, H. C., <u>Brick and Tile Engineering</u>, Structural Clay Products Institute, Washington, D.C., 1962

DISCUSSION

J. M. Melander[1] (written discussion) – The author presents some interesting data. However, the emphasis on the flexural bond strength of the masonry veneer to solve problems of the steel stud/masonry veneer system has inherent limitations. I would ask the author to comment on the following issues that such an approach fails to recognize:
1. Water penetration of masonry walls is not necessarily related to the bond strength of mortar to units.
2. Increasing the strength of mortars reduces the workability, board life, and water retentivity of plastic mortar and reduces the extensibility of the hardened mortar. While the measures taken may maximize the bond strength of test panels and prisms, they could reduce the overall level of performance of the masonry under jobsite construction conditions.
3. Both laboratory tests and field experience indicate that a four inch masonry veneer does not provide an effective barrier to water penetration.
4. As noted, variability of bond strength test results is high. The precision and bias of the test methods have not been established. Without such information, one cannot effectively distinguish between testing variability and materials performance variability.
5. The basic design concept of a masonry veneer drainage wall is that loads on the veneer are transferred to the back-up. Requiring the stiffness of the back-up to be compatible with the veneer would seem to be a more workable approach than designing the masonry veneer as a structural flexural member.

---

[1]Masonry Specialist, Engineered Structures and Codes, Portland Cement Association, Skokie, IL.

Krauklis, A. T. (author's closure)

The author would like to thank Mr. Melander for his interest, comments and discussion.

The purpose of the paper was to discuss one aspect of SS/BV construction, that appears to be neglected by the "cook book" design methods of selecting a steel backup system. Selecting a steel back-up system based on a deflection criteria that is linear (L/???), when deflection for uniform loads is a function of L^4, appears to be irrational. The author never intended to imply that this paper addresses all of the issues of SS/BV.

The original outline for this paper included the observations made during permeance testing of the various mortar mixes. In order to reduce the length of the paper, this area was deleted. The author agrees that water penetration of masonry walls is not related to bond strength. The permeance test results for the first set of samples were relatively good, while the bond strengths were relatively poor.

It is commonly understood that water penetrates a 4-inch masonry veneer and, therefore, requires a complete waterproofing and flashing system to control moisture infiltration. This waterproofing system must also provide protection of the backup system. In the case of steel studs, the susceptibility to water damage and corrosion must be addressed in the waterproofing details.

The author was careful to remain practical in selection of the criteria for construction. The fluidity of the mortar was increased only to the point were the masons remained comfortable. Spot checking the fluidity was easily verified by the vertical trowel test. The lime reduction appears to have been offset by the fineness of the locally available sand, and presoaking of the brick resulted in a few rumblings from the laborers. The only complaints from the bricklayers were noted during the cooler months of construction, as to their hands being cold. The overall production was reduced, not so much by the items indicated above, but by the rigidly enforced requirement to keep the cavity clean of mortar droppings.

It is commonly know that test results will vary from one laboratory to another, and even from one technician to another. However, the dramatic increase in bond strength between the first two test groups and the third is significantly more than the bias of the test method. These tests were conducted using the same equipment, the same procedures and the same technicians. This not to say that all of the test bias has been eliminated, but it has been reduced enough to rely on the trends indicated by the results.

Although the design of SS/BV was not the focus of this paper, the structural mechanism was discussed in general terms. The brick veneer stress requirements are related to the backup, ties, and masonry support end conditions and, therefore, are affected by them. By understanding these relationships and providing details that reflect this understanding, a designer can properly size each of the members. The selection of backup material in combination with tie selection, and the location of control joints can effectively reduce the stresses in the masonry. Just providing a stiffer backup may not reduce the masonry veneer stresses as effectively as a well-placed control joint or wall tie.

DISCUSSION

J. Gregg Borchelt[1]   This paper presents valuable data related to flexural tensile stress of various brick and mortar combinations. The number of variables is rather daunting and makes drawing appropriate conclusions rather difficult. I do not agree with all of the variables (such as wetting all of the brick) or all of the test methods (using the pieces from the E 518 test for C1072 specimens) but I choose not to discuss these. There are major concerns though, that must be mentioned:

1.   Mortars used in Pilot Tests #1 and #2 are within the proportions of Type N and Type O, not Type S and Type N. Lime contents, based on a density of 40 lb/ft^3 (640 kg/m^3), range from 0.6625 to 1.325. Pilot Test #3 does use a Type S mortar by proportions. Although mortar is proportioned in the field by dividing 50 lb (23 kg) bags of lime this practice is a violation of C 270.--Does the author feel that this practice should be stopped or the requirements of C 270 changed?

2.   The differences in mortar type change the allowable flexural stress used to establish the project requirements. Brent Gabby's paper on Flexural Bond Stress in ASTM STP 1063 reports an ASTM E 72 air bag test, on which building code allowable flexural tensile stresses are based, with average values of 131 psi (903 kPa) for Type S mortar and 102 psi (704 kPa) for Type N mortar.--Does the author feel that using these values to establish project requirements is more valid?

3.   Correlation between the various test methods used to evaluate flexural bond strength has not been established. Gabby's paper provides a comparison of average values which indicates that reported C 1072 averages are 96% and 92% of E 72 air bag averages. But this is a global comparison and it is not verified by comparative testing.--Does the author feel it is appropriate to relate C 1072 values to allowable flexural bond stresses?

---

[1]Director, Engineering and Research, Brick Institute of America, Reston, VA.

J. Gregg Borchelt[1]...continued

4.   The premise for this paper is its application to brick veneer with steel stud backing. Further, the author states that the brick veneer wythe should be treated as a structural element. In my opinion neither of these assumptions are correct. The exterior wythe of a veneer wall is not required to meet the allowable flexural tensile stresses of engineered masonry. The backing material, whatever it is, can be designed to limit cracking in the veneer wythe. Load distribution means component end conditions and the performance must be known in order to provide an adequate analysis. The wall system can be designed to prevent cracking in the veneer wythe if the cracking stress is known. An additional piece of information about this research will help supply part of that information (assuming item 3 can be satisfactorily answered).--What curing condition was used for the masonry prisms? Please include those prisms cut from the water penetration specimens and those cured for 1600+ days.

Krauklis, A. T. (author's closure)

This author would like to thank Mr. Borchelt for his interest, comments and discussions.

The two items mentioned in his introductory paragraph need to be addressed. It is this author's belief that the presoaking of bricks is part of the solution to increase masonry bond strength. It may be as simple as washing the brick to remove contaminants (dirt and dust) or it may be more. The test results indicate a significant bond strength increase between the soaked and non-soaked samples, for bricks with a variety of IRAs. More research is required to confirm just what the effect of presoaking is.

The author took precautions in order to limit the damage to the surviving portions of the ASTM E 518 tests, including the placement of a foam cushion under the test specimens. Damage would result in a conservative strength capacity in the ASTM C 1072 test. Although a formal comparison of the results of the two test methods was not conducted, the results appear to indicate capacities in the same range.

1.   The author believes that there is a problem with the traditional method of specifications, and construction of mortar. A specified Type S mortar, may not be a Type S mortar when proportioned and mixed by the traditional methods of masonry construction.

In the first set of tests, the Architect/Engineer specified mortar types by volume, and the Contractor mixed the mortar by two methods, the volume box and the bag method. In both cases the Architect/Engineer and the Contractor thought they were getting a specific type of mortar (N, N/S, and S), but actually, they were getting something less.

The second set of tests continued the study of the common practice of mixing mortar by the "Bag Method". The resulting mortars were less than the Type S as specified. Both sets of tests indicate that if you depend on traditional specification and construction methods you can get less than the type mortar you specified, and it may not have the structural properties required.

In the third set of tests, those mortars mixed with colored cement also may or may not have met the Type "S" requirements of ASTM 270 as the color pigments reduces the amount of Portland Cement in the 94 lb. bag. The purpose of this paper was not to challenge the traditional construction practices but to understand them. The results of these studies indicate that when reliable bond strength is required, the traditional methods of specification and construction may not be satisfactory, and compatibility studies may be a better approach.

2.& 3.   A correlation to the "Air Bag" method, as described in ASTM E 72, was not conducted. The author would like to respond to these two discussion points concerning the comparison of average test strength results. A wide variance of results was observed in the two test methods used in this paper and, therefore, the average test results alone are meaningless. One must account for both the average test results and the amount of variance. A comparison of the reliable test results at a specific confidence level for the various test methods may prove more useful. This topic has the potential for future studies.

4.   All building elements, which are subject to internal and/or external loadings, have structural strength requirements. Brick veneer is no exception. Therefore, this author disagrees with the comment that:

> "it is not required to meet the allowable flexural tensile stresses.."

Limiting cracking is not the solution to durable brick construction, when an uncracked wall can be achieved.

Brick veneer is an element commonly subjected to loading from wind, which results in stresses that must be resisted by the element's structural strength capacity. The brick veneer stress requirements are related to the backup, ties, and masonry support end conditions and, therefore, are affected by them. By understanding these relations and providing details that reflect this understanding, a designer can properly size each of the members. The selection of backup material in combination with the selection of ties and the location of control joints can affect the brick stresses.

The method of curing is an age old question. The object of this study was not to produce a laboratory strength but to produce a strength that was representative of the expectations for this repair project. Samples for Pilot Tests #1, #2, and #3 were built in May, June, & August, respectively. All three tests were cured outdoors in an uncontrolled environment, similar to job-site conditions. The 14 and 28-day test prisms were constructed outside where they remained until testing. The tops of the samples were protected from rain.

The long-term test samples were also made, cut into prisms, and stored outside, until approximately 2 or 3 days prior to testing, at which time the samples were moved into the laboratory. The long-term prism samples were subjected to the weather conditions of the Chicago area.

John M. Melander[1], Satyendra K. Ghosh[2], Val S. Dubovoy[3], Edwin G. Hedstrom[4], and Richard E. Klingner[5]

FLEXURAL BOND STRENGTH OF CONCRETE MASONRY PRISMS USING MASONRY CEMENT MORTARS

REFERENCE: Melander, J. M., Ghosh, S. K.; Dubovoy, V. S., Hedstrom, E. G., and Klingner, R. E., "**Flexural Bond Strength of Concrete Masonry Prisms Using Masonry Cement Mortars,**" _Masonry: Design and Construction, Problems and Repair, ASTM STP 1180_, J. M. Melander and L. R. Lauersdorf, Eds., American Society for Testing and Materials, Philadelphia, 1993.

ABSTRACT: Results of a test program to determine flexural bond strengths associated with Type M, S, and N portland cement/lime mortars under controlled test conditions were reported in the February 1991 issue of _The Masonry Society Journal_. Subsequent to that investigation on portland cement/lime mortars, similar tests were performed using masonry cement mortars. In these tests, flexural bond strengths were determined for Type S and Type N masonry cement mortars, using special standard concrete brick. The laboratory tests were designed to control materials, prism fabrication, curing, and testing. The mortar materials included Type S and Type N single-bag masonry cements and standard Ottawa sand. Results of tests conducted at Construction Technology Laboratories, Inc. (CTL), at the research laboratory of the National Concrete Masonry Association (NCMA), and at the University of Texas at Austin (UTA) are summarized and discussed in this paper.

For each of the mortar mixes, the initial flow, water retention, air content, and cube compressive strengths at 7 and 28 days were determined. Six prisms, each consisting of six standard concrete brick and five mortar joints, were fabricated and tested for flexural bond strength at 28 days according to ASTM C 1072-86, Method for Measurement

---

[1]Masonry Specialist, Engineered Structures & Codes, Portland Cement Association, Skokie IL 60077.

[2]Director, Engineered Structures & Codes, Portland Cement Association, Skokie, IL 60077.

[3]Senior Masonry Engineer, Construction Technology Laboratories, Skokie, IL 60077.

[4]Director of Research, National Concrete Masonry Association, Herndon, VA 22070.

[5]Phil M. Ferguson Professor in Civil Engineering, The University of Texas, Austin, TX 78712.

of Masonry Flexural Bond Strength. Tests of some mortars were
duplicated in different laboratories. The mean, standard deviation, and
coefficient of variation were calculated for the 30 joints tested in
each test series. Results provide a general indication of the flexural
bond strength of masonry cement mortars under the test conditions of the
program.

KEYWORDS: bond wrench testing, concrete masonry units, concrete
brick, masonry, mortar, flexural bond strength.

<hr>

     At the request of the International Conference of Building
Officials (ICBO), a task group established by the Masonry Industry Code
Committee (MICC) coordinated a testing program to determine the
statistical variation of the flexural bond strength of masonry cement
mortars currently used in the United States. This program was intended
as a sequel to a similar testing program conducted on portland
cement/lime mortars [1]. In this and in the previous testing program,
an effort was made to eliminate as many variables as possible in the
mortar properties, the prism construction, and the testing procedures.
     As a result of the testing program described here, MICC has
recommended and ICBO has adopted ratios of allowable tensile bond
stresses between masonry cement mortars and portland cement/lime
mortars. The results of this test program could also be used for
comparative purposes in other research activities with regard to
flexural bond. Results do not reflect actual bond strength values in
structures.

TESTING PROGRAM

<u>Testing Laboratories</u>

     The test program was carried out by the research laboratories of
the National Concrete Masonry Association (NCMA), Construction
Technology Laboratories, Inc. (CTL), and the University of Texas at
Austin (UTA). The Portland Cement Association sponsored the testing at
NCMA and CTL. The National Lime Association sponsored the testing at
UTA.

<u>Materials Tested</u>

     Materials tested included masonry cement, sand, and units.

     <u>Masonry cement</u>--All mortars were made from single-bag masonry
cements, with no added portland cement. At UT Austin, 6 Type S and 2
Type N masonry cement mortars were tested. The Type S mortars were made
using 5 masonry cements, selected at random as described below, and one
Type S masonry cement which had been tested previously by Chemstar Lime.
The Type N masonry mortars were made using two masonry cements
previously tested by Chemstar Lime. At NCMA, mortars were made using 5
Type S and 7 Type N masonry cements. At CTL, mortars were made using 6
Type S and 4 Type N masonry cements. Selection of all masonry cements
was not coordinated between laboratories, although CTL did duplicate

some of the NCMA tests. Some of the data from the NCMA and CTL tests were obtained in conjunction with other test programs.

Standard Ottawa silica sand--In addition to masonry cement and water, mortar included a 50/50 blend by weight of standard graded and 20/30 Ottawa silica sand.

Standard concrete brick--As in the first MICC study [1], units were the standard concrete brick produced by NCMA in accordance with the requirements outlined in the Uniform Building Code Standard Test Method for Flexural Bond Strength of Mortar Cement (UBC 24-30) [2].

Procedure used to select Type S masonry cements for test at UT Austin--

1)   A list was compiled of all Type S masonry cements currently marketed in bag form in the United States. In compiling the list, identical brands produced at different plants were considered as separate cements. A total of 61 cements were identified.

2)   The 61 Type S masonry cements were assigned numbers from 1 to 61, and 8 numbers were selected with a random number generator.

3)   The first 5 cements selected were ordered from retail points of supply.

4)   In addition, one Type S masonry cement and two Type N masonry cements previously tested in Chemstar Lime's laboratory (Henderson, NV), were selected for verification testing. The Type S masonry cement had been excluded from the above list of 61.

Mortar designations--At UT Austin, the 6 Type S and 2 Type N masonry cements were numbered from 1 to 8 (Type S cements are numbered 1 through 6; the Type N cements, 7 and 8). Each masonry cement - sand combination was designated by the cement designation number, followed by an "N" or an "S" as appropriate. For example, the nomenclature "1S" denotes the mortar made with Masonry Cement 1 (Type S). The mortar designations at UTA did not correspond with those used at the other laboratories, and none of the masonry cements tested at UTA was tested at the other two laboratories.
At NCMA, the Type S mortars were designated S-2, S-3, S-5, S-6, and S-7; the Type N mortars were numbered N-1 through N-7. At CTL, the Type S mortars were designated S-7, 8051, 8052, 8053, 8054, and 8055; the Type N mortars were designated N-1 through N-4. The numbering systems used at NCMA and at CTL do not correspond completely; however, those mortars with the same designations are the same. Thus, one Type S masonry cement (S-7) and four Type N masonry cements (N-1, N-2, N-3, and N-4) were tested in duplicate at NCMA and at CTL.
At NCMA, tests were also conducted on two other Type S mortars, designated S-1 and S-4. Because the data from those two series were not available when the MICC group considered the overall test results, they were not included in the set of data that formed the basis for the MICC recommendations, and they are not discussed further in this paper. Those interested in obtaining these data should contact the first author.

Testing Procedures

Mortar batching, mixing, and standard material tests--At all three laboratories, the different combinations of Type S and N single-bag masonry cements were proportioned by Uniform Building Code Standard on Mortar for Unit Masonry and Reinforced Masonry Other Than Gypsum (UBC 24-20) [3] (identical to ASTM C 270, Specification for Mortar for Unit Masonry), using a ratio of one part (by volume) of masonry cement to three parts (by volume) of blended Ottawa silica sand. The blended standard Ottawa silica sand used in all the mortars was determined to have a unit weight of 100 lb/ft^3(1600 kg/m^3). A batch size of 1/2 cubic foot (0.014 cubic meters), requiring 50 lb (22.7 kg) of blended sand, was selected to assure sufficient mortar to perform the required mortar tests and fabricate the prisms. Mixing was carried out in accordance with ASTM C 780, Test Method for Preconstruction and Construction Evaluation of Mortars for Plain and Reinforced Unit Masonry, with an initial flow of 125 ± 5%. As specified by UBC 24-30 [2], the following properties were determined for each mortar: initial flow, water retention, air content, initial cone penetration, cone penetration at 15-minute intervals, and cube compressive strength at 7 and 28 days.

    <u>Prism fabrication</u>--Each mortar was used to fabricate 6 prisms, each consisting of six standard concrete brick and five mortar joints. In order to reduce the effects of workmanship in fabricating the prisms, construction followed the procedure of UBC 24-30 [2], using a jig, template form, and drop hammer to reduce the human factor in laying the concrete brick.
    Cone penetrometer readings were taken at 15-minute intervals during prism fabrication, and mortar that had lost over 20% of its initial cone penetration was not used. Prism fabrication time never exceeded 45 minutes, and none of the mortars had less than 80% of the initial cone penetration at the completion of fabrication.

    The prisms were fabricated on 1/2-in. (12.7-mm) plywood pallets using alignment jigs, shown in Fig. 1. The six bolts on each of the three angles were aligned true and plumb.

    The mortar for each joint was placed using a template form as shown in Fig. 2. The template form was made of 1/4-in. (6.4-mm) Plexiglass having inside dimensions slightly larger than the standard concrete unit. Its depth was 1 in. (25.4 mm), and two screws on each end were positioned to

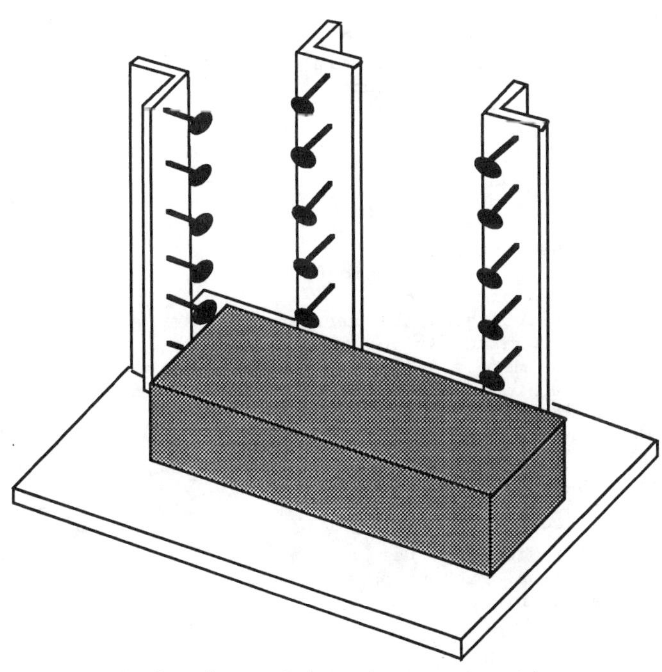

FIG. 1--Jig used in prism construction.

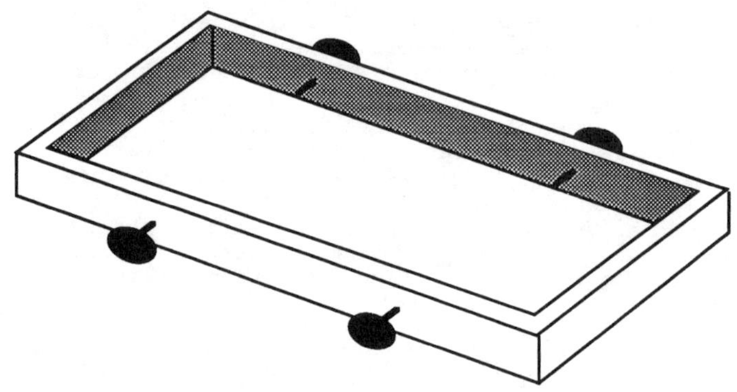

FIG. 2--Mortar joint template used in prism construction.

give a 1/2-in.-thick (12.7-mm-thick) mortar joint prior to compaction.
The template form was placed on the concrete unit. Mortar was placed in the template and struck off with a straight edge. The template was then removed.
Immediately after the removal of the template, the next concrete unit was placed on the mortar bed in contact with the three alignment bolts for that course. A bulls-eye level was used on top of each unit during placement to assure that the initial thickness of the joint was uniform.

Each bed joint was then consolidated using a drop hammer, shown in Fig. 3. The hammer was placed on top of the concrete unit, and its 4-lb (1.8-kg) weight was dropped once from a height of 1.5 in. (38.1 mm). The compacted mortar joint was approximately 3/8 in. (9.5 mm) thick. The six prisms were constructed continuously, one course at a time, until the six courses were completed. Mortar joints were then cut flush with no tooling.

Immediately after construction, each prism was covered with a plastic bag,

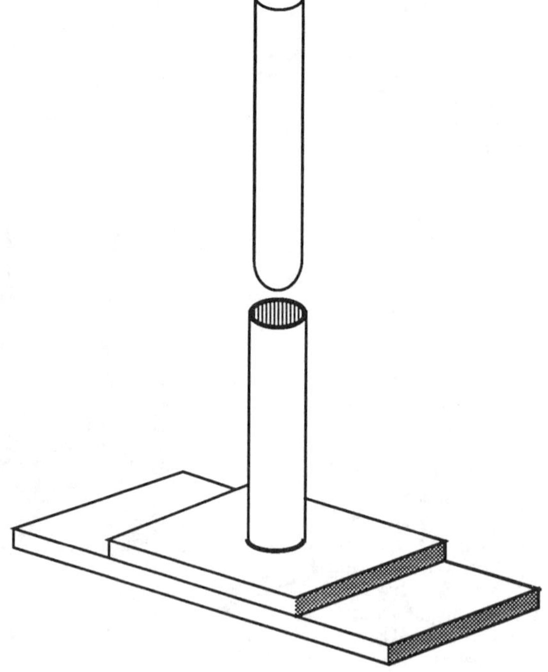

FIG. 3--Drop hammer used in prism construction.

TABLE 1--<u>Summary of physical properties and bond wrench data,
Type S masonry cement mortars.</u>

Mortar Mix ID	Added[A] Water, lb	Init. Flow, %	Water Ret., %	Cone Penet., mm	Air Cont., %	Unit[B] Wt., lb/ft^3	Cube[C] Strength 28 Day, psi	Bond Strength 28 Day		
								Mean, psi	Std Dev, psi	COV, %
NCMA										
S-2	7.44	120	89	70	25.3	106.9	1730	128	18	14
S-3	7.52	130	92	74	18.0	116.7	2490	148	37	25
S-5	7.12	124	87	66	18.7	116.7	2150	103	23	22
S-6	7.30	121	88	69	17.6	118.1	1720	130	25	19
S-7	7.70	127	88	70	15.3	119.5	1880	132	26	20
CTL										
S-7	7.59	124	84	64	15.3	119.7	1800	114	20	17
8051	6.60	120	73	50	17.6	119.2	1940	99	19	20
8052	8.20	122	79	48	17.8	117.6	3290	110	22	20
8053	7.10	127	75	48	18.1	116.7	1960	105	20	19
8054	6.70	120	83	46	20.1	114.7	2660	126	22	17
8055	6.80	128	75	45	15.0	121.3	2880	166	26	16
UT Austin										
1S	8.58	127	84	59	16.3	115.4	1263	54	11	20
2S	10.28	130	88	68	20.0	113.6	1401	96	19	20
3S	9.34	123	80	61	14.5	121.6	2052	92	18	19
4S	9.98	124	87	73	13.0	116.4	1521	103	22	21
5S	10.20	128	91	72	12.8	123.1	1482	102	17	17
6S	9.68	121	81	63	17.0	125.0	2073	87	18	21

[A] kg = lb X 0.4535
[B] kg/m^3 = lb/ft^3 X 16.02
[C] MPa = psi X 0.006895

temporarily sealed to the working surface, and left in place on its pallet for 20 ± 4 hours. The next day, each prism was uncovered and removed from its jig, placed and sealed in the same plastic bag, and moved to a storage area in the laboratory for bond wrench testing at 28 days.

<u>Prism testing</u>--Flexural bond strength was determined by bond wrench testing according to UBC 24-30 [2] and ASTM C 1072-86. For every mortar, flexural bond strength was determined for each of 30 joints; the mean, standard deviation, and coefficient of variation (COV) were calculated.

TEST RESULTS

Tables 1 and 2 present a summary of the physical properties and bond wrench data (mean flexural bond strength, standard deviation and coefficient of variation) for all mortars tested by the three participating labs. Flexural bond strength results for all mortar joints (nominally 30 joints per mortar combination) tested by the three labs are presented in Appendix A. The five joints of each prism are grouped together, and are given from top to bottom. The calculated values for mean, standard deviation, and coefficient of variation in Tables 1 and 2 were determined from the test data carried to two decimal places rather than from the rounded values shown in Appendix A.

SIGNIFICANCE OF TEST RESULTS

The immediate objective of these tests was to establish characteristic flexural bond strength values for masonry cement mortars, and to compare those values with corresponding characteristic flexural bond strength values for portland cement/lime mortars, obtained in the previous MICC study [1]. By agreement within MICC, the mortar strengths were to be compared in terms of lower characteristic values (values set at the limit below 90% of the test data). Ratios of those lower characteristic strengths were to be used to set corresponding ratios of allowable tensile bond stress. For a normal distribution, such a 10% fractile is computed as the mean minus 1.28 standard deviations.

Using all results from Appendix A, the following lower characteristic bond strength values were obtained for the Type S and Type N masonry cement mortars tested in this study:

Type S masonry cement mortar: 69 psi (0.48 MPa)
Type N masonry cement mortar: 44 psi (0.30 MPa)

TABLE 2--_Summary of physical properties and bond wrench data,_
Type N masonry cement mortars

Mortar Mix ID	Added[A] Water, lb	Init. Flow, %	Water Ret., %	Cone Penet., mm	Air Cont., %	Unit[B] Wt., lb/ft^3	Cube[C] Strength 28 Day, psi	Bond Strength 28 Day Mean, psi	Std Dev, psi	COV, %
NCMA										
N-1	7.18	126	91	75	21.4	112.0	690	66	13	20
N-2	7.10	122	89	70	24.5	108.0	640	50	6	12
N-3	7.24	130	94	74	18.9	115.6	740	56	10	17
N-4	7.46	127	91	72	16.1	119.2	1490	65	13	20
N-5	7.30	124	92	71	18.7	115.6	770	60	9	15
N-6	7.84	126	94	75	15.2	119.7	870	71	13	18
N-7	7.04	121	92	70	21.2	112.8	1030	69	13	19
CTL										
N-1	7.10	130	96	69	19.8	112.7	600	77	15	20
N-2	7.32	123	93	70	25.2	106.5	600	49	18	36
N-3	7.15	125	90	67	22.2	111.3	780	65	15	23
N-4	7.65	124	91	72	16.3	118.5	1440	75	19	25
UT Austin										
7N	8.60	129	83	65	13.8	121.2	1535	92	13	14
8N	8.86	125	92	69	21.0	109.0	598	68	14	20

[A]kg = lb X 0.4535
[B]kg/m^3 = lb/ft^3 X 16.02
[C]MPa = psi X 0.006895

Using all results from the previous MICC study of portland cement/lime mortars [1], except those for one portland cement with entrained air, the following lower characteristic bond strength values had been obtained:

    Type S portland cement/lime mortar: 126 psi (0.87 MPa)
    Type N portland cement/lime mortar: 90 psi (0.62 MPa)

Those two sets of lower characteristic values were used to calculate the following ratios between the flexural bond strengths of masonry cement mortar and non-air-entrained portland cement/lime mortar:

        Type S mortar: 0.55
        Type N mortar: 0.50

After consideration of those ratios, MICC agreed to submit the following allowable bond stress ratios (for masonry cement mortar, compared to portland cement/lime mortar) to ICBO for use in the Uniform Building Code:

        Type S mortar: 0.60
        Type N mortar: 0.50

CONCLUSIONS

The immediate objective of these tests was to establish characteristic flexural bond strength values for masonry cement mortars, and to compare those values with corresponding characteristic flexural bond strength values for portland cement/lime mortars, obtained in the previous MICC study [1]. To establish those values, tests were conducted in three different laboratories to establish the statistical variation of flexural bond strengths using the bond wrench test.

Despite the carefully designed test program, which eliminated many possible sources of differences between labs, not all factors could be completely controlled. For example, the amount of water to be used for each mortar was specified as that which would provide the desired flow of 125 ± 5%. Tables 1 and 2 show that water weights varied somewhat between labs, resulting in different water/cement ratios. Also, different mixers at each laboratory may have produced different air contents. Factors such as these would be expected to create some differences in flexural bond strength values between laboratories.

However, as shown in Fig. 4, the average bond strengths for those masonry cement mortars tested in duplicate did not vary consistently between the two research laboratories (NCMA and CTL) that conducted duplicate tests. Also, the results within all three laboratories showed coefficients of variation which were usually 25% or less.

The test results given here were obtained using strictly controlled materials, construction techniques, and curing conditions. The flexural bond strengths obtained in this test program are intended only to provide a means of evaluating and comparing materials, and should not be used as design values or compared directly with allowable flexural bond stresses prescribed by masonry building codes.

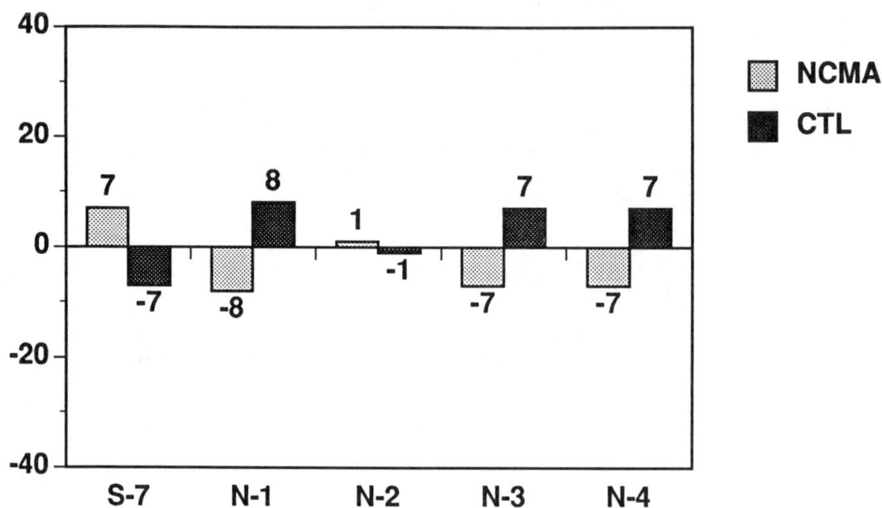

FIG. 4--Comparison of Bond Wrench Results, Type S and Type N Masonry
Cement Mortars

ACKNOWLEDGMENTS

The authors would like to thank the Portland Cement Association and
the National Lime Association for sponsoring the testing program. The
opinions and findings expressed in this paper are those of the authors,
and not necessarily those of the sponsors. The authors appreciate the
excellent work and enthusiasm of all who worked hard to complete this
project. The authors are also very appreciative of the input of the
Masonry Industry Code Committee (MICC), chaired by Stu Beavers. Testing
at UT Austin was conducted by the last author with the assistance of
Graduate Research Assistants Deepak Ahuja and Chris Higgins.

REFERENCES

[1]   Hedstrom, E. G., Tarhini, K. M., Thomas, R. D., Dubovoy, V. S.,
       Klingner, R. E., and Cook, R. A., "Flexural Bond Strength of
       Concrete Masonry Prisms using Portland Cement and Hydrated Lime
       Mortars," The Masonry Society Journal, Vol. 9, No. 2, February
       1991, pp. 8-23.

[2]   "Uniform Building Code Standard No. 24-30 (Standard Test Method
       for Flexural Bond Strength of Mortar Cement)," International
       Conference of Building Officials, Whittier, CA, 1991.

[3]   "Uniform Building Code Standard No. 24-20 (Mortar for Unit Masonry
       and Reinforced Masonry Other Than Gypsum)," International
       Conference of Building Officials, Whittier, CA, 1988.

APPENDIX A - COMPLETE BOND WRENCH RESULTS

TABLE A1--<u>Bond wrench results (psi)</u>[A] <u>from NCMA and CTL,</u>
<u>Type S masonry cement mortars.</u>

Joint No.	NCMA - Mortar Mix ID					CTL - Mortar Mix ID					
	S-2	S-3	S-5	S-6	S-7	S-7	8051	8052	8053	8054	8055
1	159	233	117	139	133	129	77	128	69	90	133
2	77	...	111	135	135	109	120	...	74	126	143
3	136	177	86	155	120	100	99	82	99	132	189
4	103	223	81	174	92	104	133	...	87	112	134
5	122	217	120	163	102	142	100	116	117	117	232
6	152	180	119	192	143	133	79	108	108	126	137
7	120	117	127	...	137	92	128	99	87	89	192
8	140	118	146	133	173	109	72	118	86	139	182
9	145	132	121	187	159	112	94	113	130	125	151
10	117	115	95	122	150	140	107	150	119	144	221
11	143	206	70	139	182	135	106	99	99	150	167
12	...	115	109	140	179	80	126	63	96	146	175
13	121	126	82	124	120	97	118	104	119	137	168
14	122	121	91	123	118	95	114	128	80	137	158
15	117	127	141	107	161	...	95	143	129	139	187
16	117	181	121	113	161	113	63	70	129	116	147
17	113	186	80	108	129	123	108	123	88	113	141
18	146	151	118	114	160	82	125	96	103	101	132
19	132	133	106	115	126	90	89	122	137	123	157
20	136	124	130	122	155	146	69	134	113	173	169
21	140	149	66	118	121	117	108	126	124	120	168
22	121	121	116	117	107	113	96	80	100	106	158
23	123	111	68	109	110	93	121	99	104	114	150
24	116	123	118	10	100	109	83	94	126	115	165
25	159	144	105	115	122	156	99	94	98	145	204
26	112	175	111	106	133	117	86	123	92	157	150
27	151	125	65	117	96	120	107	111	88	144	144
28	114	131	100	109	104	114	98	97	87	73	204
29	117	112	66	108	117	95	63	143	111	120	142
30	142	109	117	157	101	128	91	118	152	142	165
Mean	128	148	103	130	132	114	99	110	105	126	166
Std Dev	18	38	23	25	26	19	19	22	20	22	26
COV, %	14	25	22	19	20	17	20	20	19	17	16

[A]MPa = psi X 0.006895

TABLE A2--Bond wrench test results (psi)[A] from UT Austin, Type S masonry cement mortars.

Joint No.	UT Austin - Mortar Mix ID					
	1S	2S	3S	4S	5S	6S
1	63	124	98	91	83	103
2	34	78	81	88	74	86
3	71	94	71	90	102	77
4	77	84	96	88	103	101
5	59	87	134	122	99	124
6	49	113	79	77	85	94
7	38	95	78	82	86	89
8	77	95	76	79	93	98
9	54	79	110	96	104	118
10	47	102	83	127	119	92
11	39	63	67	84	88	65
12	55	102		65	79	73
13	49	134	90	71	87	114
14	49	102	103	103	110	69
15	52	132	97	100	131	97
16	48	116	56	133	91	102
17	61	95	100	119	84	51
18	68	88	90	99	86	73
19	53	89	77	84	129	100
20	72	96	94	96	110	68
21	57	86	88	100	124	61
22	38	100	97	113	87	105
23	65	81	93	135	99	85
24	41	112	99	126	112	90
25	48	134	144	150	124	100
26	48	74	88	117	104	82
27	54	60	89	87	126	52
28	50	88	91	124	120	84
29	58	76	101	126	86	80
30	58	96	92	131	132	85
Mean	54	96	92	104	102	87
Std Dev	11	19	18	22	18	18
COV, %	20	20	19	21	17	21

[A] MPa = psi X 0.006895

TABLE A3--<u>Bond wrench test results (psi)[A] from NCMA,
Type N masonry cement mortars.</u>

Joint No.	NCMA - Mortar Mix ID						
	N-1	N-2	N-3	N-4	N-5	N-6	N-7
1	72	50	69	54	50	81	63
2	61	43	48	58	56	56	60
3	53	52	45	63	58	61	74
4	76	48	42	80	56	54	71
5	105	48	38	62	62	60	54
6	82	46	55	79	64	63	92
7	51	56	46	67		82	80
8	86	41	43	45	59	49	59
9	67	52	51	51	40	49	60
10	62	44	60	85	56	65	65
11	53	40	65	57	71	80	81
12	50	48	60	51	62	69	75
13	63	57	61	51	58	74	69
14	65	48	53	85	67	73	92
15	64	43	61	74	77	00	07
16	60	49	59	53	58	50	62
17	51	41	58	80	50	71	57
18	75	58	62	57	59	78	48
19	69	54	68	62	74	73	53
20	46	53	54	52	57	85	76
21	80	43	79	88		69	75
22	61	54	68	76	76	55	90
23	56	62	72	55	57	74	55
24	61	51	47	74	52	77	51
25	67	63	55	58	60	88	78
26	84	57	52	41	47	92	70
27	72	54	49	67	55	83	86
28	60	51	51	65	75	86	64
29	62	45	49	79	60	83	84
30	53	51	59	78	70		52
Mean	66	50	56	65	60	71	70
Std Dev	13	6	10	13	9	13	13
COV, %	20	12	17	20	15	18	19

[A]MPa = psi X 0.006895

TABLE A4--<u>Bond wrench test results (psi)[A] from CTL and UT Austin,
Type N masonry cement mortars.</u>

Joint No.	CTL - Mortar Mix ID				UT Austin - Mortar Mix ID	
	N-1	N-2	N-3	N-4	7N	8N
1	74	84	61	74	103	65
2	72	87	52	60	75	56
3	79	36	70	51	99	47
4	67	...	80	34	92	57
5	62	...	63	50	102	49
6	110	52	76	100	112	82
7	76	41	63	57	80	82
8	73	39	44	81	105	74
9	65	34	45	74	95	78
10	68	72	72	94	70	80
11	83	48	83	106	110	64
12	71	64	59	54	93	84
13	71	42	51	75	97	69
14	71	17	30	73	113	57
15	74	56	75	113	102	102
16	83	55	64	81	108	54
17	73	46	48	60	79	69
18	55	30	68	52	102	61
19	89	51	54	71	91	45
20	112	66	95	91	84	69
21	117	65	60	70	68	64
22	64	40	44	79	86	84
23	85	64	70	86	88	73
24	81	40	75	81	77	55
25	87	18	76	104	69	80
26	73	43	65	93	86	63
27	65	49	73	60	89	83
28	73	35	76	80	86	70
29	82	24	74	80	91	51
30	47	63	93	58	95	52
Mean	77	49	65	75	92	67
Std Dev	15	18	15	19	13	14
COV, %	20	36	23	25	14	20

[A] MPa = psi X 0.006895

Brian E. Trimble[1] and Bohdan T. Kulakowski[2]

FRICTIONAL CHARACTERISTICS OF CLAY BRICK PAVERS

REFERENCE: Trimble, B.E., Kulakowski, B.T., "Frictional Characteristics of Clay Brick Pavers", _Masonry: Design and Construction, Problems and Repair, ASTM STP 1180_, J. M. Melander and L.R. Lauersdorf, Eds., American Society for Testing and Materials, Philadelphia, 1993.

ABSTRACT: The use of brick as a paving material has a long history. Over the past few years brick has again been specified in vehicular areas. Engineers are now requesting information on the skid resistance properties of brick. A research project was undertaken to determine brick's skid resistance, as well as to compare brick's frictional characteristics with those of other paving materials. It was also desired to determine if skid resistance correlated with other physical properties of brick.

Tests using the British Pendulum Tester confirmed that these brick have excellent frictional characteristics. It was also determined that skid resistance does not correlate with brick's compressive strength, abrasion index nor cold water absorption.

KEYWORDS: brick, friction, paving, skid resistance.

INTRODUCTION

Brick has been used successfully in vehicular applications in the past. In the early part of the 20th century brick roads were the norm. At the time brick provided a stable, durable paving surface which was fairly safe. As concrete and bituminous roadways became more popular, the use of brick as a major surfacing material dropped off. It wasn't until recently that brick streets became desirable again because of their performance, unique look and their connection with the past in many places. The design of modern roadways is very complex, and designers require testing on all facets of a roadway. This includes testing the skid resistance of paving materials. There has been little data in this country to support the use of brick in vehicular applications. A research program was undertaken to determine the frictional characteristics of clay brick pavers [1]. The objective of the research was to determine the skid resistance of new clay brick pavers as well as to investigate a correlation between skid resistance and other physical properties of brick.

Skid resistance is measured by the dynamic coefficient of friction. It is related to the tire/roadway friction and is therefore applicable to vehicular traffic. Slip resistance is related to pedestrian traffic and is measured by the static coefficient of friction. The test methods used

---

[1] Senior Engineer, Technical Services, Brick Institute of America, Reston, A 22091.

[2] Professor of Mechanical Engineering, Pennsylvania State University, niversity Park, PA 16802.

in this research measure the dynamic coefficient of friction. Although dynamic coefficient of friction is primarily thought to relate to skid resistance, one author believes that these frictional tests are also appropriate for slip resistance [2]. More research needs to be conducted to verify this theory.

SCOPE

The research program consisted of testing a set of 10 brick pavers from 14 different manufacturers for skid resistance, compressive strength, 24 hour cold water absorption and 5 hour boiling water absorption. The brick were tested as received. Three basic categories were identified by method of manufacturing: extruded, molded and drypressed. The pavers were selected to represent the entire range of the types of clay brick pavers commonly available. Therefore, the results obtained would be applicable to the majority of clay brick pavers within their categories.

The pavers, identified by code, are shown in Table 1 with their manufacturing type. The size of the units was the same for all brick, typically 3 5/8 in. by 2 1/4 in. by 7 5/8 in. (92.1 mm by 57.2 mm by 193.7 mm). Two exceptions are Pavers 742 and 610B. Pavers 742 and 610B were 3 in. (76.2 mm) wide which is narrower than the width required by the friction test method. In addition, paver 610B was 3/8 in. (10 mm) thick.

TABLE 1
Types and Codes of Brick Pavers

Extruded – Wirecut	712A
Extruded – Wirecut	300A
Extruded – Wirecut	735
Extruded – Wirecut	1208
Extruded – Wirecut	610A
Extruded – Wirecut	640
Extruded – Smooth	742
Extruded – Smooth	610B
Molded (Water Struck)	106
Molded (Sand Struck)	450
Molded (Sand Struck)	406
Molded (Sand Struck)	712B
Repressed	300B
Drypressed	1165

TEST PROCEDURES

Frictional properties were tested according to ASTM E 303-83 Standard Method for Measuring Surface Frictional Properties Using the British Pendulum Tester. The British Pendulum Tester is a device widely used around the world to measure skid resistance of materials. The value measured by the tester, or British Pendulum Number (BPN), represents the amount of kinetic energy lost when a rubber slider attached to the end of the pendulum arm is propelled over the test surface. The results produced by the British Pendulum tests depend primarily on the surface microtexture, i.e., on the surface deviations from the planar surface with characteristic dimensions of wavelength and amplitude less than 0.02 in. (0.5 mm). In general, microtexture determines the frictional resistance of a dry surface. The effectiveness of the microtexture in generating friction between vehicle tires and pavement surface when the surface is wet depends on the surface drainage ability. The drainage ability is determined in part by the surface macrotexture, which is defined as the deviations of the pavement surface from true planar surface with characteristic dimensions of wavelength and amplitude from 0.02 in. (0.5 mm) up to those that no longer affect tire-pavement interaction.

Measurement of macrotexture, as described by the ASTM E 965-87 Standard Test Method for Measuring Surface Macrotexture Depth Using a Volumetric Technique, requires a test area larger than the area of an individual paver. These measurements should be conducted in the future on actual brick pavements.

Physical properties of brick were tested according to ASTM C 67-89a Standard Test Method of Sampling and Testing Brick and Structural Clay Tile. Compressive strength, 24 hour cold water absorption and 5 hour boiling water absorption tests were conducted on at least five of each type of brick.

RESULTS OF SKID RESISTANCE TESTS

The results of the skid resistance tests are shown in Table 2. The pavers are listed in order of decreasing British Pendulum Number (BPN).

In analyzing the data it appears that sand-struck molded brick tend to have higher skid resistance values. This may be due to the uneven surfaces and sand coating on the molded brick. Also, in the BPN testing procedure, the surface of the paver is flushed with water. Molded brick, which tend to be more absorptive, may be surface dry when tested, similar to being tested in a dry condition. The result is a higher skid resistance value.

In general, the microtexture of brick provides for a skid resistant surface. A smooth surfaced brick such as a die-skin brick will have a low BPN value. Repressing or drypressing the unit will flatten out any applied texture resulting in lower BPN values. Textures such as wirecut, velour, or blade cut will increase the BPN value. The addition of a grit, such as sand, on the paver's surface can also increase the BPN value.

The results show that most new brick have adequate skid resistance. Typically, the minimum acceptable value of BPN is 55. BPN values of

TABLE 2
British Pendulum Tests Results
(Average of 6 Tests)

PAVER CODE	PAVER TYPE	MEAN BPN	STD. DEV.
712B	Molded (SS[a])	87.3	4.58
300A	Extruded-wirecut	82.7	8.25
1165	Drypressed	82.5	5.26
450	Molded (SS)	81.4	2.03
406	Molded (SS)	77.9	3.30
1208	Extruded-wirecut	75.8	5.50
640	Extruded-wirecut	68.9	6.95
712A	Extruded-wirecut	68.8	7.22
735	Extruded-wirecut	68.3	2.71
300B	Repressed	63.8	3.96
106	Molded (WS[b])	61.0	4.73
610A	Extruded-wirecut	51.4	4.58
742[c]	Extruded-smooth	50.9	2.17
610B[c]	Extruded-smooth	42.3	3.83

[a] Sand-Struck   [b] Water-Struck   [c] Narrow Paver

other road surfacing materials in-place including both asphalt and portland cement concrete pavements range from 42 to 92 [1]. The British standard for brick pavers requires brick to have a mean wet skid resistance value of 60 [3]. The Australians recommend a BPN of 50 [4].

Based on recommended values, the pavers can be classified into the three following groups:
- Very Good (712B, 300A, 1165, 450, 406, and 1208)
- Good (640, 712A, 735, 300B, 106)
- Poor (610A)

It must be emphasized that the above classification is somewhat arbitrary since there are no established safety standards for pavement skid resistance measured with the British Pendulum Tester. It should also be stressed that pavers 742 and 610B have not been classified since they were narrower than that specified in the British Pendulum Test.

Furthermore, measuring the skid resistance of a new paver is not a true measure of how the paver will perform in the field due to polishing action of traffic. All paving materials are subject to polishing to some degree. Brick pavers will polish over time reaching an equilibrium point within a year after installation [5].

RESULTS OF BRICK PHYSICAL PROPERTIES TESTS

The compressive strength, 24 hour cold water absorption, and 5 hour boiling water absorption test results are shown in Tables 3 and 4, respectively. The abrasion index was calculated for all pavers in accordance with ASTM C 902-91a Specification for Pedestrian and Light Traffic Paving Brick. These values are shown in Table 5.

TABLE 3
Compressive Strength Results
Average of 5 Tests

PAVER CODE	MEAN COMPRESSIVE STRENGTH, psi (MPa)	STANDARD DEVIATION psi (MPa)
640	16,900 (116.6)	590 (4.1)
735	16,500 (113.8)[a]	650 (4.5)
406	16,070 (110.8)[a]	1,390 (9.6)
300A	14,890 (102.7)	970 (6.7)
712A	14,570 (100.5)[a]	2,040 (14.1)
300B	14,500 (100.0)[a]	830 (5.7)
610A	13,890 (95.8)	840 (5.8)
610B	11,310 (78.0)	1,160 (8.0)
742	10,420 (71.9)[a]	1,240 (8.6)
450	9,480 (65.4)	1,500 (10.3)
1208	8,290 (57.2)[a]	730 (5.0)
106	4,480 (30.9)	940 (6.5)
1165	4,400 (30.3)	760 (5.2)
712B	3,690 (25.4)	440 (3.0)

[a] Average of 4 tests

TABLE 4
Absorption and Saturation Coefficient Results
Avg of 6 Tests

PAVER CODE	24 HR COLD WATER ABSORPTION (%)	5 HR BOILING WATER ABSORPTION (%)	SATURATION COEFFICIENT
106	0.238	1.021	0.23
712A	2.891	4.623	0.63
406	3.314	4.572	0.73
640	4.140	4.492	0.92
450	4.370	6.246	0.70
610A	4.545	5.686	0.80
735	4.748	5.901	0.80
742	5.138	7.113	0.72
300A	5.577	7.080	0.79
300B	6.114	7.635	0.80
1208	6.421	8.508	0.75
610B	7.222	7.560	0.96
712B	7.875	12.261	0.64
1165	17.831	19.746	0.90

TABLE 5
Abrasion Index

PAVER CODE	ABRASION INDEX
106	0.005
712A	0.020
406	0.021
640	0.025
735	0.029
610A	0.033
300A	0.037
300B	0.042
450	0.046
742	0.049
610B	0.064
1208	0.077
712B	0.214
1165	0.405

CORRELATIONS

One purpose for this research was to identify any correlation between skid resistance of brick pavers and brick's physical properties. It was hoped that if skid resistance correlated with any physical property which is regularly tested, a separate test for frictional characteristics of brick would not have to be conducted.

Coefficients of correlation were calculated for combinations of two parameters; skid resistance as one parameter and compressive strength, cold water absorption, boiling water absorption and saturation coefficient as the second parameter. Table 6 shows the correlation coefficients of the pavers. A linear correlation exists when the coefficient is equal to +1.0 or -1.0. Lower values indicate less correlation. The two narrow pavers, 742 and 610B were excluded from the analysis because the uncertainty associated with their BPN values.

The results indicate that little correlation exists between BPN and the abrasion index, compressive strength, cold water absorption or saturation coefficient of brick pavers.

TABLE 6
Correlation Analysis vs. BPN

Abrasion Index	0.510
Compressive Strength	-0.334
Cold Water Absorption	0.460
Saturation Coefficient	0.590

CONCLUSIONS

The results of these tests indicate that brick pavers have good initial values for skid resistance. The texture and the absorption characteristics of the units provide for a rough microtexture which leads to a fairly skid resistant surface. Although the initial skid resistance values of these new brick pavers are good, the skid resistance in place and over time must be evaluated. The wearing action of traffic is known to polish paving materials while in use. Also, the effect of joints and chamfers will increase the skid resistance of the pavement. The joints will also channel water away which reduces the possibility of hydroplaning.

The results of the correlation analysis showed that no statistically significant relationship could be established between skid resistance and the physical properties of brick. These physical properties include compressive strength, cold water absorption, abrasion index and saturation coefficient.

REFERENCES

[1] Kulakowski, B.T., _Evaluation of the Frictional Characteristics of Brick Pavers_, Final Report submitted to the Brick Institute of America, Pennsylvania Transportation Institute, University Park, PA, Nov. 1991.

[2] James, D.I., "Assessing the Pedestrian Slip Resistance of Clay Pavers", _Clay Paving Bricks_, Proceedings No. 44, The Institute of Ceramics, Stoke-on-Trent, United Kingdom, Nov. 1989.

[3]   Clay and Calcium Silicate Pavers for Flexible Pavements, Part 1: Specification for Pavers, BS 6677:1986, British Standards Institution, United Kingdom, 1986.

[4]   Specifying and Laying Clay Pavers, Clay Brick and Paver Institute, Wentworthville, Australia, Aug. 1989.

[5]   Walsh, I. D., "The Use of Clay Pavers in the Highway", Clay Paving Bricks, Proceedings No. 44, The Institute of Ceramics, Stoke-on-Trent, United Kingdom, Nov. 1989.

Robert J. Kudder[1] and John E. Slater [2]

LABORATORY STUDY OF THE CORROSION OF STEEL STUDS AND SCREWS USED IN MASONRY WALLS

REFERENCE: Kudder, R. J., and Slater, J. E., "Laboratory Study of the Corrosion of Steel Studs and Screws Used in Masonry Walls," Masonry: Design and Construction, Problems and Repair, ASTM STP 1180, John M. Melander and Lynn R. Lauersdorf, Eds., American Society for Testing and Materials, Philadelphia, 1993.

ABSTRACT: Steel-stud/brick-veneer walls utilizes metal studs as the backup for brick masonry veneers. Screws are used to secure a tie or anchor to the stud. This system is unique in masonry construction because the engagement of the threaded shank of the screw into a cold-formed steel stud is relied upon to laterally secure the exterior masonry wythe. The generally successful use for the traditional cavity wall ties and anchors mays not necessarily apply to steel-stud/brick-veneer walls because of this reliance on the screw threads. The environment in a masonry cavity may support corrosion. The effects of corrosion on the load capacity of the screw is a concern associated with this masonry system. This paper discusses an laboratory study of the tensile capacity of screws in metal studs, and the change in capacity over time when these metals are in a warm, humid environment.

KEYWORDS: brick veneer, corrosion tests, corrosion, galvanizing, pull-out tests, screws, steel studs

Steel ties, anchors and joint reinforcement are customary components of cavity wall construction. With proper coatings, detailing and construction, steel components have a history of generally successful use in masonry walls. Their function is to laterally secure the masonry wythe to a stiff backup. Ties and anchors are normally embedded in the bed joint of the exterior wythe, and are secured to the backup by one of several methods including embedding them in the bed joints, locking them in dovetail slots, or hooking them through rods welded to a structural steel member.

---

[1]Principal, Raths, Raths and Johnson, Inc., 835 Midway Drive, Willowbrook, Illinois 60521.

[2]Principal, Invetech, Inc. Engineering and Materials Consultants, 10645 Richmond, Suite 130, Houston, Texas 77042.

In steel-stud/brick-veneer walls, metal studs are used as the backup, and screws are used to secure the tie or anchor to the stud. Suggested alternatives such as clips which engage onto or are screwed into the side of the stud are impractical because they interfere with the installation of sheathing. Screws installed into the stud through the sheathing are the most common configuration. To provide lateral resistance, the screws must resist loads in direct tension. This system is unique in masonry construction because the engagement of the threaded shank of the screw into a cold-formed steel stud is relied upon for the tensile resistance. The history of successful use for the traditional cavity wall ties and anchors may not necessarily apply to steel-stud/brick-veneer walls because of the reliance on this connection. The environment in a masonry cavity will at times be warm and have a high humidity. These conditions can initiate and support corrosion, with the potential to compromise the tensile strength of the connection.

The potential effects of corrosion over time on the load capacity of the screw is a concern associated with this masonry system. Concern has fostered a controversial battle in the construction press, including imagery of masonry walls being held by one thin thread [1,2]. Issues of back-up stiffness and cracking behavior aside, the question remains: How will the screws perform, and how will they fail?

TEST PROGRAM

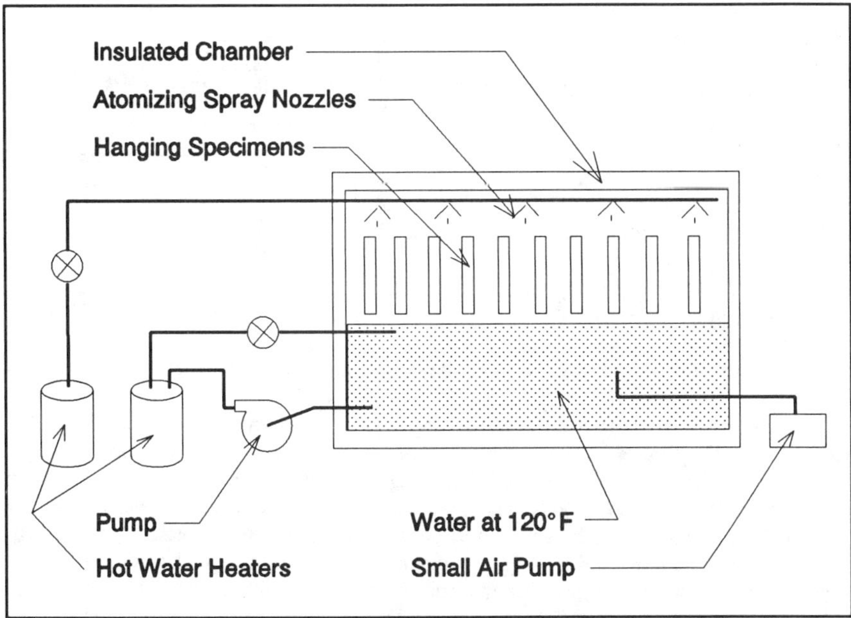

FIG 1 - Schematic Diagram of Test Chamber Construction

A testing program was developed to investigate the behavior of screws under conditions which can be expected in steel-stud/brick-veneer walls. Since corrosion rate is a function of both time of wetness of the corroding surfaces, and the temperature [3], both were maximized in this study to minimize testing time. A test chamber was designed and built to accelerate the rate of corrosion of screws in steel studs by maintaining 100% relative humidity and a 120°F (49°C) temperature. The temperature was maintained by insulating the chamber walls and maintaining 2-foot (0.61 m) depth of water at 120°F (49°C) using a circulating pump and a small domestic hot water heater. The humidity was maintained by sealing the chamber, and periodically spraying the interior with hot water through atomizing nozzles. To prevent oxygen depletion in the closed system and a resulting decrease in corrosion rate, a small diaphragm air pump was used to bubble fresh air through the water and into the chamber. The chamber configuration is shown in Figure 1.

The specimens were either 14 gage and 20 gage galvanized steel studs with a strip of gypsum sheathing attached, as shown in Figure 2. Bugle-head drywall screws and hex-head self-drilling screws were included in the program. For the 14 gage stud specimens, both #6 bugle-head and #10-16 hex-head screws with S-12 type tips were used. For the 20 gage stud specimens, the screws were #6 bugle-headed with an S type tip. The bugle-head screws had a black oxide finish typical of drywall screws. The hex-head screws had an integral washer, and had a zinc plating combined with a polymer protective finish. These screws are considered representative of the types currently in use for brick veneer/metal stud construction. The hex-head screws were installed in the stud specimens after the test program began because of a delay in receiving them. Therefore, the exposure times for hex-head screws are less than the exposure times for the bugle-head screws in each specimen.

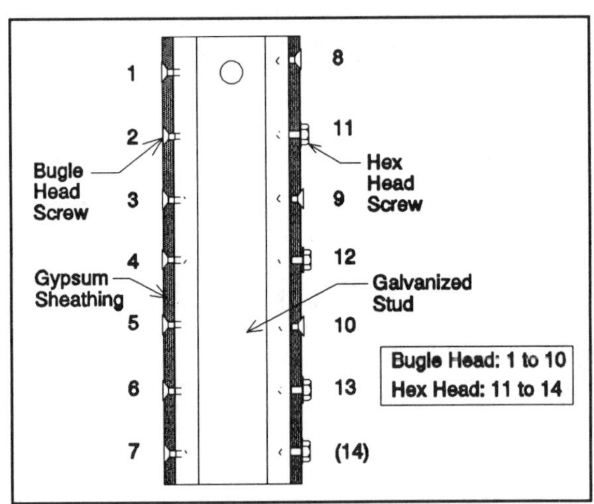

FIG. 2 - Typical 14 gage test specimen.

At intervals of exposure, the tensile capacity of the screws was measured by pull-out tests, and metallurgical examinations of the specimens were performed. Specimens were selected for testing in a random order. The program started with the 14 gage specimens in the test chamber. As each 14 gage specimen was removed and tested, it was replaced with a 20 gage specimen. The 20 gage specimens were tested at the end of the program. Metallurgical evaluations of the 14 gage specimens included: cross-sectional metallographic analysis of the screw-to-stud junction; measurement of screw dimensions under the head, in the shank; and measurement of remaining wall thickness of the stud in the area of contact with the gypsum sheathing. Measurements were preceded by cathodic cleaning in warm inhibited sulfuric acid to remove corrosion product.

VISUAL OBSERVATIONS OF CORROSION BEHAVIOR

During the planning of the test program, the selection and installation of the gypsum sheathing was assumed to be relatively unimportant to the purpose. This assumption proved to be incorrect. It became clear early in the program that the sheathing type and installation were more important than originally expected. The greatest rate of corrosion of the studs resulted at locations where the sheathing was in intimate contact with the stud surface. Similarly, the highest rate of corrosion on the screws occurred where the shank and head of the screw was surrounded by the gypsum sheathing. The sheathing material used with the 14 gage specimens had treated facing papers and a treated core. When the 20 gage specimens were fabricated, this material was not readily available. A substitute gypsum sheathing with treated facing paper but with a white, untreated core was used. This simple expedient caused problems. In the chamber environment, the sheathing with the untreated core deteriorated very quickly. It lost its body and fell off of the specimens approximately one-third of the way through the program. Once the sheathing fell off, the corrosion rate of both the studs and the screws changed markedly. The effect of freeing the metals from the sheathing was so significant that the 20 gage specimens were not tested after this occurred. In an actual wall, the masonry tie itself would keep the sheathing in contact with both the studs and the screws, and therefore expose them to the more severe condition simulated in the test chamber for samples with the sheathing attached. If this test were repeated, it is suggested that the screws be installed through a strip of sheet metal or large washers to secure the sheathing in place.

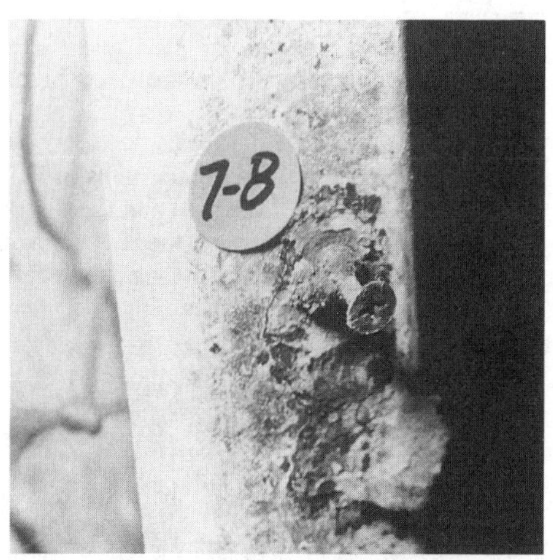

FIG. 3 - Representative bugle-head screw after 2409 hours of exposure.

FIG. 4 - Representative example of a bugle-head screw after 19805 hours of exposure, showing a disintegrated head.

The studs typically exhibited a steady loss of the galvanizing coating, eventually resulting in the exposure of steel. With further exposure in the chamber, the steel corroded. The area of maximum corrosion was centered on the screw holes, but the corrosion was not necessarily most severe immediately adjacent to the screw hole. Apparently, penetration of the galvanizing when the screw is installed will initiate the corrosion process. However, intimate contact between the stud and the sheathing resulted in a region of essentially uniform corrosion centered on the screw hole.

The representative photographs in Figures 3 and 4 show the widening area of corrosion on the surface of the stud and the deterioration of the screw heads with time. Figure 5 shows cross-sections through the stud/screw interface. It is clear that the major corrosion attack is <u>outside</u> the engaged threads, and is concentrated within the area of the screw surrounded by the gypsum sheathing.

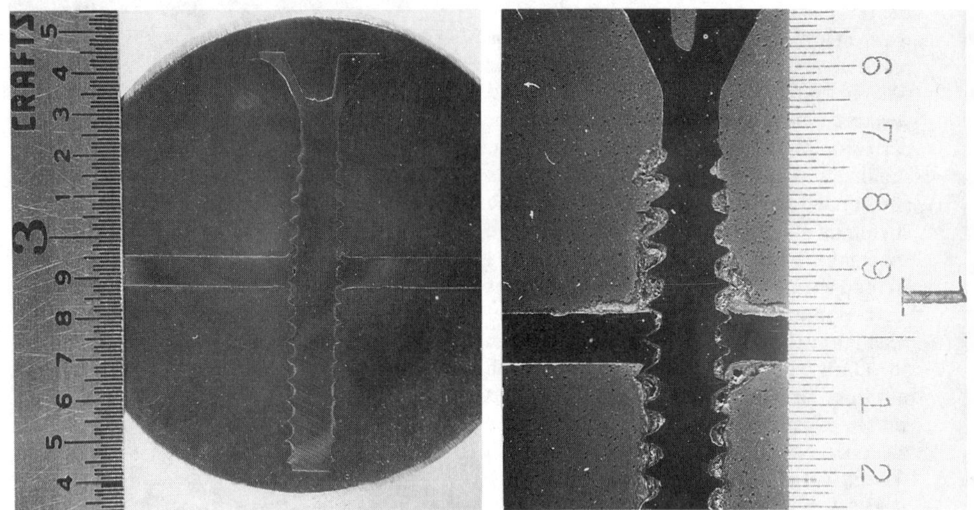

FIG. 5 - Cross-section through stud/screw interface.

PULL-OUT TESTS

The pull-out capacity of the screws was determined using the apparatus shown in Figure 6. It consisted of a steel frame with a hydraulic ram to apply load, and a load-cell to measure the load. A custom shaft was made which cradled the underside of the screw head during the test. Because of the use of the load-cell, any loading due to eccentricity of the specimen in the frame was automatically accounted for in the load data. The sheathing was removed and the sample was allowed to dry for at least four hours. The shank of the screws was then cleaned of gypsum and most of the corrosion product that still adhered so that it would fit inside the jaw of the apparatus. If the threads slipped and then regrabbed during a pull, failure was considered to occur at the first slip.

The results of the pull-out tests are given in Tables 1, 2, and 3. Typically, three screws of each type on each sample were tested. The remaining screws were left in place for metallurgical evaluation. To the extent possible, the range of conditions that the screws exhibited were tested. Towards the end of the program, some screw heads completely disin-

| TABLE 1 - PULL-OUT CAPACITY TEST RESULTS, 14GAGE STUD, HEX-HEAD SCREW | | | | | |
| SAMPLE | | SCREW NUMBER | | | | AVERAGE |
NUMBER	HOURS	11	12	13	14	
2	306	1080 {1} (4804)	1110 (4938)	1090 (4649)		1093 (4863)
3	516	1130 (5026)	1080 (4804)	1170 (5204)		1127 (5012)
4	845	1170 (5204)	1032 (4591)	1164 (5178)		1122 (4991)
5	1173	1044 (4644)	1144 (5089)	1028 (4573)		1072 (4768)
6	1719	1120 (4982)	1110 (4938)	1125 (5004)		1118 (4975)
7	2049	912 (4057)	992 (4413)	1016 (4519)		973 (4330)
8	3268	1065 (4737)	1093 (4862)	1126 (5009)		1095 (4869)
9	4696	1151 (5120)	1075 (4782)	1068 (4751)		1098 (4884)
10	6393		1180 (5249)	1198 (5329)		1189 (5289)
11	8907		1103 (4906)	1047 (4657)		1075 (4782)
12	9888		797 (3545)	1088 (4840)		943 (4192)
13	10577	1250 (5560)	1237 (5502)			1244 (5531)
14	12244	963 (4284)	1127 (5013)	1074 (4777)		1055 (4691)
15	12982	832 (3701)	907 (4035)			870 (3868)
16	14970	980 (4359)	0	1175 (5227)		718 (3195)
17	17814	1084 (4822)	685 (3047)	1012 (4502)		927 (4123)

TABLE 1 - PULL-OUT CAPACITY TEST RESULTS, 14GAGE STUD, HEX-HEAD SCREW					
SAMPLE		SCREW NUMBER				AVERAGE
NUMBER	HOURS	11	12	13	14	
18	19444	375 (1668)	703 (3127)	1022 (4546)		700 (3114)
19	21764	977 (4435)	1114 (4955)	1099 (4889)		439 (1953)
20		880 (3914)	880 (3914)	346 (1539)		
21	24085	848 (3772)	678 (3016)	758 (3372)		761 (3387)
22	25231	966 (4297)	1004 (4466)	1098 (4884)		413 (1835)
23		884 (3932)	884 (3932)		706 (3140)	
24		0	0	974 (4333)	909 (4043)	

{1} Units Pounds
(Newtons)

TABLE 2 - PULL-OUT CAPACITY TEST RESULTS, 14 GAGE STUDS, BUGLE HEAD SCREWS

SAMPLE		SCREW NUMBER										AVERAGE
NUMBER	HOURS	1	2	3	4	5	6	7	8	9	10	
1	343	{1}	916 (4075)		1040 (4626)		1075 (4782)					1010 (4494)
2	667		980 (4359)		1070 (4760)		930 (4137)					993 (4419)
3	887			1003 (4462)	1050 (4671)	940 (4181)						998 (4438)
4	1206		940 (4181)		920 (4092)		966 (4297)					942 (4190)
5	1534		932 (4146)		1038 (4617)		1060 (4715)					1010 (4493)
6	2044		1020 (4537)		1040 (4626)		1000 (4448)					1020 (4537)
7	2409		972 (4324)		945 (4204)		1014 (4510)					977 (4346)
8	3647		903 (4017)		952 (4235)		924 (4110)					926 (4121)
9	5056		956 (4252)		969 (4310)		1022 (4546)					982 (4370)
10	6754		175 (778)		859 (3821)		1016 (4519)					683 (3040)
11	9268	636 (2829)	662 (2945)	0	0	0	0				0	185 (825)
12	10249	908 (4039)	108 (480)	0	0		127 (565)		0			191 (847)

TABLE 2 - PULL-OUT CAPACITY TEST RESULTS, 14 GAGE STUDS, BUGLE HEAD SCREWS

SAMPLE		SCREW NUMBER										AVERAGE
NUMBER	HOURS	1	2	3	4	5	6	7	8	9	10	
13	10935	{1}	770 (3425)		257 (1143)							517 (2284)
14	12605		279 (1241)		115 (512)		264 (1174)					219 (976)
15	13342		270 (1201)		688 (3060)		554 (2464)					504 (2242)
16	15331		759 (3376)		703 (3127)	0	840 (3737)					576 (2560)
17	18175		0		901 (4008)		143 (636)		0	0		209 (929)
18	19805	856 (3808)	0	0	0	0	0	360 (1601)	0	0	695 (3092)	191 (850)
19	22126		998 (4439)		981 (4364)		913 (4061)					200 (892)
20		0	0	174 (774)	0	0	0	791 (3519)	857 (3812)	0	0	
21	24446			938 (4172)	961 (4275)	748 (3327)		0		542 (2411)		638 (2837)
22	25592	712 (3167)	0	0	807 (3590)	0	0		595 (2647)		0	113 (504)
23			911 (4052)		678 (3016)		107 (476)		0	0	0	
24		0	0	0	0	0	43 (191)		0	193 (859)	827 (3679)	

{1}: Units Pounds (Newtons)

TABLE 3 - PULL-OUT TEST CAPACITY RESULTS, 20 GAGE STUD, BUGLE HEAD SCREWS

| SAMPLE | | SCREW NUMBER | | | | | | | | | AVERAGE |
NUMBER	HOURS	1	2	3	4	5	6	7	10	11	
43	1146	{1}	357 (1588)		368 (1637)		372 (1655)			370 (1646)	367 (1631)
42	3488		462 (2055)				453 (2015)			441 (1962)	452 (2011)
41	5787		354 (1676)		351 (1561)			355 (1679)		352 (1588)	353 (1670)
40	7417		345 (1535)		321 (1428)		334 (1486)			354 (1575)	339 (1506)
39	10261		305 (1357)			0	306 (1361)	231 (1028)		294 (1308)	227 (1011)
38	12250		311 (1383)			327 (1455)		286 (1272)			308 (1370)
37	12987			324 (1441)			321 (1428)	290 (1290)		298 (1326)	308 (1371)
36	14657	334 (1486)	0	0	0	0	322 (1432)	54 (240)	363 (1615)		134 (597)

{1} Units Pounds
(Newtons)

tegrated in the chamber, and no tests were possible. A value of "0" is given when a test was not possible, and the "0" values are included in the averages reported. There is a clear loss of pull-out capacity over time. The average pull-out capacity test results are shown in Figure 7.

QUANTITATIVE CORROSION ASSESSMENT

Measurements on shank immediately below the head, initial thread area between the head and the face of the stud, and threads engaged in the stud of cleaned bugle-head screws from the 14 gage specimens were made. Figure 8 shows these measurements as a function of chamber time. Major metal loss occurs primarily in the shank and initial thread areas, and significant loss occurs after approximately 10,000 hours exposure. Measurements at the threads engaged in the stud did not reveal any significant loss at exposure times up to 16,000 hours.

FIG. 6 - Pull-out Test Apparatus

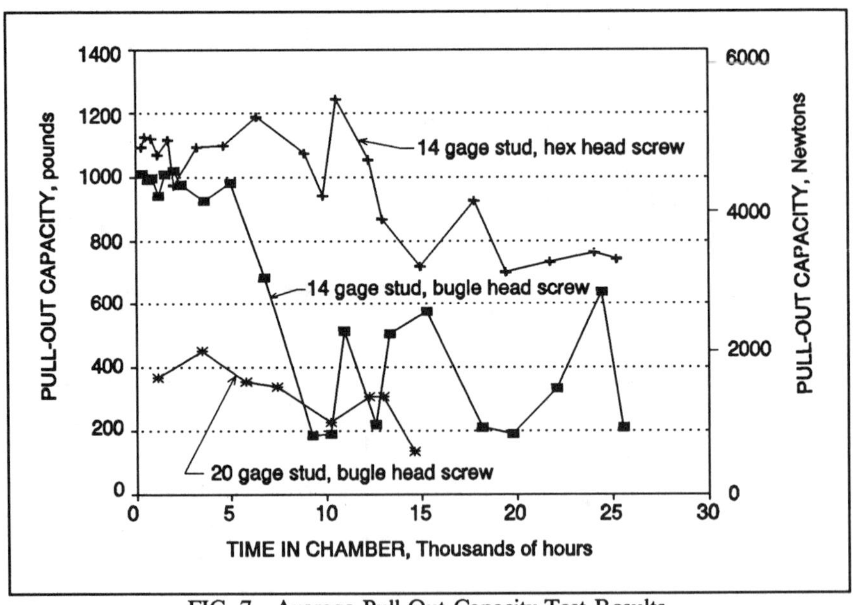

FIG. 7 - Average Pull-Out Capacity Test Results.

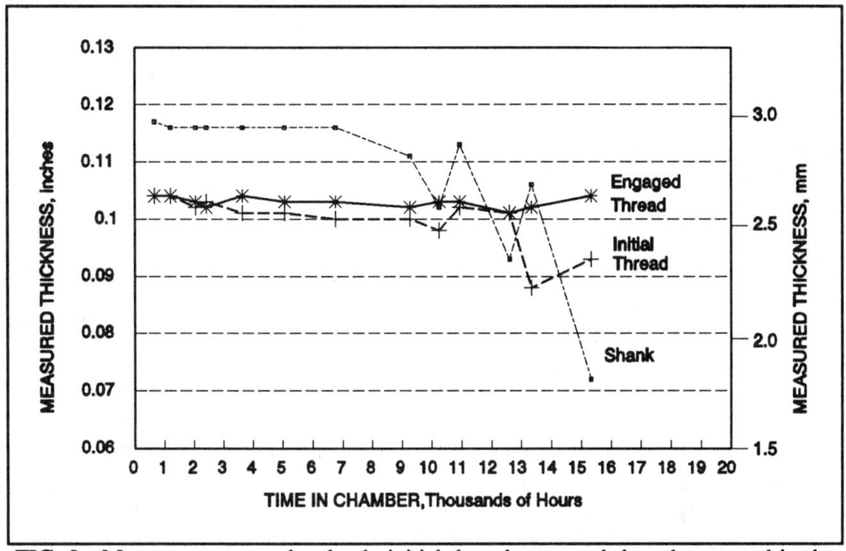

FIG. 8 - Measurements on the shank, initial thread area, and threads engaged in the stud of cleaned bugle-head screws.

Figure 9 shows data on metal loss in corroded areas of the 14 gage studs as a function of exposure time up to 10,000 hours. Three sections of the stud were evaluated and the average minimum thickness over the three sections was calculated. As expected, galvanizing provided protection for some time - to approximately 4,000 hours - before significant corrosion loss occurred.

DISCUSSION

Concern for steel-stud/brick-veneer systems may be motivated by the image of masonry being laterally restrained by the engagement of one or two screw threads in a thin piece of sheet metal, in an environment

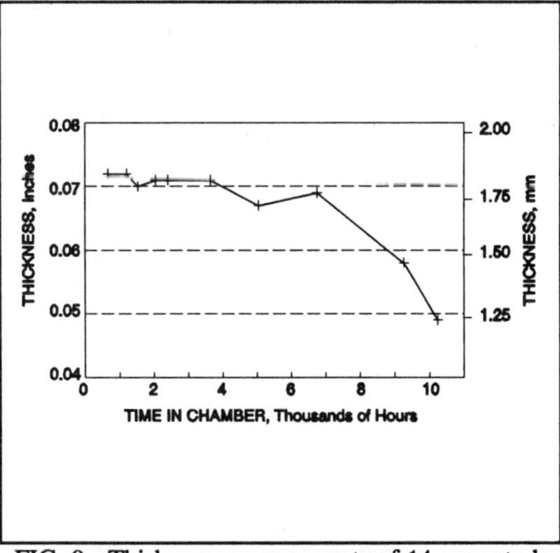

FIG. 9 - Thickness measurements of 14 gage studs as a function of time in the test chamber.

susceptible to corrosion problems. A 14 gage stud is a substantial structural element. The testing program indicates that pull-out failures occur at the threads only early in the life of the specimens, when they are performing their best. The early stages of the corrosion process do not weaken the threads or the thread engagement, and may strengthen the engagement because of expansion of the corrosion products. Certainly, the corrosion process will reduce the pull-out capacity of the screw, but not by the mechanism customarily suspected.

If corrosion progresses sufficiently to cause significant rusting of the steel stud, the process does appear to compromise the engagement of the threads. When a specimen reaches this stage, it is far more likely that the screw itself will fail at the base of the head. For the bugle head screw, the ratio of the perimeter to the cross-sectional area is a maximum at the base of the head, as shown in Figure 10. The large surface area relative to the cross-sectional area results from the geometry of the Phillips driver slots. The base of the head is therefore the location most suscepti-

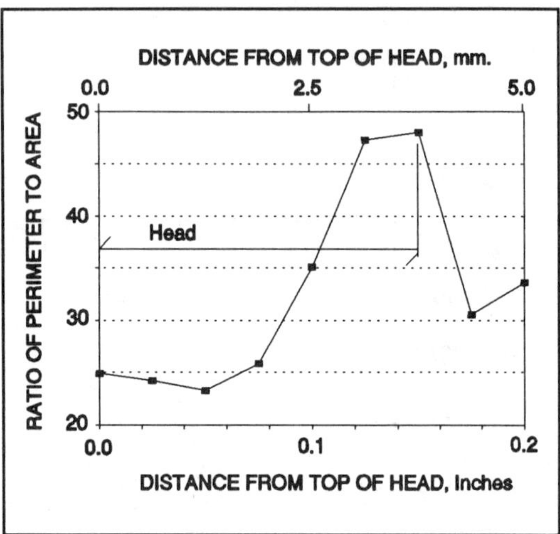

FIG. 10 - Ratio of Perimeter to Cross-Sectional Area Through the Head of a #6 Bugle-Head Screw.

ble to damage from corrosion. A similar observation was also made for the hex head screws, but the cause is probably related to the geometry of the screw in a different way. The manufacturing process which forms a right angle between the base of the head, or its built-in washer, and the shank of the screw may result in an inherently weak plane.

APPLYING TEST RESULTS TO A BUILDING EVALUATION

The testing program was intended to achieve an effective acceleration of the corrosion process. The time of wetness can be assumed to be 100%, and the temperature was held at 120°F (49°C). Assuming wetness, the corrosion rate is a function of temperature [4]. The usual "rule of thumb" is that the corrosion rate doubles for a temperature rise of 18°F (10.0°C) [5]. Thus, if in-wall temperatures are known, the results of this study can be used to estimate corrosion behavior and loss of screw capacity. The effective time of wetness can be estimated with a knowledge of relative humidity within the wall. The "effective" corrosion rate based on time of wetness and temperature within the wall can be theoretically estimated. This rate can then be related to a tensile capacity rating based on the results of the pull-out tests.

Since the environment in a building wall is seldom constant, using the theoretical relationship to the chamber environment directly may not be practical. It might be more useful to use the observations reported above to determine an "equivalent time in the chamber" for an actual building. For example, measurements of actual screw thickness can be compared with the data shown in Figure 8. The stud thickness can be compared with the data shown in Figure 9. Alternatively, the visual condition of the screws can be compared to the photographs in such as those shown in Figures 3 and 4[1]. These comparisons can be used to place a building on the time scale of the test program.

CONCLUSIONS

The exposure environment and the materials immediately surrounding a screw are critical to the longevity of the screw. The rate of corrosion has been found to be most severe at locations where gypsum sheathing remains in intimate contact with the surface of the stud and surrounds the shank of the screw. This situation is most likely to occur at a masonry tie, particularly one which incorporates a base. As expected, a combined coating of zinc and a polymer appears to offer better protection against corrosion than a simple black oxide coating. However, the protection is not absolute, and deterioration and loss of tensile capacity is only delayed by the coatings and not prevented.

For heavy gage studs such as the 14 gage specimens in the test program, it is more likely that corrosion will reduce the tensile capacity of a screw by reducing the strength and condition of the head than by damage to the thread engagement. For the lighter gage studs such as the 20 gage specimens in the test program, corrosion can cause reduction of tensile capacity by both deterioration of the head and failure of the thread engagement.

Material selection alone will not prevent corrosion damage and a reduction of tensile capacity. The maximum longevity of a steel-stud/brick-veneer wall can be achieved only if its design and construction include carefully selected materials and well-executed details to control moisture penetration and to allow ventilation to remove moisture. The corrosion process is inevitable, and at advanced stages, will reduce the pull-out capacity of screws used to anchor masonry.

REFERENCES

[1] Grimm, Clayford T., "Brick Veneer: A Second Opinion", Construction Specifier, April, 1984.

[2] Heslip, John A., "Design Alert #3: A Critical Review-Update", Masonry Advisory Council, Park Ridge, Illinois, 1983.

[3] Sereda, P. J., "Weather Factors Affecting Corrosion of Metals", Corrosion in Natural Environments, ASTM STP 558, American Society for Testing and Materials, 1974, pp. 7-22.

[4] Grossman, P. R., "Investigation of Atmospheric Exposure Factors that Determine Time-of-Wetness of Outdoor Structures", Atmospheric Factors Affecting the Corrosion of Engineering Metals, ASTM STP 646, S. K. Coburn, Ed., American Society for Testing and Materials, 1978, pp. 5-16.

[5] Corrosion Basics, An Introduction, National Association of Corrosion Engineers, Houston, Texas, 1984, p. 348.

---

[1] Charts with color photographs showing the progression of corrosion are available from the authors.

T.W. Bremner,[1] David Rae,[2]

Influence of Aggregate Microstructure on the Volume Stability of Lightweight Concrete Masonry

---

REFERENCE: Bremner, T.W. and Rae, David, "Influence of Aggregate Microstructure on the Volume Stability of Lightweight Concrete Masonry", _Masonry: Design and Construction, Problems and Repair, ASTM STP 1180_, John M. Melander and Lynn R. Lauersdorf, Eds., American Society for Testing and Materials, Philadelphia, 1993.

ABSTRACT: The microstructure of four types of manufactured lightweight aggregate was studied using scanning electron microscopy and the results were used to provide insight into the dimensional stability of concretes made from these aggregates. Dimensional stability was determined according to the Test Method for Length Change of Hardened Hydraulic-Cement Mortar and Concrete (ASTM C 157) as modified by the procedures covered in the Specification for Lightweight Aggregates for Concrete Masonry Units (ASTM C 331).

Aggregate types studied were rotary kiln produced expanded shale, sintered fly ash, pelletized cold bonded fly ash and expanded glass. Scanning electron microscopy revealed the nature of the aggregate pore structure and the extent to which the vesicular structure, typical of most lightweight aggregates, is interconnected. When used in concrete, the three aggregates produced at high temperature met the ASTM C 331 shrinkage requirements while the one made by cold-bonding did not. After 100 days of drying the aggregates were immersed in a lime saturated water for an additional 251 days, followed by air drying during which length measurements were taken periodically. The microstructure was shown to have a pronounced effect on the volume stability of the aggregate.

KEYWORDS: concrete, drying, lightweight masonry, microstructure, shrinkage, volume stability, wetting

---

Cracking in concrete masonry walls occurs for many reasons with drying shrinkage of the individual masonry units being one of the most significant. Masonry units normally are made with a limited binder volume and a large volume fraction of aggregate, with the result that aggregates assume an important role in controlling the volume changes of the cement paste due to drying. To learn more about the process of drying shrinkage the microstructures of four types of manufactured lightweight aggregates were studied using a scanning electron microscope (SEM). The observations made were used to explain the results of drying shrinkage tests performed on concretes made from the same types of aggregates.

---

[1]Professor, Department of Civil Engineering, University of New Brunswick, Fredericton, NB, Canada, E3B 5A3

[2]Electron Microscopy Unit, University of New Brunswick, Fredericton, NB, Canada, E3B 5A3

The four aggregates tested were a sintered fly ash, a rotary kiln expanded shale, expanded glass, and a pelletized cold bonded fly ash. When examined under the SEM, considerable similarity was noted between the expanded shale, sintered fly ash and the expanded glass with the main difference being the amount and size of the vesicles inside the aggregate particles.

The measured shrinkage also showed trends that could be related to microstructure. Concrete made with the lightest aggregate (expanded glass) produced more shrinkage than those with the sintered fly ash or the expanded shale. The shrinkage of concrete made from the fourth aggregate (cold bonded fly ash) was substantially greater than the other three. The reason for this appears to be that cold bonded fly ash is not an inert aggregate but is apparently composed of incompletely bonded fly ash particles that have the potential to expand and contract in the same way as the cementitious binder.

AGGREGATES USED IN THE STUDY

Sintered Fly Ash

The sintered fly ash was produced by pelletizing fly ash in a pan pelletizer with the possible addition of pulverized coal to form a green pellet. This is placed on a sinter strand and passed under a flame hood that ignites the bed, developing a temperature in excess of 1200°C. At this temperature the fly ash pellets sinter into a solid particle with interstitial voids between the fly ash particles.

Expanded Shale

The expanded shale was produced using a rotary kiln in which pre-sized shale was subjected to a temperature of about 1200°C. At this temperature gases form in the shale particles, causing them to expand. This expansion is retained upon cooling.

Expanded Glass

The expanded glass was made by heating a glass cullet.

Cold Bonded Fly Ash

The cold bonded fly ash aggregate was produced by pelletizing fly ash with 3 to 6% lime plus proprietary additives, then hardening the pellets at 70°C to 90°C in a humid atmosphere. Curing time is between 12 to 16 hours.

MICROSTRUCTURE OF THE AGGREGATE

Specimen preparation entailed examining the aggregates visually, selecting ten typical 10 to 20 mm size particles, breaking them in two using a pair of wire cutters and examining them with a stereo microscope to identify typical microstructure. Three aggregates which were judged typical were coated with gold and examined using the scanning electron microscope. Representative micrographs were taken of the fractured aggregate surfaces.

Sintered Fly Ash

The vesicular structure typical of sintered fly ash can be seen in Figures 1(a) and 2(a). Note that the vesicles do not appear to be interconnected and, even under very high magnification, conduits or cracks leading from one vesicle to another are rare. The walls surrounding the vesicles appear to be formed of a very dense ceramic-like material which effectively isolates one void from another, creating a particle with

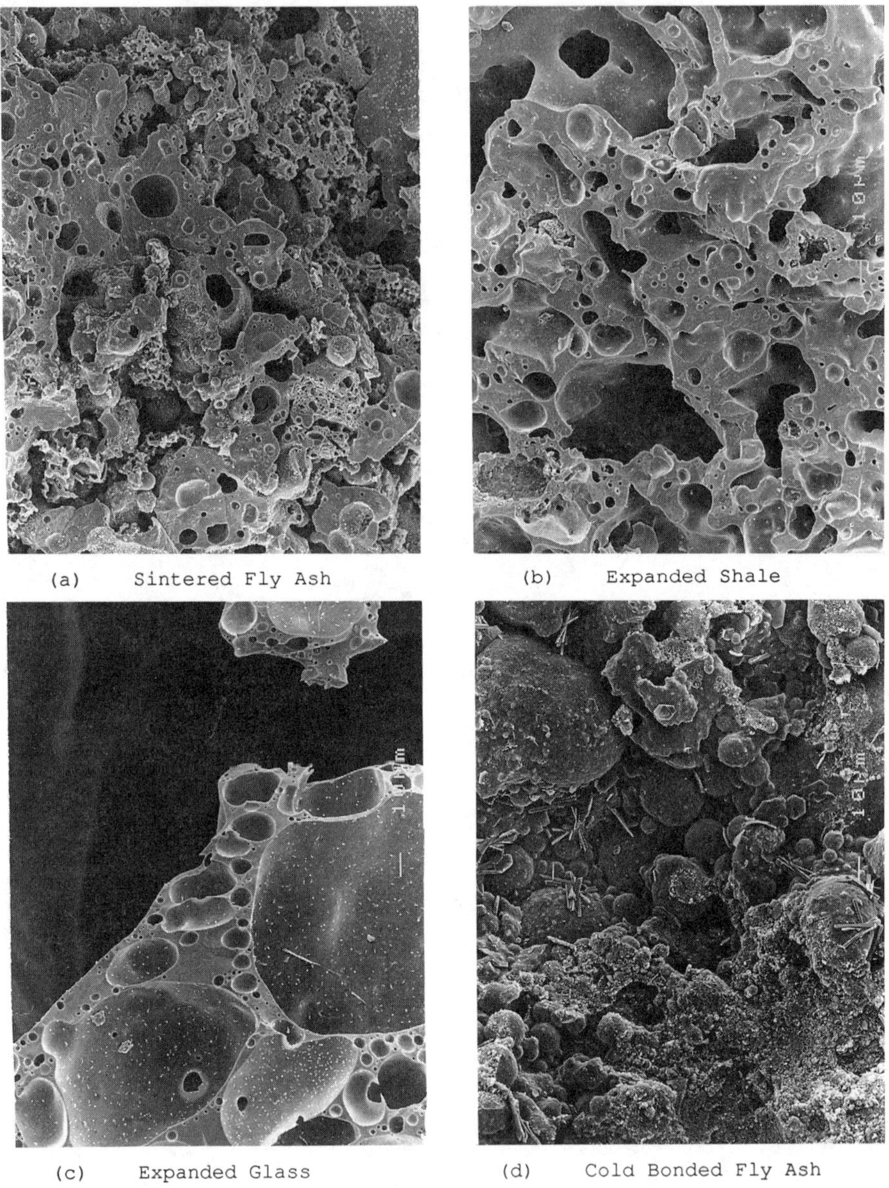

(a) Sintered Fly Ash	(b) Expanded Shale
(c) Expanded Glass	(d) Cold Bonded Fly Ash

Figure 1.   Microstructure of Fracture Surfaces of Lightweight Aggregates.

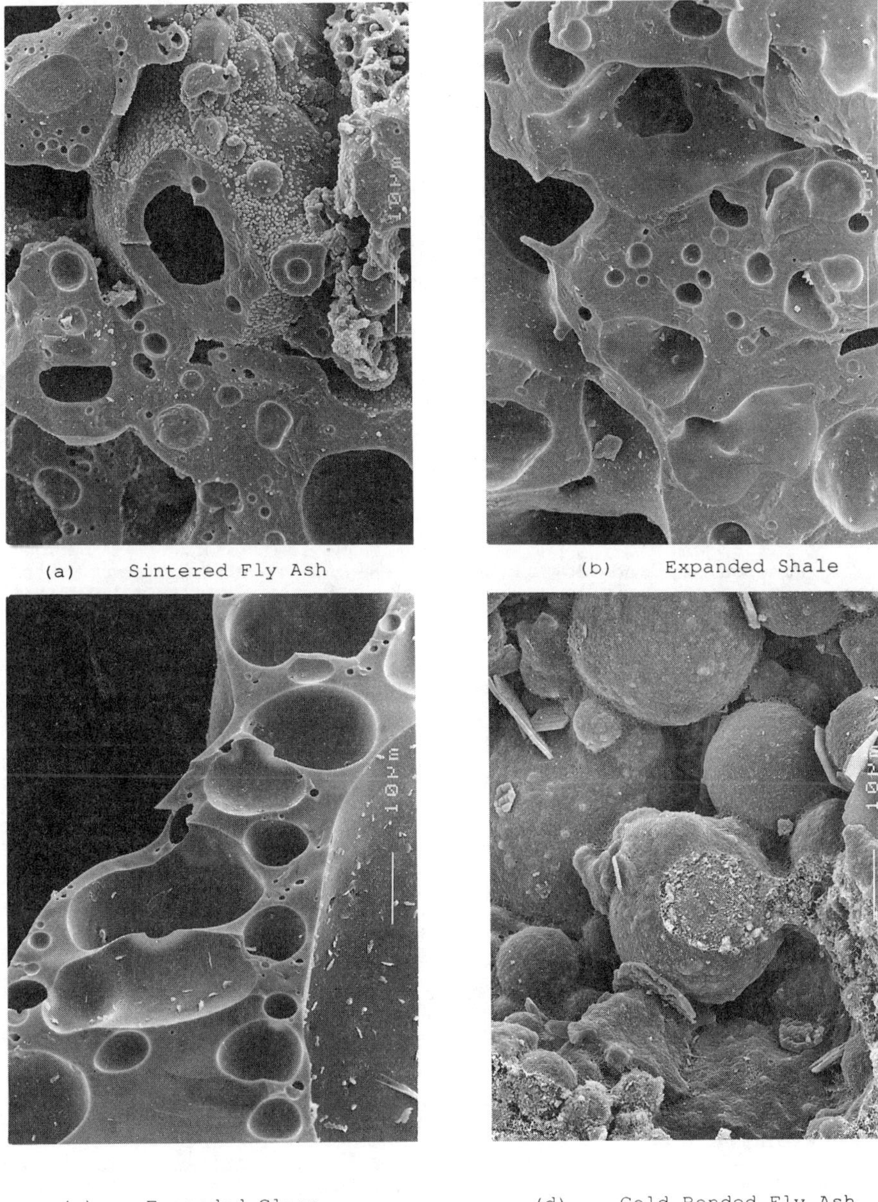

Figure 2.   Microstructure of Fracture Surfaces of Lightweight Aggregates.

limited permeability to liquids and gases. These aggregates may have been subjected to a slightly higher sintering temperature than that examined by Swamy et al. [1] who reported vestiges of fly ash particles indicating that the sintering process had been done at a lower temperature. For the aggregates used in this study the particles were well rounded and had a vesicular structure with thick walls and limited interconnections.

Expanded Shale

The microstructure of the expanded shale, shown in Figures 1(b) and 2(b), reveals a slightly higher degree of expansion than the sintered fly ash. As with the sintered fly ash a dense ceramic matrix surrounds the individual pores with limited interconnection between the pores. Although the other aggregate types were essentially spherical, the expanded shale had a rounded cubical shape.

Expanded Glass

The expanded glass has a microstructure similar to sintered fly ash and expanded shale with the exception that the degree of expansion (ratio of voids to total gross volume of the particle) is about 80% compared to 40% for the expanded shale. The walls of the vesicles in the expanded glass are correspondingly thinner and it appears that there are some inter-connections between the voids. (see Figures 1(c) and 2(c)).

Cold Bonded Fly Ash

Figures (1d) and (2d) indicate that a multitude of fly ash particles have been bonded as a mortar with a lack of complete adhesion between the particles. Where this lack of bonding occurs, water can be expected to move into and out of the particle (see Figure 1(d)). The low density apparently arises from the fact that there is significant space between particles.

Preparation and Testing of Concrete Prisms for Shrinkage and Expansion

The standard dimensional stability test for length change of hardened mortar and concrete (ASTM C 157), as modified by the provisions covered in the standard specifications for lightweight aggregates for concrete masonry units (ASTM C 331), was performed on concretes made from each of the four aggregates. All aggregates were oven dried, sieved and had their grading adjusted in an attempt to meet the ASTM C 331 grading limits (see Table 1) for 3/8 inch (9.5 mm) to 0 size fraction. All aggregates used met the grading requirements with the exception of the expanded glass. This material changes quite significantly when crushed and therefore it was decided not to manufacture finer particles by crushing larger ones.

As required by section 8.6.1 of ASTM C 331 the concrete mix consisted of one part of Portland cement to six parts of combined fine and coarse aggregates, measured by dry loose volume. The water was adjusted to produce a slump of 2 to 3 inches and the concrete consolidated using tamping followed by very low frequency vibration. This method was necessary to consolidate the expanded glass concrete and was used on the other three types of concrete as well. All specimens except those made with expanded glass were free from honeycomb. The top surface of the expanded glass specimen showed some signs of segregation in the top 5 mm layer of concrete, most likely as a result of aggregate floatation. The results of aggregate bulk density tests and concrete density (fresh) are given in Table 2. The aggregate particle relative density was obtained from literature supplied by the manufacturers.

After the required 7 days moist curing at 73.4 $\pm$ 3°F (23 $\pm$ 1.7°C) and a relative humidity of not less than 95%, the initial lengths of the prisms were measured. Subsequent storage (in air) was at 73.4 $\pm$ 3°F (23 $\pm$ 1.7°C) and relative humidity of 50 $\pm$ 5% for a period of 100 days. Prism

Table 1
Grading of Recombined Aggregates With ASTM C 331 Limits
for 3/8 in. (9.5 mm to 0) Combined Fine and Coarse Aggregates

Sieve Size	ASTM Limit	Sintered Fly ash	Passive Expanded Shale	Expanded Glass	% Cold Bonded Fly Ash
1/2 in.	100	100	100	100	100
3/8 in	90-100	95	97	90	100
No. 4	65-90	77	71	80	78
8	35-65	50	49	53	49
16	-	-	34	25	13
50	10-25	17	19	0	10
100	5-15	10	12	0	9

Table 2
Aggregate and Concrete Properties

Aggregate Type	Aggregate Bulk Density kg/m^3 (pcf)	Aggregate Particle Relative Density g/cm^3 (S.G.)	Concrete Density (fresh) kg/m^3 (pcf)	Concrete Dynamic Modulus GPa (ksi)
Sintered Fly Ash	691 (43)	1.7	1939 (121)	7.8 (1131)
Expanded Shale	947 (59)	1.6	1693 (106)	6.1 (885)
Expanded Glass	261 (16)	0.3	633 (40)	2.0 (290)
Cold Bonded Fly Ash	1073 (67)	1.8	1722 (108)	7.8 (1131)

lengths were measured periodically in the controlled atmosphere room. After the end of this 100 day period the specimens were placed in a lime saturated solution and stored in a room at 73 $\pm$ 2°F. After a further 251 days in solution the concrete was returned to air at 73 $\pm$ 3°F. No effort was made to control the humidity at this stage, which ranged from 35 to 42%. After 181 days air drying, the concrete was re-immersed in the lime saturated solution for a further 154 days. Finally, the prisms were air dried, with length measurements taken periodically. The results of the shrinkage and expansion tests are given in Table 3.

RESULTS OF SHRINKAGE AND EXPANSION TESTS ON CONCRETE PRISMS

All but one of the aggregates met the 0.1% shrinkage requirements set out by ASTM C 157 for lightweight concrete (see Table 3 and Figure 3). Of the three aggregates that met the specifications, those aggregates with the lowest particle densities produced concretes with the lowest unit weight and had the highest shrinkage values. Lower density concretes also tended to have higher weight loss on a percent mass basis (see Figure 4). Calculations of percentage change in moisture are based on a datum determined at the start of initial drying. The cold bonded fly ash concrete failed to meet standard requirements by a significant margin during the first 100 day drying period. Apparently the aggregate is contributing to, rather than restraining, the shrinkage of the cement matrix during the drying process. Haller [2] indicates that this level of shrinkage would normally be associated with a cement paste with a water to cement ratio of 0.5.

The shrinkage measurements for the subsequent wetting and drying cycles (Figure 3) indicate that both the sintered fly ash and expanded shale prisms continue to give results well within the 0.1% limit of acceptability for lightweight concrete. The expanded glass suffered more shrinkage during the second air-drying cycle and, at the lower humidity, failed to meet the 0.1% limitation. It is possible that at a higher relative humidity of 50% (similar to the first air-drying cycle) this material would have met the standard requirement of 0.1%. The cold-bonded fly ash prisms showed an improvement in performance in subsequent cycles but continued to be outside of acceptable limits.

DISCUSSION OF RESULTS

Aggregates are normally considered as the stiff inclusions in a volumetrically unstable matrix of cement paste. Upon wetting and drying the matrix expands and contracts due to its capillary porous nature. The aggregates can be considered as restraining elements that hinder this free contraction and expansion of the cement paste matrix. With increasing degrees of aggregate expansion the stiffness of the aggregate particles is reduced and consequently they become less effective in restraining the cement paste matrix.

The test results presented here support this hypothesis. Considering first the three aggregates expanded at high temperatures, it is clear that the greater the aggregate expansion the greater the shrinkage in the concrete prisms. The sintered fly ash and expanded shale concretes have substantially less shrinkage than the expanded glass concretes.

The concrete made from the cold bonded fly ash behaved differently from the other three in that it had a much greater shrinkage. The dynamic modulus was essentially the same for both the cold bonded fly ash and the sintered fly ash which, according to the above reasoning, should have produced similar shrinkage. This apparent dichotomy can be explained by comparing the microstructural aspects of the cold bonded fly ash with the other three aggregates. Although the poorly bonded fly ash particles provide significant restraint to the shrinkage of the cement paste matrix, in itself it is a product subject to volumetric changes similar to that which the matrix experiences and as a result the restraint mechanism is significantly reduced.

Table 3

Shrinkage and Expansion of the Concretes (as a %). Measured at the End of
a Particular Cycle of Drying in Air or Immersion in Lime solution. All
Readings at the Beginning of Each Cycle are Taken as Zero.

Aggregate	Shrinkage		Expansion	
	First Drying (after 100 days)	Second Drying (after 181 days)	First Wetting (after 251 days)	Second Wetting (after 154 days)
Sintered Fly Ash	1 .031	.049	.012	.030
Expanded Shale	1. .043 2. .047 3. .047	.054 .056 .060	.014 .016 .018	.023 .026 .026
Expanded Glass	1. .064 2. .069 3. .065 4. .062	.109 .117 .113 .102	.011 .007 .013 .007	.059 .064 .067 .056
Cold Bonded Fly Ash	1. .238 2. .234 3. .228	.224 .216 .204	.113 .108 .099	.113 .105 .094

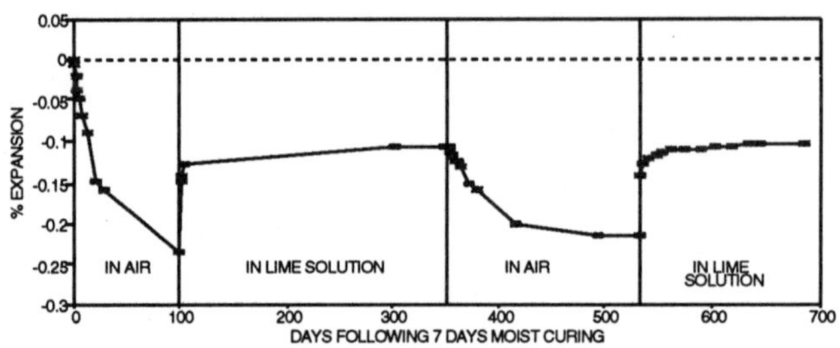

A) Cold Bonded Fly Ash

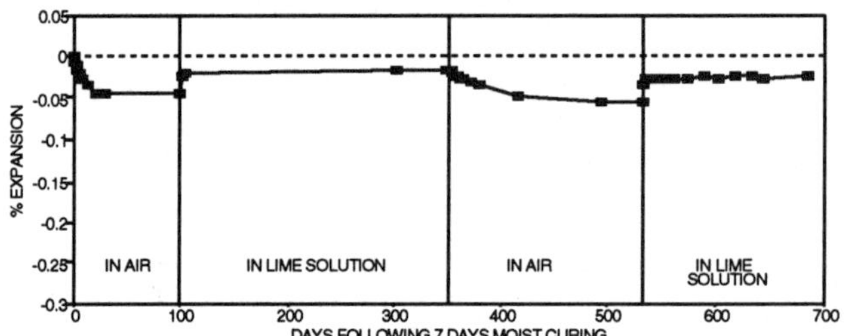

B) Expanded Shale

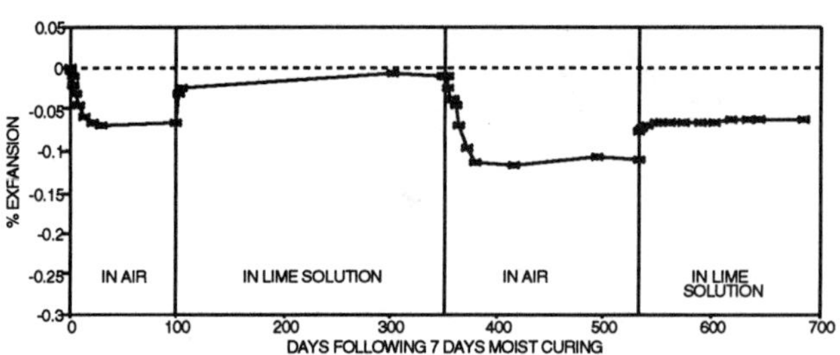

C) Expanded Glass

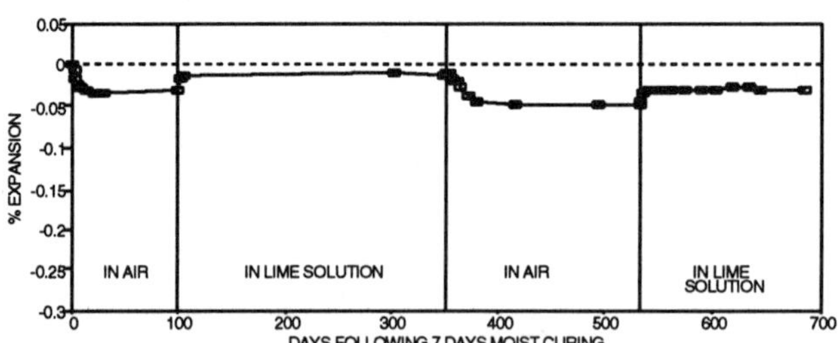

D) Sintered Fly Ash

FIGURE 3: Shrinkage and Expansion of Lightweight Aggregate Concrete

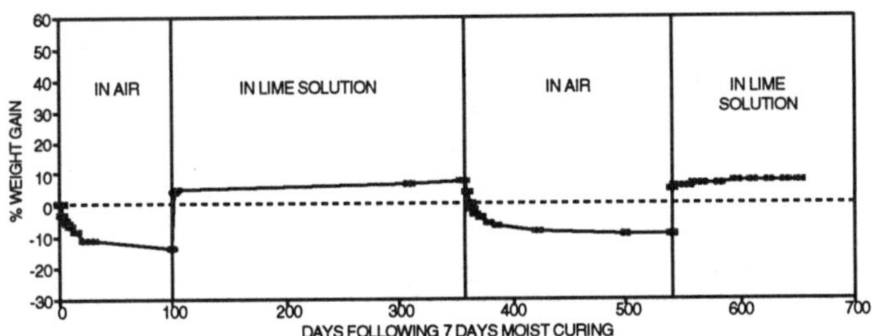

A) Cold Bonded Fly Ash

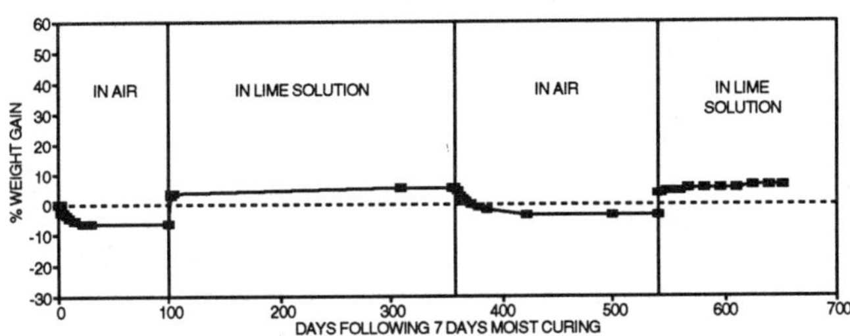

B) Expanded Shale

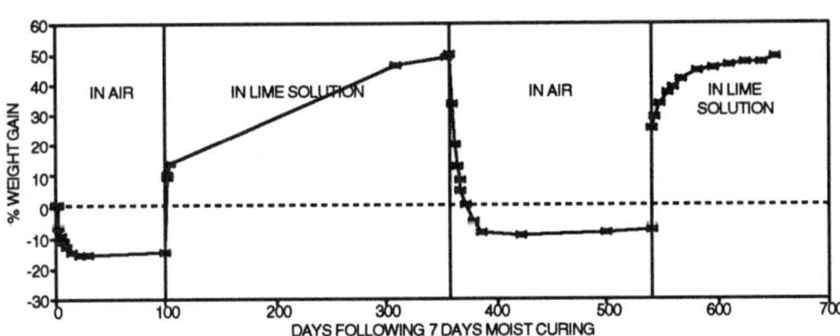

C) Expanded Glass

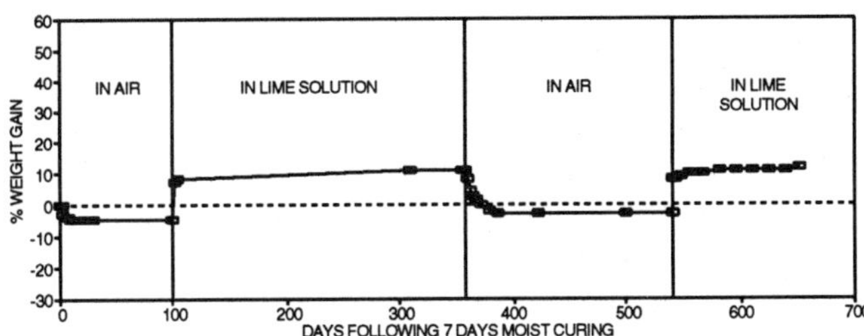

D) Sintered Fly Ash

FIGURE 4: Moisture Gain or Loss Calculated on a Percent Mass Basis
    for Lightweight Aggregate Concrete

CONCLUSION

Concrete prisms made from aggregates produced at high temperature experienced shrinkage rates within the ASTM C 331 limit of 0.1%, whereas the prism made from cold-bonded fly ash aggregate exceeded this limit. These results can be explained by the observed differences in microstructure. The aggregates produced at high temperature have an essentially non-interconnected pore structure, in marked contrast to the more permeable nature of the cold-bonded aggregate. Differences between the three vesicular aggregates can also be related to their microstructure, with the aggregate with the largest vesicles and thinnest pore walls exhibiting the greatest shrinkage.

ACKNOWLEDGEMENT

The financial support of National Science and Engineering Research Council of Canada and the Expanded Shale, Clay and Slate Institute is acknowledged.

REFERENCES

[1]   Swamy, R.M. and Lambert, G.H., "The Microstructure of Lytag Aggregate", _International Journal of Cement Composites and Lightweight Concrete_, Vol. 3, No. 4., November 1981, pp. 273-282.

[2]   Haller, P., "Shrinkage and Creep of Mortar and Concrete", _Diskussionbericht_, No. 124, EMPA, Zurich, 1940 (in German).

Bruce T. Wright,[1] Rick D. Wilkins,[1] and George W. John[2]

VARIABLES AFFECTING THE STRENGTH OF MASONRY MORTARS

REFERENCE: Wright, B. T., Wilkins, R. D., and John, G. W., "Variables Affecting the Strength of Masonry Mortars," Masonry: Design and Construction, Problems and Repair, ASTM STP 1180, John M. Melander and Lynn R. Lauersdorf, Eds., American Society for Testing and Materials, Philadelphia, 1993.

ABSTRACT: Masonry cement mortars with additives which entrain various levels of air as well as mortars containing lime were evaluated. A mathematical model was developed describing brick-to-mortar bond strength. Lime mortars, prepared as specified by ASTM with higher proportions of portland cement, provided higher bond strengths and higher compressive strengths compared to masonry cement mortars. When prepared with the level of portland cement typical of masonry cement mortars, lime mortars had comparable bond strengths but weaker compressive strengths at similar air content. A one percent increase in portland cement content increased the bond strength four times as much as reducing the air content by 1%. Lime did not provide increased bond strength. The higher bond strengths achieved with Type S portland cement/lime mortars over Type N are a direct result of higher portland cement content.

KEYWORDS: masonry cement, portland cement/lime, air entrainment, bond strength, compressive strength

---

Two general cement mortar types are typically used in masonry construction. One mortar type is portland cement/lime (PC/L) and the other is masonry cement. It is known that PC/L mortars with air contents between 4 and 6%, following ASTM Specification for Mortar for Unit Masonry (C 270) proportions, generally have higher bond and compressive strengths compared to masonry cement mortars. It is also accepted that masonry cement mortars that typically entrain between 14 and 18% air content have superior workability, freeze-thaw durability, and reduced drying shrinkage compared to PC/L mortars.

Proponents of lime claim that lime enhances mortar-to-brick bond strength by a pozzolanic reaction mechanism [1]. It is also claimed that higher levels of air, typical of masonry cement mortars, reduce bond strength. The goal of this project was to investigate the influence of air, lime, and portland cement content on bond strength.

A comparison of masonry cement mortars with PC/L mortars was performed. PC/L mortars usually contain higher levels of portland cement than masonry cement mortars. Therefore, a comparison of masonry cement mortars and PC/L mortars with similar portland cement content was

---

[1]Manager and senior technical specialist, respectively, Mineral Process Technology, Westvaco Corporation, Charleston Heights, SC 29415

[2]Engineering Consultant, 145 Manchester Rd., Charleston, SC 29407

made to evaluate the effect of lime on bond strength. In addition to developing data on mortar-to-brick bond strengths, compressive strengths of the various mortar types were measured for comparison of masonry cement mortars to PC/L mortars containing the two different proportions of portland cement.

RESULTS AND DISCUSSION

The Influence of Air Content and Portland Cement Level on Flexural Bond Strength

Bond strength determinations were made as described in the experimental section of this report using a modified ASTM Method for Measurement of Masonry Flexural Bond Strength (C 1072) Bond Wrench apparatus. Compressive strengths were performed according to ASTM Specification for Masonry Coment (C 91). Mortar and specimen preparations followed ASTM procedures as described in the experimental section.

Initial evaluations utilized Type N masonry cement mortar prepared with an air entraining masonry additive to determine the effect of cure time on bond strength. The cure profile was useful for predicting the final bond strength expected from a given mortar broken at shorter intervals than the 28 day standard curing. Figure 1 is a graph of the bond strength versus cure time of couplets made using Formulation B. It is seen that approximately 67% of the final bond strength was achieved after 7 days of specimen curing, approximately 85% of the final bond strength was achieved after 14 days, and the bond strength was not significantly increased beyond 28 days of curing and so could be considered "final." Using this relationship, cure times of 7 and 14 days were chosen for brick couplets prepared with mortars using masonry additives with varying levels of air contents.

The masonry additive series, Formulations A-F, was designed to entrain increasing levels of air in masonry cement mortars, as well as to function as a grinding aid in masonry cement production. Table 1 lists the data for mortars prepared with the masonry additive series and bond strengths from specimens made using the mortars which were broken after 7 and 14 days of curing. The upper portion of Table 1 is data observed using Type N masonry cement mortars which contained 49% portland cement content in the cement portion of the mortars. The lower portion of Table 1 is data obtained using Type S masonry cement mortars which contained 69% portland cement. The additives are listed in order of increasing air content. It should be noted that as the air content increases, the water required to achieve the specified mortar consistency decreases; and the workability and plasticity as gauged by the Westvaco Mason's Rating[3] increases. The general trend observed is a decrease in bond strength as the air content increases with both Type N and Type S masonry cement mortars.

Figure 2 depicts the effect of air content and portland cement level on bond strength using the data from Table 1. Figure 2 also indicates a similarity in slopes between Type N and Type S plots and shows that the bond strengths for Type S averaged 42% higher than Type N. The increased bond strength is directly proportional to the

---

[3]The Mason's Rating is a qualitative test designed by Polychemicals Department, Westvaco Corporation, Charleston Heights, South Carolina 29415. The mortar evaluation rates workability and plasticity on a scale of 1 to 5, with 5 being the best.

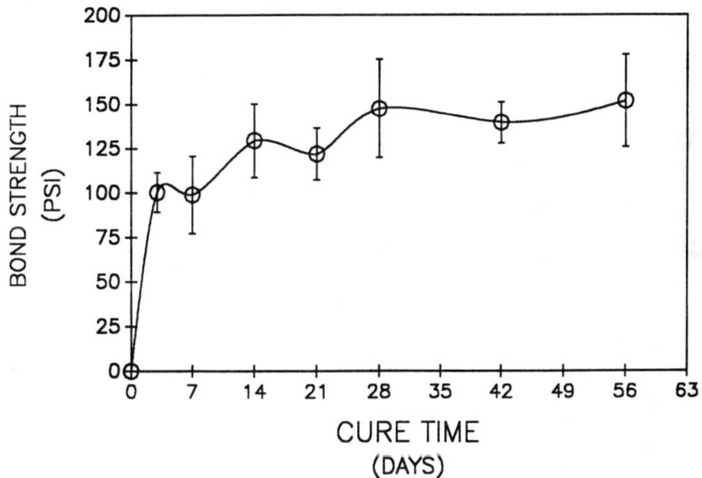

FIGURE 1--Bond strength vs. cure time (Type N masonry
cement with Formulation B.

TABLE 1--Masonry cements mortar data.

Additive	Water Level*, gms	ASTM Air, %	Mason's Rating	Bond Strength, psi** (mean ± standard deviation)	
				7 day	14 day
Type N Masonry Cement Mortar Data:			343g Type I Portland Cement ground with 350g Marl, 7g Gypsum, 0.15% Additive and mixed with 2400g Sand		
Formulation A	404	8.7	2.4	123 ± 34	119 ± 30
Formulation B	386	13.9	3.8	99 ± 20	129 ± 19
Formulation C	376	14.5	3.3	110 ± 24	75 ± 16
Formulation D	377	15.3	4.4	94 ± 21	122 ± 30
Formulation E	370	13.0	4.2	95 ± 15	90 ± 20
Formulation F	364	22.0	4.5	72 ± 16	86 ± 33
Type S Masonry Cement Mortar Data:			480g Type I Portland Cement ground with 210g Marl, 10g Gypsum, 0.15% Additive and mixed with 2400g Sand)		
Formulation A	406	8.5	2.9	218 ± 38	212 ± 50
Formulation B	372	15.7	4.0	160 ± 27	158 ± 40
Formulation C	372	18.7	4.2	142 ± 14	134 ± 28
Formulation D	366	18.5	3.8	121 ± 42	138 ± 21
Formulation E	370	24.8	4.5	107 ± 20	119 ± 40

*Water required to achieve flow of 125%.
**To convert from psi to MPa, multiply by 0.00698.

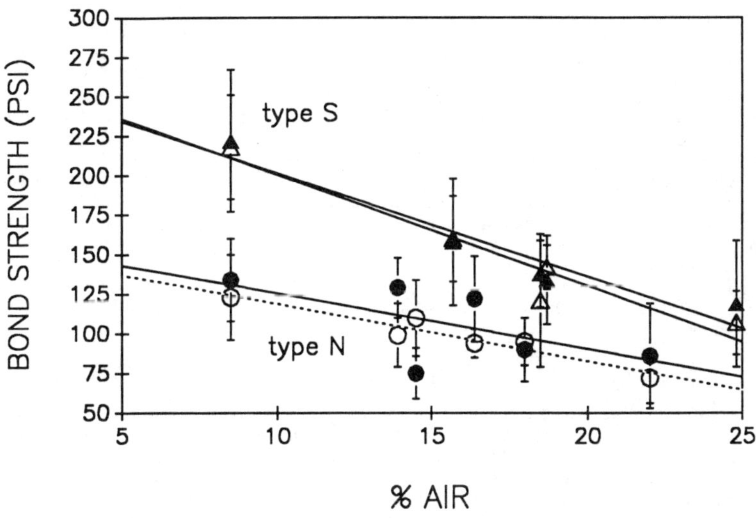

NOTE: Portland cement content is directly
        related to bond strength.

        Increasing air content lowers bond
        strength.

Figure 2--Effect of air content on bond strength.

difference in portland cement content (40%) in the two mortar types. Based on these observations, linear regression analyses of the plots were performed to derive the following mathematical equation to describe the effect of air and portland cement content on bond strength (see Appendix):

$$\text{Bond Strength (psi)} = (16.3 \times \% \text{ PC}) - (4.2 \times \% \text{ Air}) \tag{1}$$

Equation (1) shows that the impact of percentage portland cement level is four times greater than the impact of percentage air content on mortar-to-brick bond strength. This equation is believed to be characteristic of the specific masonry cement used for this investigation.

Figure 3 shows the effect of the water-to-portland cement (w/c) ratio versus bond strength. Another general trend observed in Table 1 is the increase in bond strength with an increase in the water required to achieve the desired mortar consistency. As previously mentioned, the increase in water demand parallels a decrease in mortar air content as the two factors are dependent. The observation that an increase in water demand caused an increase in bond strength was unexpected because higher levels of water are known to weaken the compressive strength [2]. This observation is believed to be a result of the initial rate of absorption (IRA) of the masonry units used for specimen preparation which absorbs moisture out of the mortar upon contact. Mortars with lower w/c ratios are believed to result in lower bond strengths due to insufficient water for proper curing after water loss to the brick. However, mortars with higher levels of water have improved flow and adequate water for hydration after water removal by the masonry unit.

A Comparison of Masonry Cement and Portland Cement/Lime Mortars

In the next phase of the investigation, masonry cement mortars were compared to portland cement/lime mortars using lower proportions as well as the higher portland cement content typically used. The masonry mortars evaluated were made using Type S Formulations A and B that entrained low and medium levels of air (8.5 and 15.7%). Table 2 lists the data for mortar properties and bond strength data for the various mortars evaluated. Specimen breaks were made at 3, 7, 14, and 28 days. The bond strengths did not increase with extended curing time using Type S mortar as they did with Type N mortar. The portland cement/lime bond strengths were again stronger than the bond strengths of masonry cement mortars. However, the levels of portland cement were also significantly higher as well, which contribute to the difference in strengths. Portland cement/lime mortars typically entrain only 4 to 6% air. Addition of an air entraining agent to a PC/L mortar to raise the air content lowered the bond strength similarly to that for Formulation A at the 14-day mark.

Portland cement/lime mortar made using "masonry cement proportions" of portland cement resulted in bond strengths similar to Formulation A treated mortars. These data show there is no bond strength enhancement contribution from the hydrated lime in the mortar (see Appendix). Figure 4 graphs the difference in bond strength between PC/L mortars and masonry cement mortars. Figure 5 plots the bond strengths of PC/L mortars with "masonry cement proportions" of portland cement and masonry cement mortars made using Formulation A.

The portland cement/lime mortar bond strength data were compared to the predicted values calculated using the previously derived bond strength equation. The results are listed in Table 3. The calculated bond values of portland cement/lime mortars made using typical proportions were approximately 10% higher than the experimental result. The PC/L mortar using masonry proportions calculated bond strength value

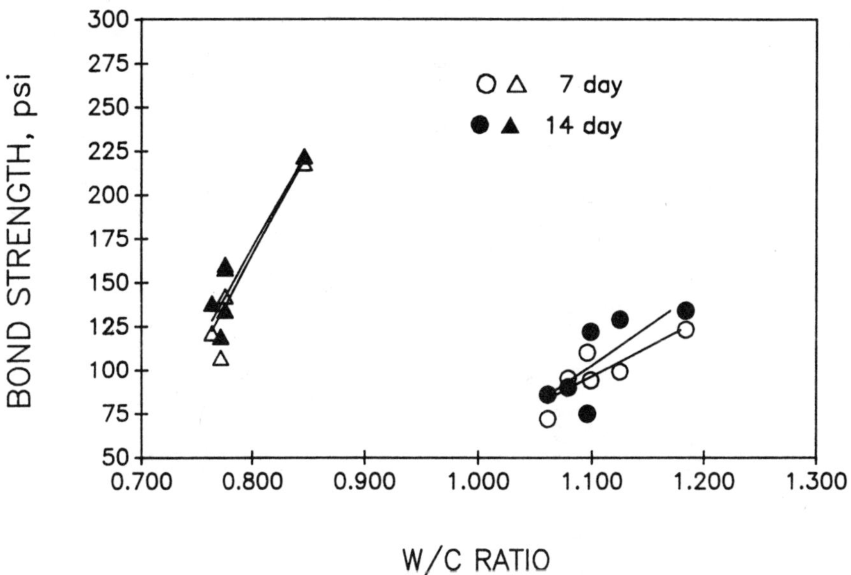

NOTE: General trend of increasing w/c
ratio increases bond strength.

FIGURE 3--Effect of w/c ratio on bond strength.

TABLE 2--Type S mortar data.

Couplet Mortar (125 Flow)

Mortar	PC in Total Mortar, %	H_2O/PC	Couplet Mortar, % Air	Mason's Rating	Bond Strength, psi* (mean ± standard deviation)			
					3 day	7 day	14 day	28 day
Formulation A	13.7	406/480	8.5	3.0	233 ± 43	218 ± 33	212 ± 45	218 ± 58
Formulation B	13.8	372/480	15.7	4.0	161 ± 26	160 ± 27	158 ± 40	166 ± 19
PC/L (Standard)	17.8	430/627	5.2	2.5	203 ± 37	244 ± 84	247 ± 51	228 ± 25
PC/L (High Air)	17.9	406/627	14.2	---	166 ± 37	157 ± 46	216 ± 31	216 ± 43
PC/L (Masonry Cement Proportions)	13.4	476/480	4.3	3.5	---	213 ± 29	228 ± 27	---

*To convert from psi to MPa, multiply by 0.00698.

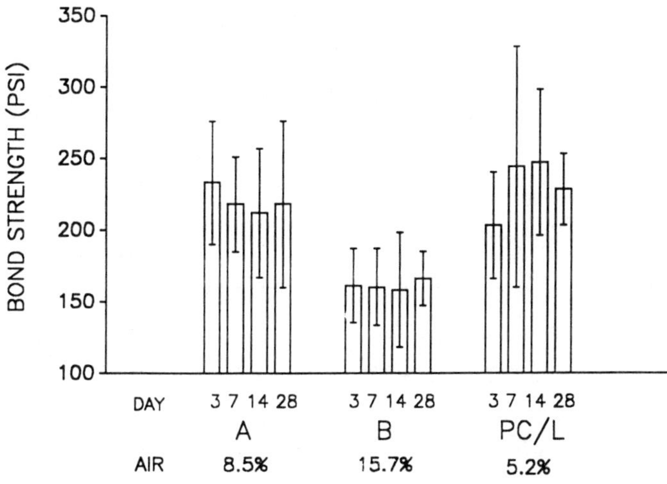

FIGURE 4--Bond strength of masonry cement vs. PC/L mortar.

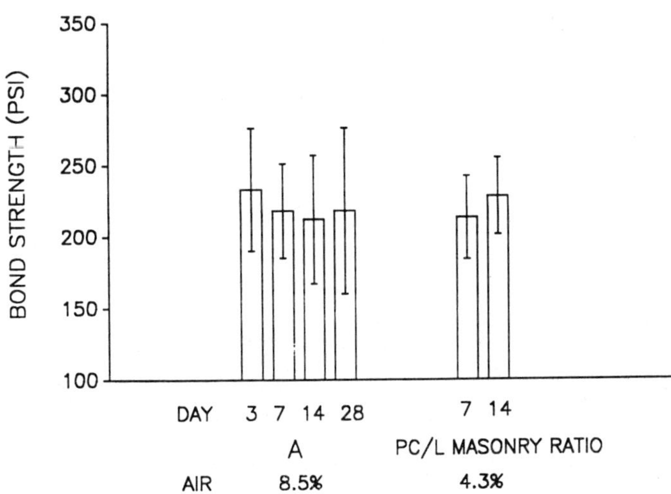

FIGURE 5--Bond strength of PC/L with masonry proportions
of portland cement.

Table 3--<u>Check of bond strength model for PC/L mortars.</u>

Mortar	Calculated, psi*	Experimental, psi*
PC/L (Normal)	268	246
PC/L (Masonry Cement Proportions)	200	222

NOTE: <u>Bond Strength Equation</u>:

Bond Strength (psi) = 16.3(% PC) − 4.2(% Air)   (1)

<u>PC/L</u> <u>Normal</u>:

16.3(17.8) − 4.2(5.2) = 268 psi

[Calculated value 9% higher than experimental
  result.]

<u>PC/L</u> <u>Masonry</u> <u>Cement</u> <u>Proportions</u>:

16.3(13.4) − 4.2(4.3) = 200 psi

[Calculated value 10% lower than experimental
  result.]

---

*To convert from psi to MPa, multiply by 0.00698.

was approximately 10% lower than the experimental bond strength. The equation, good to approximately 10% of experimental results, confirms the relationship of bond strength to portland cement content for portland cement/lime mortars. The standard deviation of the bond strength data is typically 20-25% which is large enough to account for the differences between calculated and experimental data.

Table 4 lists the compressive strength data for the same mortars evaluated previously for bond strength. Again, the compressive strength of PC/L mortar is significantly higher than the masonry cement mortars, as expected, due to the higher portland cement content.

It is interesting to note that the PC/L mortar made using masonry proportions had lower compressive strengths than the masonry cement mortars by 15% to 25%, while the PC/L mortar with high air had greater compressive strength than masonry cement mortars. These data once again indicate that cement content affects mortar strength much more than air content.

EXPERIMENTAL

Masonry Cement Preparation

Masonry cements were prepared by combining 2 940g Type I portland cement, 3 000g pre-ground marl, 60g gypsum, and 9g of additive (0.15% dosage) for Type N and 4 100g Type-1 portland cement, 1 800g pre-ground marl, 90g gypsum, and 9g of additive for Type S. These materials were ground for thirty minutes in a pilot ball mill, 27 inches long x 20 inches diameter (69 cm x 51 cm).

Specimen Preparation

Bond strength mortars were prepared by mixing 700g masonry cement with 2 400g sand (50% 20-30 and 50% graded) and sufficient water in a Hobart mixer in accordance with ASTM procedures. Mortar flows of 125 ± 5% were used for couplet preparation. Mortar was applied to standard concrete masonry units (UBC-24-30) purchased from the National Concrete Masonry Association (NCMA) using a jig for alignment [3]. The masonry units had an IRA of 61g/30in^2 (76 cm^2) as reported by NCMA. In-house testing resulted in an average IRA of 68g/in^2 (2.5cm^2). The top unit was placed on top and subjected to two 1.5in (3.75 cm) drops of a 4-pound (1.9 kg) drop hammer. The couplets were sealed in plastic bags until the desired break date.

All datum points are an average of breaks of six specimens for bond strength and three cubes for compressive strength.

The bond wrench apparatus was made according to ASTM C 1072 except that the steel thicknesses are one-eighth inch (0.3 cm) greater than specified to make the apparatus sturdier. The test apparatus was also equipped with a precision jack and pressure load cell to provide for the application of controlled force.

CONCLUSIONS

1.   Bond strength of all mortars tested at the same flow is directly dependent on portland cement content.

2.   Increased air content has a measurable effect in reducing bond strength.

TABLE 4--<u>Type S mortar compressive strength data.</u>

Mortar	PC in Total Mortar, %	H_2O/PC	ASTM Air, %	Compressive Strength, psi*			
				3 day	7 day	14 day	28 day
Formulation A	13.7	378/480	9.2	2 600	3 217	3 408	3 558
Formulation B	13.8	356/480	14.5	2 550	3 050	3 142	3 742
PC/L (Standard)	17.8	408/627	6.6	3 658	3 517	4 683	5 350
PC/L (High Air)	17.9	382/627	12.6	3 217	3 617	3 892	4 685
PC/L (Masonry Cement Proportions)	13.4	458/480	4.9	2 183	2 533	2 750	3 217

NOTE: PC/L mortar with high cement content has greater compressive strength than masonry cement mortars.

PC/L mortar with portland cement proportions used in masonry cement has lower compresive strength than masonry cement mortars.

---

*To convert from psi to MPa, multiply by 0.00698.

3.   In portland cement/lime mortars, no bond enhancement was observed
      with the addition of lime.

4.   Portland cement/lime mortars with the same portland cement content
      as Type S masonry cement had lower compressive strength.

5.   Workability of portland cement/lime mortars was improved with
      increasing lime content.

6.   Increased air content improves workability of both masonry and
      portland cement/lime mortars.

APPENDIX

<u>Bond Strength Model</u>

<u>Portland cement level in masonry cement mortars</u>

Type N: $\dfrac{343 \text{ g PC}}{3{,}480 \text{ g total mortar}}$ = 9.9%

Type S: $\dfrac{480 \text{ g PC}}{3{,}480 \text{ g total mortar}}$ = 13.8%

NOTE: Bond strength
        equation for   = 160 - 3.89(% Air)   [corr. 0.91]
        Type N

        divide the Y-intercept by % PC

        60 ÷ 9.9 = 16.2 (% PC)

        Bond strength
        equation for   = 225 - 4.6(% Air)   [corr. 0.88]
        Type S

        divide the Y intercept by % PC

        225 ÷ 13.8 = 16.3 (% PC)

<u>Type N Equation</u>

        Bond strength = 16.2 (% PC) - 3.9 (% Air)   (2)

<u>Type S Equation</u>

        Bond strength = 16.3 (% PC) - 4.6 (% Air)   (3)

<u>Averaging Type N and Type S Equations</u>

        Bond strength (psi) = (16.3 x % PC) - (4.2 x % Air)   (1)

Type N equation developed without 14-day data due to low
correlation (0.4).

Type S equation developed using both 7 and 14-day data
excluding Formulation A data.

REFERENCES

[1].   Walker, D. D., "Hydrated Lime, An Irreplaceable Mortar Plasticizer," _International Lime Conference Proceedings_, Rome, Sept. 13-14, 1990.

[2].   Lea, F. M., _The Chemistry of Cement and Concrete_, 3rd Edition, Chemical Publishing Co., Inc., New York, NY, 1971, p.392.

[3].   Uniform Building Code 24-30.

Testing and Evaluation

W. Mark McGinley[1]

FLEXURAL BOND STRENGTH TESTING - AN EVALUATION OF THE BOND WRENCH TESTING PROCEDURES

REFERENCE: McGinley, W. M., "Flexural Bond Strength Testing - An Evaluation of the Bond Wrench Testing Procedures," _Masonry: Design and Construction, Problems and Repair_, _ASTM STP 1180_, John M. Melander and Lynn R. Lauersdorf, Eds., American Society for Testing and Materials, Philadelphia, 1993.

ABSTRACT: The recent adoption of flexural tensile bond as a criterion for acceptance of materials in building construction by the Uniform Building Code has directed considerable attention to the ASTM C 1072 - 86 Standard Method for Measurement of Masonry Flexural Bond Strength and the bond wrench testing apparatus. Described is an experimental evaluation of the Bond Wrench Testing Apparatus and testing procedures. During the investigation a device for calibration of the bond wrench was developed and evaluated.

KEYWORDS: masonry, flexural bond, testing, bond wrench, evaluation, calibration

---

The recent adoption of flexural tensile bond as a criterion for acceptance of materials in building construction by the Uniform Building Code [1] has directed considerable attention to the ASTM C 1072 - 86 Standard Method for Measurement of Masonry Flexural Bond Strength and the bond wrench testing apparatus. Some tests using this method have produced highly variable results and mixed opinions on the validity of bond wrench testing [2] [3] [4].

There are a number of factors that affect the development of bond in masonry assemblies. These include, the material properties of the mortar and masonry units, workmanship, curing conditions and the testing procedures and apparatus. Variations in the mortar and masonry unit properties are inherent to the constituent materials and, while their variability can be reduced, it cannot be eliminated. The effect of differences in curing conditions and workmanship can be reduced by explicit specimen fabrication procedures, and it appears the procedures defined in the Uniform Building Code Standard No. 24-30 [1] go a long way towards

---

[1]Assistant Professor, Department of Architectural Engineering, North Carolina A & T State University, Greensboro, NC, 27411

achieving this end. The mortar mixing procedures and the standard concrete brick unit required by this standard also appear, in the most part, to reduce the effect which variations in these materials have on the measured flexural tensile bond. However, the procedures and apparatus used for testing require investigation to determine whether they significantly affect the measured flexural bond strengths and their coefficients of variation.

To address the accuracy of the bond wrench testing apparatus and the procedures described in ASTM Standard C 1072 - 86, an experimental investigation was conducted. The goals of the investigation were to: 1) determine whether the bond wrench apparatus and testing procedures actually produce the assumed linear stress distribution on masonry prism specimens; 2) develop a calibration procedure for the bond wrench apparatus; 3) determine what factors have an effect on the applied stress distribution; and 4) recommend changes in the testing procedures and bond wrench description that will reduce the variability of test results.

This report describes the testing program, summarizes the experimental results and evaluates the validity of the bond wrench and the current testing procedures.

EXPERIMENTAL   PROGRAM

<u>Phase 1</u>

During the first phase of the experimental program, a block appliance was fabricated to evaluate the stresses applied to masonry specimens within the bond wrench testing apparatus. This block was fabricated using an epoxy material with an elastic modulus that approximated the elastic modulus of masonry. As shown in Figure 1, the epoxy block was rectangular in shape with approximately the same depth and width as a masonry prism specimen fabricated with standard concrete brick units. The height of the unit was approximately 280 mm. A total of sixteen strain gauges were mounted to the four faces, centered 62 mm from its top. When the block was placed in the bond wrench, these gauges were located so that they would fall at the same elevation as the top mortar joint of a masonry prism specimen. Two additional strain gauges were mounted on each of sides 1 and 2 to measure the block strains below the lower clamping bracket of the bond wrench.

To determine the exact elastic modulus of the epoxy material and evaluate the performance of the strain gauges, the block was placed in a compression testing machine and an axial load was applied to the unit at eccentricities of 0 mm, 19 mm, and 38 mm (See Figure 2). Each load test was repeated three times.

Each strain gauge was shunt calibrated before testing began, and this calibration was repeated after completion of Phase 1.

EPOXY CALIBRATION BLOCK FOR THE ASTM C 1072 BOND WRENCH

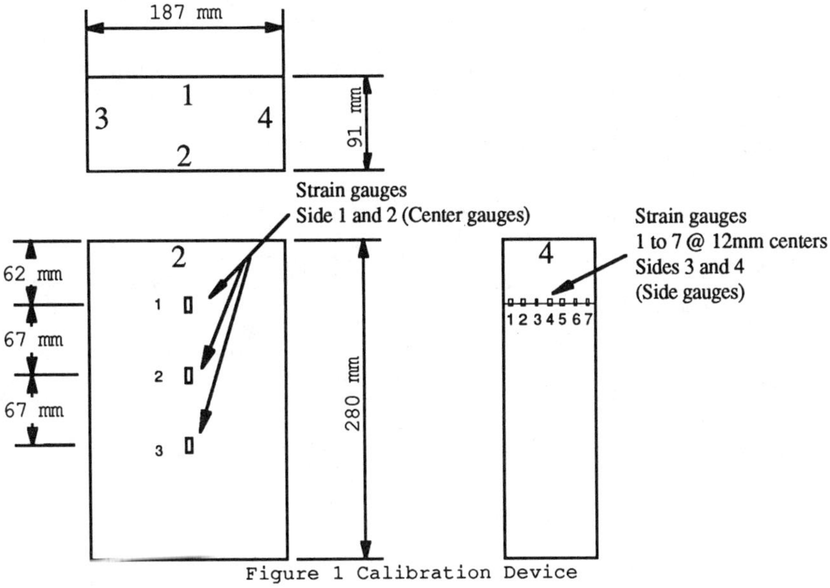

Figure 1 Calibration Device

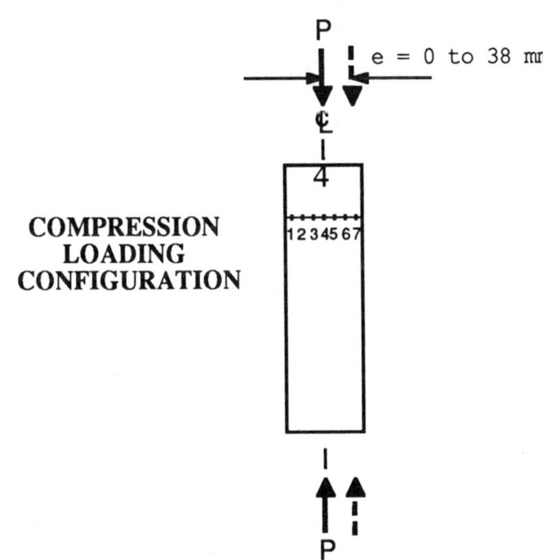

**COMPRESSION
LOADING
CONFIGURATION**

Figure 2 Configuration of Compression Loading on the Epoxy Block

<u>Phase 2</u>

During the second phase of the investigation, the calibration block fabricated during Phase 1 was used to evaluate the performance of the bond wrench testing apparatus at North Carolina A & T State University. This bond wrench was fabricated using the plans presented in ASTM C 1072 - 86, which are identical to those shown in UBC Standard 24-30 [1]. In addition, the apparatus had also been modified to incorporate changes currently under consideration by ASTM Task Group C 15.04.13.

The calibration block was placed into the lower clamping bracket and carefully aligned to the center of the testing apparatus. The center of the upper level of gauges was located at approximately 12 mm (0.5 in) above the top of the lower clamping bracket. A spacer of compressible foam was placed between the bottom of the block and the prism base support. The gauges were balanced and monitored while the upper clamping bracket was attached to the block and the clamping bolts tightened. An eccentric load was applied to the upper clamping bracket by advancing a nut along a threaded rod. The load was monitored by a load cell placed in-line with the threaded rod. All signals from the strain gauges and load cell were monitored and recorded with a computerized data acquisition system. Figure 3 shows the calibration block and its configuration within the bond wrench.

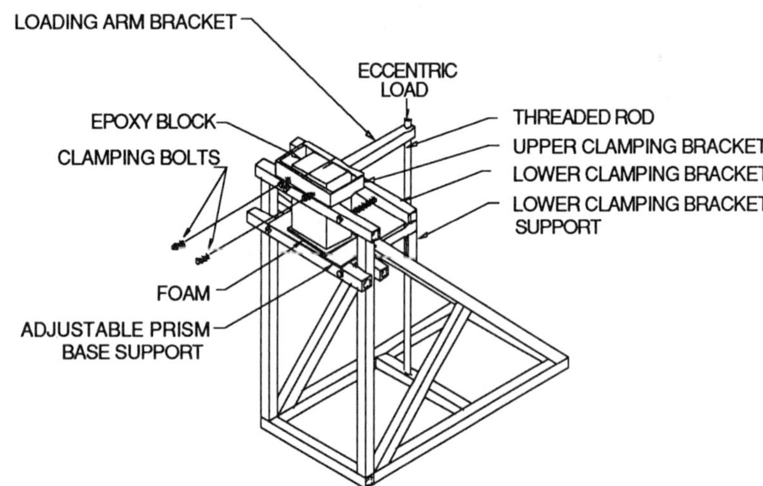

Figure 3 Calibration Block in the Bond Wrench

Three different operators were used to apply loads up to approximately 2670 N (600 lb) to the loading arm bracket. Three identical tests were conducted.

During testing, what appeared to be significant stress concentration effects were observed. To help determine the causes of these effects, additional tests were run with the upper level of gauges located at approximately 38 mm (1.5 in) above the top of the lower clamping bracket. These tests were repeated three times.

One of the apparatus modifications described above included the replacement of the solid plate bearing surfaces in the lower and upper

clamping brackets with individual bearing pads (see Figure 4). Since these
pads may have contributed to the stress concentrations, the pads on the
upper clamping bracket were replaced with solid bearing surfaces. Bond
wrench tests were repeated for a gauge height of approximately 12 mm (0.5
in) above the lower clamping bracket. Three tests were run.

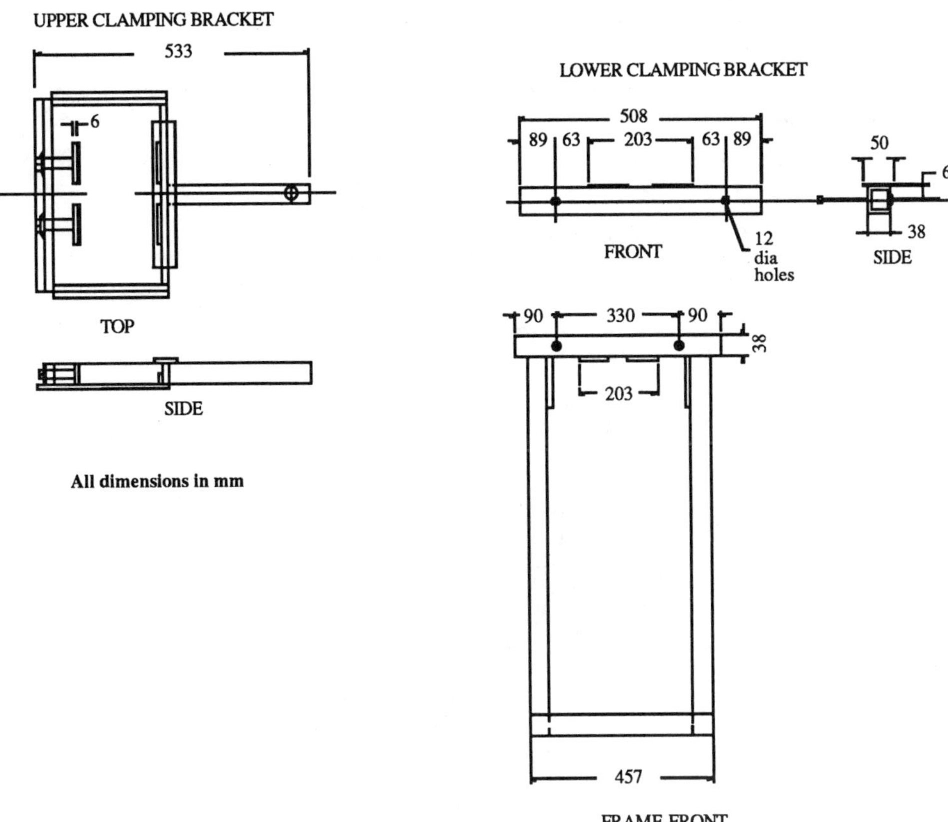

Figure 4 Upper and Lower Clamping Bracket Modified Configurations

<u>Phase 3</u>

     The fact that loading rate affects measured strengths has been
generally accepted for many years. Just how significant this effect is on
the flexural bond strengths measured by the bond wrench must be evaluated
since the current description of loading rate may not be adequate to ensure
consistent results. Phase three of this investigation evaluated the
effects of three loading rates on the bond strengths measured for standard
concrete masonry prisms.

     In an attempt to further reduce the influence which prism fabrication
procedures may have on measured bond strength, two-high conrete brick
prisms (couplets) were fabricated and tested during this phase of the
investigation. This specimen configuration also eliminates any pre-loading

on the lower joints of the specimen in the bond wrench. A total of sixty couplets were fabricated using the following procedures:

1. Prepackaged type S masonry cement, sand and water were mixed in the proportions and to the flow defined in UBC Specification 24-30 [1] using a laboratory mixer and the procedures outlined in ASTM Standard C 305 - 87 Mechanical Mixing of Hydraulic Cement Pastes and Mortars of Plastic Consistency. The proportions of each constituent are summarized in Table 1.
2. Two concrete units were placed in the prism fabricating jig as shown in Figure 5 [5] and the horizontal angles were adjusted level to the top of the units.
3 Two, 12 mm (0.5 in) thick plexiglass forms were placed on the top of the angles so that the dowels fit within alignment holes in the horizontal angles, forming a 12 mm deep mortar joint form around the masonry units.
4 A metal spoon was used to place mortar into one of the forms.
5 The mixing container was recovered and a metal straight edge was used to screed off the excess mortar using a single sawing motion.

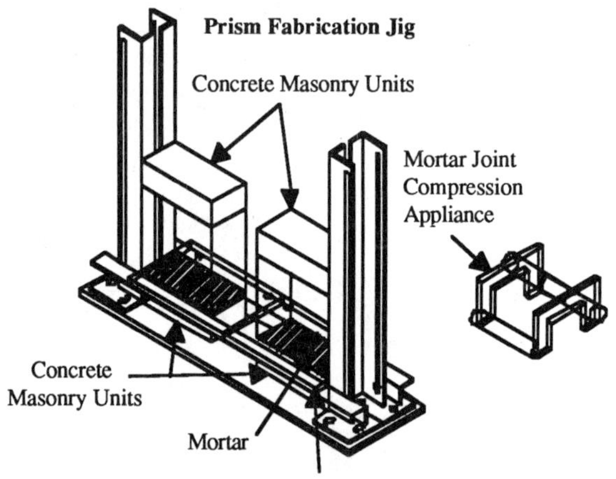

Figure 5 Prism Fabrication Jig

6 The plexiglass form was removed and a masonry unit was placed on top of the mortar.
7 The mortar joint was compressed to a uniform 9.5 mm (3/8 in) thickness using the mortar joint compression appliance. This device was placed on the top masonry unit and a downward force was applied until the edges of the appliance touched the horizontal angles.
8 Steps 4 through 7 were repeated for the remaining unit in the jig.
9. The two prism units were removed from the jig and the excess mortar was struck off using a metal spatula and a single non sawing motion on each face.
10 Each unit was placed in a plastic bag that was sealed after removing the majority of the air.
11 Steps 2 through 10 were repeated. Four prisms were made from each mortar batch and the total the time between completion of mortar mixing and the fabrication of the final prism never exceeded seven minutes.

For each mortar batch, a cone penetrometer test was conducted as described in ASTM Standard C 780 - 87 Preconstruction and Construction Evaluation of Mortars for Plain and Reinforced Unit Masonry before prism fabrication began. In addition, mortar was sampled from three batches and a total of three flow tests and three air entrainment tests were conducted as described in ASTM Standards C 91 Standard Specification for Masonry Cement and C 780 - 87, respectively. Six mortar cubes were also constructed and tested for compressive strength at an age of 28 days, as described in ASTM Standard C 780 - 87.

Table 1 Mortar Mix Proportions

Sand (mm^3)	Masonry Cement (Type S) (mm^3)	Water (ml)
2100	699	515

The prism specimens were cured for twenty-eight days at an average temperature of 23° C ± 3°. They were then removed from the plastic bags, allowed to dry for a minimum of one hour and tested in the bond wrench using the procedures specified in ASTM Standard C 1072 - 86. Twenty specimens were tested at an applied loading rate of 130 N/min (30 LB/min), twenty specimens were tested at an applied loading rate of 270 N/min (60 LB/min) and twenty specimens were tested at an applied loading rate of 400 N/min (90 LB/min). These loading rates corresponded to maximum tensile stress rates of 160 kPa/min, 320 kPa/min and 480 kPa/min, respectively.

EXPERIMENTAL RESULTS AND DISCUSSION

Phase 1

Figure 6 shows a typical strain distribution within the axially loaded block. It appears that there are strain concentration effects caused by misalignments within the testing machine and imperfections in the bearing surfaces, especially at higher strain levels. However, a linear regression performed on the average stress and strain values for all three tests yielded an elastic modulus of 3,330 MPa (480,000 psi), with an error, R^2, of 0.998.

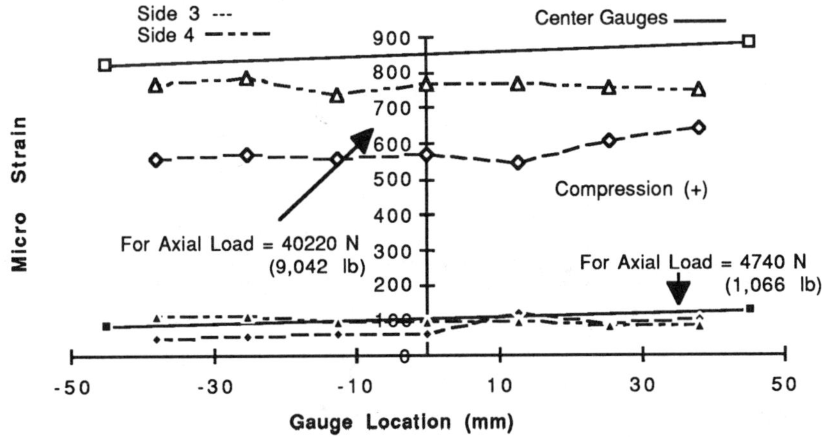

Figure 6 Strain Distribution Across the Depth of the Axially Loaded Block

Furthermore, it appears that when an eccentricity is introduced in the loading, better performance of the system results. Figure 7 shows good agreement between strains measured on block faces 3 and 4, for an eccentricity of 19 mm (0.75 in). As shown in Figure 8, there is also good agreement

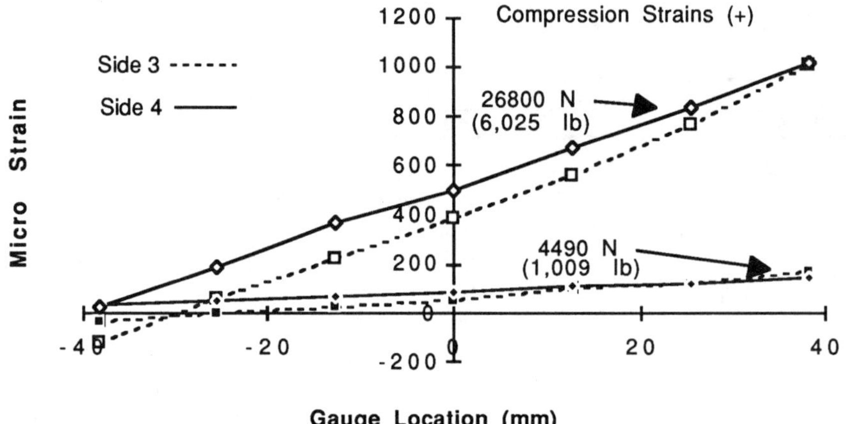

Figure 7   Block Strains on Faces 3 and 4 for an Axial Load and an Eccentricity of 19 mm (0.75 in)

between the predicted strains and the average of the strains measured on sides 3 and 4. This agreement between predicted and average measured strain was also observed for an eccentricity of 38 mm (1.5 in), as shown in Figure 9. Predicted strains were calculated using the average measured elastic modulus.

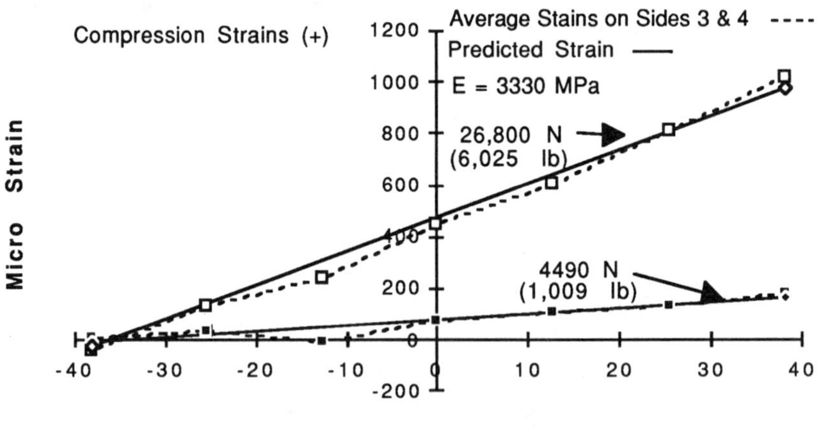

Figure 8 A Comparison of Average Strains on Sides 3 & 4 to Predicted Strains, e = 19 mm (0.75 in)

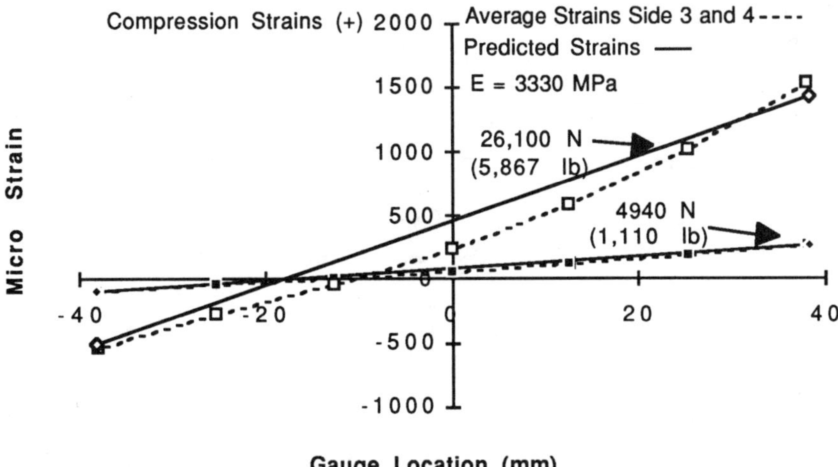

Figure 9 A Comparison of Average Strains on Sides 3 & 4 to Predicted
Strains, e = 38 mm (1.5 in)

It appears from the previous Figures that the strains, at least on
sides 3 and 4, model the expected behavior of the block under an eccentric
load reasonably well. However, further investigation of the load
concentration effects and possible retesting of the block unit under axial
load with a different bearing surface may be required before the block can
be confidently used as a calibration device. These investigations are
currently underway and the results will be reported at a later date.

Phase 2

Evaluation of the strains produced in the calibration block during
the first set of tests in this phase of the investigation suggests that the
difference in operators had little effect on the test results. This
evaluation also found that clamping torque's of up to 5.65 N.m (50 in lb)
had no significant affect on the measured strains.

Figure 10 shows a typical plot of strains in the block unit for
applied loads of 596 N (134 lb) and 2680 N (602 lb). These strains were
measured with the gauges located at approximately 12 mm (0.5 in) above the
lower clamping bracket, using the upper clamping bracket with separate
bearing pads. At lower strain levels, reasonable agreement is shown
between the strains measured at gauges on the sides and in the center.
However, as strain levels increased, the difference between the side gauges
and center gages increased. The center gauges measured significantly less
strain than the side gauges at higher strain levels. This behavior
indicates that the separate bearing pads of the modified clamping brackets
can cause stress concentrations in the block. It appears that the bond
wrench may not apply a linear stress field across the width of the specimen
at higher strain levels if separate bearing pads are used.

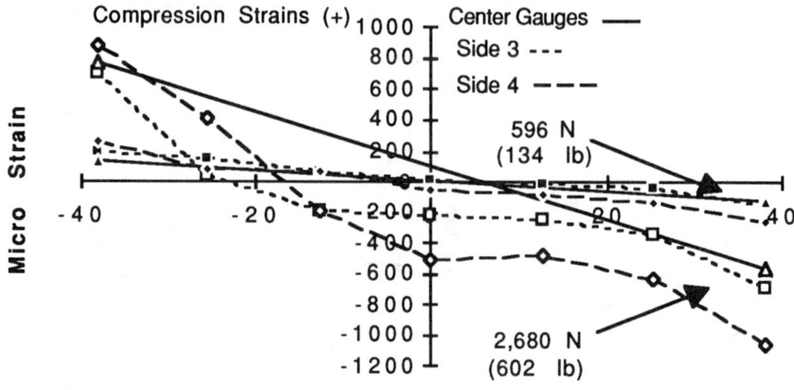

Figure 10 Block Strains in Bond Wrench using Separate Bearing Pads and a
Gauge Height Above the Lower Clamping Bracket of 12 mm (0.5 in).

Figure 11 shows a typical comparison of the average of strains measured on side 3 and 4 and the predicted strains. The predicted strains were calculated using the elastic stress formulas described in ASTM C 1072 – 86 and an elastic modulus of 3,330 MPa.

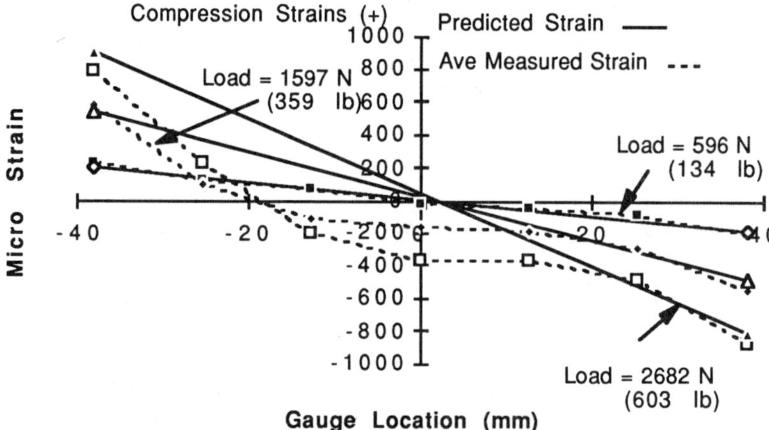

Figure 11 Predicted Versus Average Measured Strains on side 3 and 4 Using
Separate Bearing Pads and a Gauge Height Above the Lower
Clamping Bracket of 12 mm (0.5 in).

At lower strain levels, the agreement between measured and predicted values is good. However, at higher strain levels, the strains do not vary linearly across the depth of the specimen and differ significantly from the predicted values. Significant tensile strains are present across most of the specimen depth. The strain values on the right hand side of the tension zone do, however, appear to be reasonably close to the predicted values, especially near the critical outer fibers.

Figure 12 shows the average of strains measured on sides 3 and 4 when the gauge elevation is increased to 38 mm (1.5 in) above the lower clamping bracket. A significantly more linear strain distribution and better agreement with predicted strains were observed, especially in the compression zone of the specimen. These results suggest that the lower clamping bracket caused a significant portion of the stress concentration effects and better test results may be obtained by defining a minimum distance between the lower clamping bracket and the mortar joint being tested. It also appears that the configuration of the lower clamping bracket must be re-evaluated.

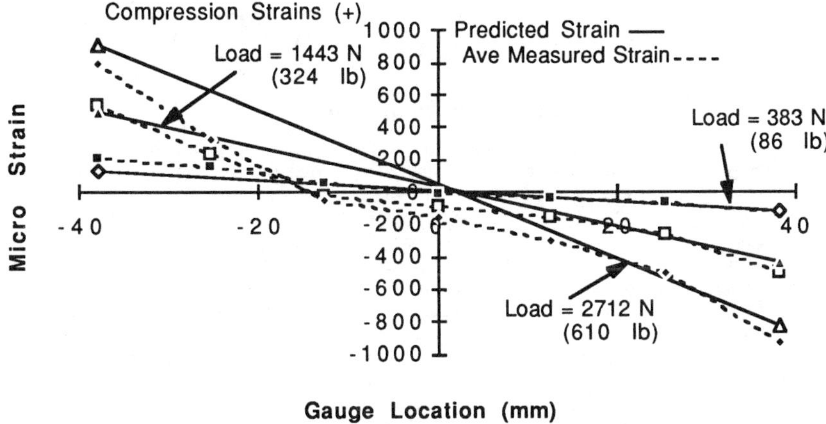

Figure 12 Predicted Versus Average Measured Strains on Side 3 and 4 Using Separate Bearing Pads and a Gauge Height Above the Lower Clamping Bracket of 38 mm (1.5 in).

As shown in Figure 13, little improvement of the strain distribution across the depth of the specimen was observed when solid bearing surfaces were used on the upper clamping bracket. This suggests that the lower clamping bracket and the flexibility of both clamps had a greater effect on the strain distribution than the bearing area of the upper clamp.

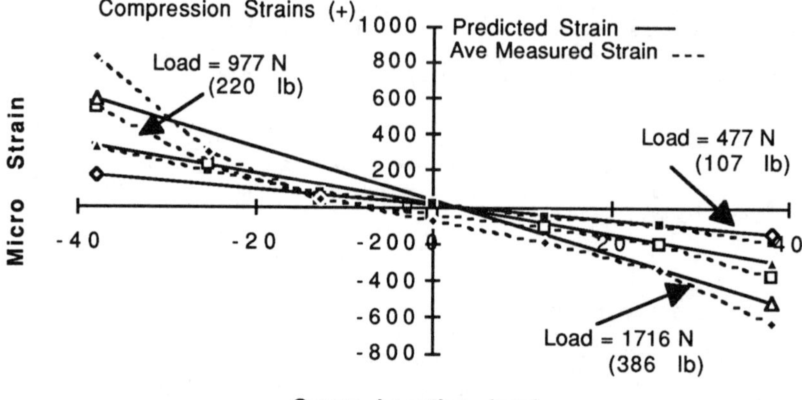

Figure 13 Predicted Versus Average Measured Strains on Side 3 and 4 Using
Solid Bearing Surfaces and a Gauge Height Above the Lower
Clamping Bracket of 38 mm (1.5 in)

Figure 14 shows the strains measured on side 2 for a gauge height
above the lower clamping bracket of 12 mm (0.5 in), separate bearing pads
on the upper clamping bracket and an applied load of approximately 1,713 N
(385 lb). Figure 15 shows the strains measured on side 2 for a gauge
height above the lower clamping bracket of 12 mm (0.5 in), solid bearing
surfaces on the upper clamping bracket and an applied load of approximately
1,557 N (350 lb). For the separate bearing and solid bearing
configurations, a comparison of the strains measured on side 2 to the
predicted strains (Figures 14 and 15) indicates that the presence of solid
bearing plates did improve the strain distribution across the width of the
specimen. Similar strain distributions were observed on Side 1 for both
testing configurations.

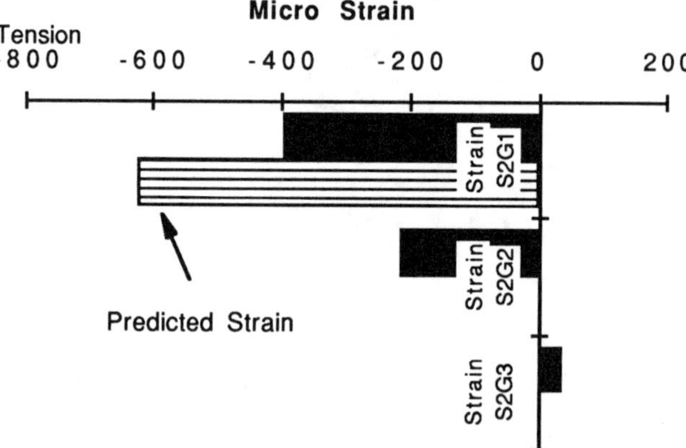

Figure 14 Strain Distribution of Center Gauges over height of Calibration
Block, Side 2, Separate Bearing Surfaces and a Height Above
Lower Clamping Bracket of 12 mm (0.5 in)

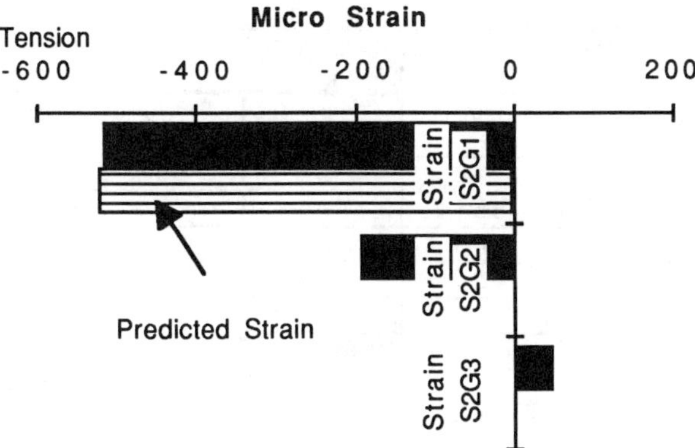

Figure 15   Strain Distribution of Center Gauges over Height of Calibration
Block, Side 2, Solid Bearing Surfaces and a Height Above Lower
Clamping Bracket of 12 mm (0.5 in)

It should be noted that strains measured by the center gauges, for
the separate bearing surface configuration and a gauge height of 38 mm (1.5
in), are also very close to the predicted values. Thus, it appears that
the strain variation across the width of the specimen can be reduced by
either incorporating solid bearing surfaces or by increasing the mortar
joint height above the lower clamping bracket. Since the latter
configuration appears to also improve the strain distribution across depth
of the specimen, improvements in the performance of the bond wrench
apparatus may be best directed to this area.

Figures 14 and 15 also show that there were significant tension
strains present in the specimen below the lower clamping bracket. The
compressible material at the base of the prism does not appear to relieve
all the strains and it may be necessary to require the lower support plate
to be lowered away from the specimen before testing.

The results from this phase of the investigation suggest that the
bond wrench testing apparatus may produce strain, and therefore stress,
distributions that significantly differ from the assumed linear
distribution. This difference may cause some of the observed variations in
experimental results. However, it is also clear that minor modifications
of the bond wrench and testing procedures should reduce these distribution
variations and therefore reduce the variation of the flexural bond
strengths measured by the bond wrench testing apparatus. These
modifications are under investigation and will be reported at a later date.

<u>Phase 3</u>

The results of the couplet prism tests are summarized in Table 2.
Also included in this table are the penetrometer test results for each
mortar batch. The average of the six mortar cube compressive tests was
11.18 MPa, with a coefficient of variation of 12.1 %. The average of the
three mortar flow tests was 128, with a coefficient of variation of 2%,
and the average of the three air entrainment tests was 17.3 %, with a
coefficient of variation of 8%.

Table 2 Mortar Test and Bond Wrench Results

Mortar Batch	Penetrometer (mm)	Flexural Strength (kPa)				
		Prism 1	Prism 2	Prism 3	Prism 4	Load Rate (N/min)
1	84	742	738	686	700	133
2	74	584	562	735	765	133
3	78	843	744	623	697	133
4	78	834	581	396	679	133
5	76	782	775	555	432	133
		Average Strength =		673	COV=	0.182
				(97.5	psi)	
6	80	382	453	606	537	267
7	80	691	658	540	717	267
8	80	779	692	754	675	267
9	75	554	545	735	713	267
10	73	1095	775	534	565	267
		Average Strength =		650	COV=	0.234
				(94.3	psi)	
11	79	921	699	773	500	400
12	74	801	495	470	396	400
13	73	720	1064	858	855	400
14	76	766	765	746	495	400
15	77	833	774	575	732	400
		Average Strength =		712	COV=	0.242
				(103.3	psi)	

Examination of the average prism flexural strengths shows no consistent trend across the three loading rates, although the average flexural bond strength measured at a rate of 400 N/min (90 lb/min) is the largest of all three averages. This suggests that there is some small increase in measured strength with loading rate. However, the strength increase is small and does not increase consistently with loading rate.

There is, however, a consistent trend in the coefficients of variation of the measured bond strengths. It appears that the lowest loading rate had a significantly smaller coefficient of variation than the two higher rates. This suggests that more consistent results could be obtained with the bond wrench if a low and uniform loading rate is specified in the testing procedures. Exactly what this loading rate should be and the tolerances that are required needs further investigation

SUMMARY AND CONCLUSIONS

An experimental investigation of the bond wrench testing apparatus and the procedures outlined in ASTM Standard C 1072 - 86 was undertaken. During the investigation a calibration device was fabricated, evaluated and used to measure the strains induced in a typical masonry specimen tested as required in the above standard. In addition, a total of sixty standard masonry specimens were fabricated and tested to evaluate proposed new fabrication procedures and determine whether loading rate has an effect on the flexural bond strengths measured using the bond wrench.

It was clear from the results that the bond wrench apparatus may produce strain, and therefore stress, distributions in masonry specimens that differ significantly from the linearly varying distribution that is assumed. These nonlinear strain variations may be partially to blame for a portion of the observed variation in the bond strengths measured using this apparatus. The test results also suggest that the differences between the expected strain distributions and the measured distribution may be drastically reduced by changes in the testing procedures and minor modifications of the bond wrench apparatus.

The results of the investigation further suggest that smaller coefficients of variation may be obtained with the bond wrench testing apparatus if a low and uniform loading rate is used during testing.

ACKNOWLEDGMENTS

The work described in this paper was sponsored by the Portland Cement Association (Project 82-02a). The opinions and the findings expressed in this paper are those of the author who is responsible for the conduct and accuracy of the program. They are not necessarily those of the Portland Cement Association.

REFERENCES

[1]   Uniform Building Code Standard No. 24-30, _Standard Test Method For Flexural Bond Strength of Mortar Cement_, International Conference of Building Officials, Whitter, CA, 1988.

[2]   Hedstrom, E. D., Tarhini, K. M., Thomas, R. D., Dubovoy, V. S., Klinger, R. E., Cook, R. A., "Flexural Bond Strength of Concrete Masonry Prisms Using Portland Cement and Hydrated Lime Mortars," _The Masonry Society Journal_, Vol. 9, No. 2, February , 1991

[3]   Ghosh, S. K., "Flexural Bond Strength of Masonry: An Experimental Review," _The Masonry Society Journal_, Vol. 9, No. 2, February , 1991

[4]   American Society for Testing and Materials, _Unpublished Report of C01.11.02 Round Robin Testing of Proposed Tensile Bond Test Method for Masonry Cement,_ 1989.

[5]   Atkinson- Noland and Associates, _Construction of Masonry Prisms,_ Engineering Report, Bolder, CO, 1985.

[6]   Uniform Building Code Standard No. 24-19, _Mortar Cement,_ International Conference of Building Officials, Whitter, CA, 1988.

Mary E. Driscoll,[1] Robert E. Gates,[1]

A COMPARATIVE REVIEW OF VARIOUS TEST METHODS FOR EVALUATING THE WATER PENETRATION RESISTANCE OF CONCRETE MASONRY WALL UNITS

REFERENCE: Driscoll, M. E., Gates, R. E., "A Comparative Review of Various Test Methods for Evaluating the Water Penetration Resistance of Concrete Masonry Wall Units," <u>Masonry: Design and Construction, Problems and Repair, ASTM STP 1180,</u> John M. Melander and Lynn R. Lauersdorf, Eds., American Society For Testing and Materials, Philadelphia, 1993.

ABSTRACT: Water damage to masonry either directly or indirectly as a result of freeze-thaw action has long been a concern to the construction industry. Over the years, the industry has developed a number of tests to measure leakage in masonry walls. However, little attention has been given to the correlation between these tests and the factors that contribute to water penetration and leakage. The forces that compel liquid water from the outside of a masonry wall into the interior of the building are:

<u>capillary forces</u>	– the propensity for water to "wick" in porous materials or through hairline cracks
<u>kinetic forces</u>	– the kinetic energy of a wind-driven rain will force it into the depth of the wall
<u>pressure differential</u>	– net pressure differentials caused by ventilation and air conditioning systems may cause the water to pass through small defects in the wall
<u>gravity</u>	– under gravity water can drip in through imperfections in flashing and parapet walls
<u>surface tension</u>	– water will tend to follow an easily wet surface, even turning around corners and edges such as in soffits, shelf angles, and loose laid metal flashing.

This paper reviews the effectiveness of existing water penetration and leakage tests such as the ASTM E 514 Test Method for Water Permeance of Masonry, RILEM tube test, AAMA 501.2-83 and other similar tests in predicting the resistance of masonry walls to water penetration caused by one or more of these forces. The authors further suggest a simple test method to complement the existing ones. A theoretical treatment of these forces and their effects upon water penetration is discussed in a separate paper.

KEYWORDS: water resistance, water penetration, masonry test method

---

[1] Technical Supervisor and Operations Manager, respectively, W. R. Grace & Company, 62 Whittemore Ave., Cambridge, MA 02140.

Exterior walls constitute a major component of a building envelope. The resistance of exterior walls to water penetration is a subject of great interest in the design of wall systems as well as the selection of proper construction materials. Over the years, masonry wall systems have proven to be high performance, durable systems that can be designed and built with excellent resistance to water damage. Although masonry wall systems can outperform many other exterior wall systems, the subject of potential water damage to masonry, either directly or indirectly, has been a concern to the construction industry. Over the years, the industry has developed a number of tests to measure the water resistance of masonry materials and walls. Unfortunately there is often much variability in the results of one type of test or between the results of different test methods.

For example, the water absorption of masonry units can be measured by ASTM Method of Sampling and Testing Brick and Structural Clay Tile (C67). However, variable results for absorption may be obtained, depending on the depth to which the masonry unit is immersed in water. The major reason for the variability is that the results are not only affected by the absorptivity of the masonry unit, but also by the pressure of the head of water above the masonry unit. This paper is intended to examine not only the inherent properties of masonry which affects its water resistance, but also the impact of various test conditions on the type and severity of fundamental forces that act to drive water into the test specimen. In addition, to adequately assess the results of these tests, careful consideration should be given to quantifying all water movement, especially water accumulation, within the test sample.

Keeping these things in mind, in any small or large scale test, the test conditions should be chosen to relate meaningfully to the same fundamental forces encountered in the real world, yet in a controllable and measureable fashion. First, let us examine the fundamental forces that act to drive water into any wall, including masonry, and the inherent material properties of masonry that may resist these forces. A detailed model describing the relationship between these properties and the driving forces of our environment is presented in another paper of this symposium. (1)

Briefly summarized the forces that drive water are:

Kinetic

This force, originating from the velocity of wind-driven rain hitting a wall, hammers moisture into the pores and cracks. Since this force is proportional to the square of the wind's velocity, the kinetic force quadruples as the velocity doubles. This is often the driving force for water penetration through cracks, voids, poorly constructed joints or improperly designed details.

Capillary

Small pores or hairline cracks may wick water powerfully in relation to a material's affinity for water. Hydrophilic materials such as concrete or clay masonry are subject to much higher capillary forces than hydrophobic materials such as organic polymers. Capillary forces act to wick water present at the exterior surface into the depth of the wall and even the interior of the building. It can also augment other avenues of water migration, carrying water further into the wall. In areas with high water tables, capillary action can wick the ground water up the wall, resulting in damage to the masonry courses near the ground level.

Gravity

Under the influence of gravity, water can drip in through imperfections in roof flashing, parapet walls, headers, and window sills. Once water is within a wall, gravity will broaden the area affected by drawing it down the structure, creating an expanding cone of wet or damp material.

Pressure Differentials

Any pressure gradients across a wall caused by ventilation and air conditioning systems where the interior air pressure in a portion of the building is lower than the exterior atmospheric pressure contribute to the migration of water into the building by sucking surface water through cracks, voids, etc. Another form of this type of force is seen in the pressure exerted by a column of water. For example, in horizontal applications of masonry such as brick or concrete pavers, ponded water exerts a pressure driving force proportional to the depth of the pond. In vertical applications, if the cores of concrete masonry units or cavities of double wythe walls are filled by water (e.g., from a roof or window leak) then the standing head of water can exert a pressure, driving water into the interior and exterior of the building. It is important to note that only a 5 cm (two inch) high column of water exerts a pressure equal to a kinetic driving force of a 100 km/h (62 mph) wind.

Surface Tension

Closely related to capillary force, surface tension will also drive water across a hydrophilic surface, even turning around corners and edges such as in soffits, shelf angles and loose laid metal flashing. Often, because of surface tension, water can work its way under loose laid metal flashing and enter the building, a problem that can be avoided by the use of fully adhered flashing and an effective drip edge.

Most importantly, the magnitude of these individual forces can be additive, thus quickly amplifying the overall driving force for water migration.

While proper building design and construction can help protect masonry against moisture penetration, two fundamental material properties also influence its resistance to moisture penetration.

The first is the permeability of masonry. An intrinsic property of porous materials, permeability is defined by Darcy's equation as a measure of a material's bulk resistance to the flow of fluids such as water. The nature of a material's pores (e.g. their size, tortuosity, connectivity, etc.), determines its permeability ($\underline{2}$). In other words, a tight dense material is more likely to resist water migration than a looser material that bears more resemblance to an interconnected network of small pipes. Obviously, discontinuities like cracks or voids strongly compromise the resistance of the masonry unit or assembly to water penetration. The hydrophilicity of the masonry material is the other major determining factor. By significantly changing the hydrophilic nature of clay or concrete to a hydrophobic one capillary action and its resulting "wicking" may be entirely defeated.

Finally, in order to get meaningful results from these test methods, it is important to quantify $\underline{all}$ water migration in a test sample. At any point in time a mass balance may be written for a wall segment in the form of:

(Total weight of water   − (Total weight of   = (Net accumulated
   put into a wall)   water that has   weight of water)
                                  flowed out)

or, alternatively:

(Rate of water   − (Rate of water   = (Rate of
   flowing in)   flowing out)   accumulation)

The data required to solve these equations are important measurements to make during a test since accumulated water, even if visibly hidden within the wall, can still pose significant threats to the building (e.g., mildew, freeze/thaw damage, etc.).

Any test method for quantifying masonry's vulnerability to water migration can and should be analyzed in terms of these fundamentals, so that we can choose test methods which give us useful information about the water resistance of masonry units.

RILEM TUBE TEST

Similar to a conventional laboratory method for determining permeability, the RILEM tube test is often applied to masonry as a measure of its resistance to moisture penetration ($\underline{3}$).

It is easily applicable to both individual masonry units as well as a wall assembly with a minimum of apparatus and time. In this test, a calibrated L-shaped tube is used to exert a starting pressure head against a small portion of the sample (see Figure 1). The drop in the height of water in the tube over time, is used as an indication of the material's vulnerability to water penetration.

Despite its apparent simplicity, the test is rather complex. First, it is highly dynamic. The test begins with a large external driving force on the block. The driving force is applied in the form

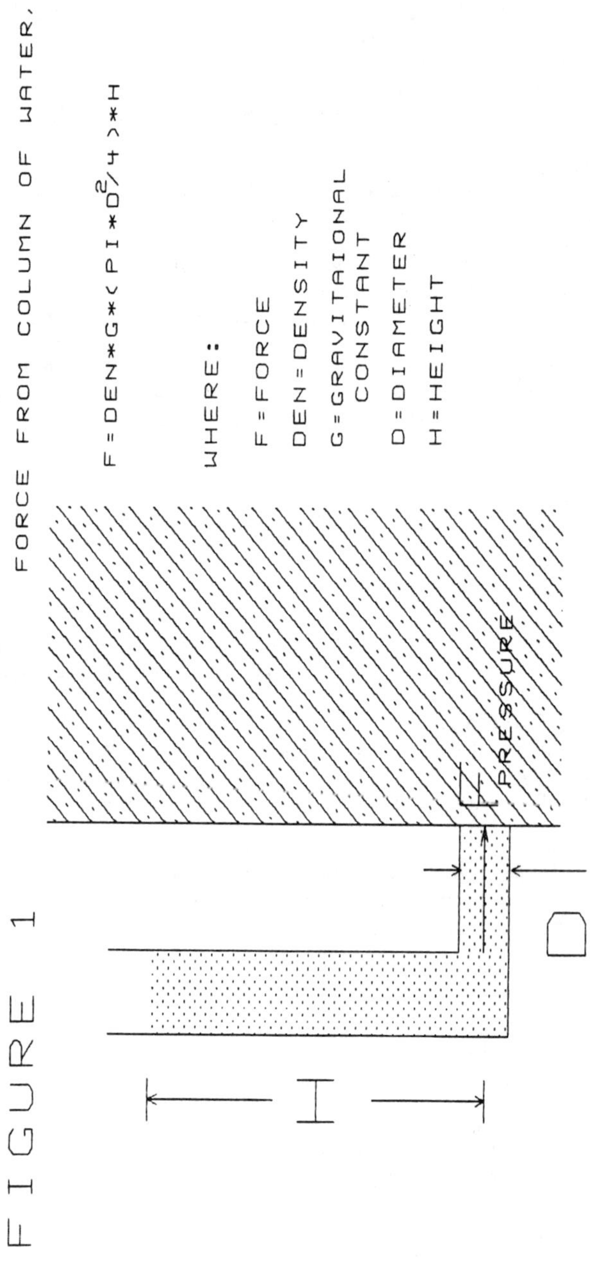
FIGURE 1
FORCE FROM COLUMN OF WATER,
F = DEN*G*(PI*D²/4)*H
WHERE:
F = FORCE
DEN = DENSITY
G = GRAVITAIONAL CONSTANT
D = DIAMETER
H = HEIGHT
H
D
PRESSURE
F

of water pressure, making use of industry conventions that the force of
wind driven rain may be simulated by a column of essentially static
water. However, as water migrates into the block from the tube, the
driving force exerted by the head of water is diminished.

Furthermore, the dynamics of this test are such that while the
pressure differential is the dominant force initially, as time goes on
and more water enters the sample, the effects of gravity and capillary
action within the sample increase, making analysis of the test results
rather complex. For instance, in the case of a masonry unit treated
with a surface applied water repellent coating, a large enough initial
force may drive water through the surface coating and into the untreated
masonry below. Once that occurs the capillary action of the untreated
material will act to suck water into the block, thus secretly augmenting
the pressure in the tube. In addition, once water saturates a portion
of the sample, gravity will also conspire to pull water in. And yet
this type of failure may not occur in a test where moderate forces are
used.

Another disadvantage of this test is the very small amount of area
tested. Material variations within a masonry unit as well as variations
between units cannot be adequately characterized without conducting many
tests.

Although the test is simple to conduct, analysis of the results to
gain insight about the permeability and water repellency of masonry
materials is quite complicated. Without proper analysis, the
conclusions drawn can be misleading.

ABSORPTION TESTS

A procedure for quantifying the water absorption of masonry units
is described in ASTM C 67 and ASTM Method of Sampling and Testing
Concrete Masonry Units (C 140). Both tests base their results on the
amount of water absorbed after the sample has been "submerged" (or
"immersed" in the text of C 140) in water. Thus the tests subject the
samples to a pressure driving force, generated by the surrounding
water. Unfortunately, the magnitude of this force is likely to be
highly variable since the total depth of the pool of water is not
specified. Thus the driving force may be different at different points
in the sample. In the case of a typical masonry block, the force near
the bottom would be several times larger than the force near the top of
the block, especially if it was immersed vertically instead of
horizontally.

Contrasting this method with the RILEM tube test, the tube test
initially exerts a singular pressure to the sample but that force may
change drastically over time. These absorption tests maintain an
essentially constant pressure over time but that pressure may vary
drastically along the sample. Meaningful comparisons between tests are
nearly impossible.

SPRAY TYPE TESTS

Tests like American Architectural Manufacturers Association "Field

Check of Metal Curtain Walls for Water Leakage" (AAMA 501.2-83, "Mask and Spray") (4), garden hose spray tests, and spray-bar tests use the kinetic energy of a stream of water to simulate rain and wind.

The success and value of this simulation is dependent on the care taken in the set up and calibration of the spray. The stream's diameter, flow rate, angle of impingement on the wall, and the distance between the nozzle and the wall (just to mention the major factors) all strongly influence the net force applied to an area of the surface (see Figure 2). And while these tests may appear to evaluate a broader area than the tube tests, the magnitude of the forces being applied are harder to calibrate and control especially in a field setting. Because of this, garden hose spray tests are often wisely restricted to merely locating areas where relatively significant leaks may occur (5).

A more carefully controlled version of this kinetic based approach is referred to as a spray-bar test. In this test, usually conducted in a laboratory setting, a fixed apparatus is used to supply several streams of water directed at the masonry sample. With proper setup and instrumentation, the force of each stream can approach the desired value. However, an inherent disadvantage of this kinetic based test is that each stream acts only on a very small area. Hence, wide variations in the results due to material variations can be expected.

<u>ASTM E 514</u>

ASTM Test Method for Water Permeance of Masonry (E 514) and its associated modifications are the most involved to set up, yet provide the most comprehensive test conditions available in the industry to date. And still the text clearly and correctly acknowledges the difficulties in establishing meaningful, useful, comparable results. The test procedure is discussed in more detail in another paper in this symposium (6).

In this test a section of wall is flooded with water, as force (in the form of air pressure) is exerted on the sample's flooded surface in order to simulate wind driven rain. It resembles the tube test in its controllable (although somewhat arbitrary) differential pressure driving force but with the added advantage of testing a larger composite sample. However, this externally applied force differs from the tube test, since it is maintained at a constant value. As water penetrates the block, capillary pressure may augment the driving force depending on the relative hydrophobicity of the material. Then as the wall becomes saturated with water, gravity will act to draw the moisture down thus slowly augmenting the capillary force. Depending upon the sampling time, equilibrium may or may not be reached between these forces and the resistance of the block.

Unfortunately, the results of this test are reported in a distinctly different way from most other methods. Instead of quantifying the total amount of water that enters the wall, it quantifies only the amount of water that passes through the wall either in the form of runoff from the back side of the sample or in the form of visible dampness.

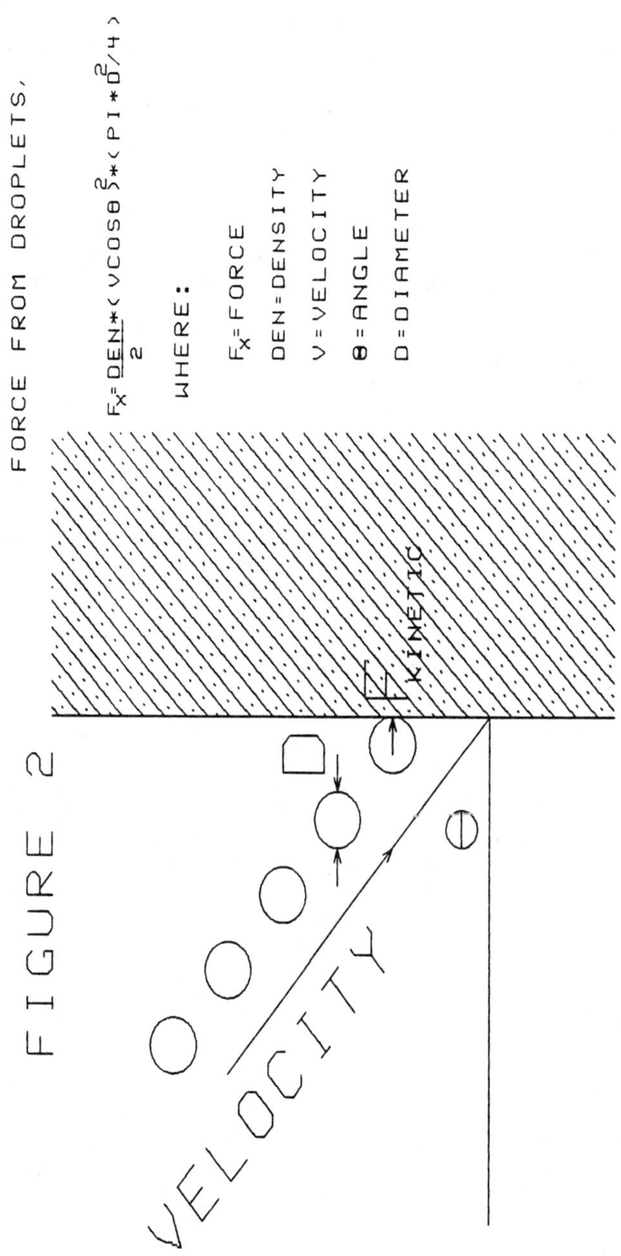

FIGURE 2
FORCE FROM DROPLETS,
Fx = DEN/2 * (VCOSθ)² * (PI*D²/4)
WHERE:
Fx = FORCE
DEN = DENSITY
V = VELOCITY
θ = ANGLE
D = DIAMETER
VELOCITY
D
F KINETIC
θ

Although an ideal wall would not permit any water penetration, in the real world water migration can be manifested in different ways:

1)   water that passes through the face of the block to the core where it runs off and is drained away from the interior.

2)   water that passes through the face of the block and continues to migrate throughout (and even up) the block under the influence of capillary force thus accumulating large quantities of water within the wall.

3)   water that passes all the way across the block and appears at the other side.

Unfortunately only this last form of water penetration is explicitly quantified in this test even though a wall that soaked up and retained a great deal of moisture would pose real concerns to a building.

A significant amount of work goes into the setup and execution of this test. Making a few more measurements to complete a water mass balance would add a great deal to the evaluation and characterization of moisture penetration in the masonry material.

THE PRISM TEST, A PRACTICAL ALTERNATIVE

One of the hidden difficulties in all these test methods is that in order to mimic the real world they typically focus on forces applied normally to a vertical wall. And, though realistic, this creates a multidimensional flow situation when one considers the influence of gravity and possible capillary forces. To simplify this problem in an attempt to gain meaningful data that can be analyzed with reasonable theoretical models, we have developed an alternative test that can be applied to a single building unit or a small composite (see Figure 3). In this test a head of water is applied to the entire face of a wall unit but from the bottom so that the forces of gravity and pressure act in the same plane. The advantage to this test is it allows us to quantify the magnitude of these forces over a broad area, with focussed attention to capillary forces, since eliminating this action is the goal of water repellent treatments.

The test is conducted by resting the sample on the gasketed lip of a tub filled completely with water. A tube, connected to the tub can be elevated above the face of the sample to exert any desired pressure head on essentially the full face of the sample. Keeping the tube filled with water keeps a constant pressure force on the sample until the forces of pressure, gravity, and capillarity reach equilibrium. A water balance around the sample can be made by measuring the tube's makeup water, any runoff water, and the sample's weight accumulation over time. Conversely, if one only needs a practical indicator of performance in the field, the tube may be kept full by gently dribbling water from a hose onto its edge.

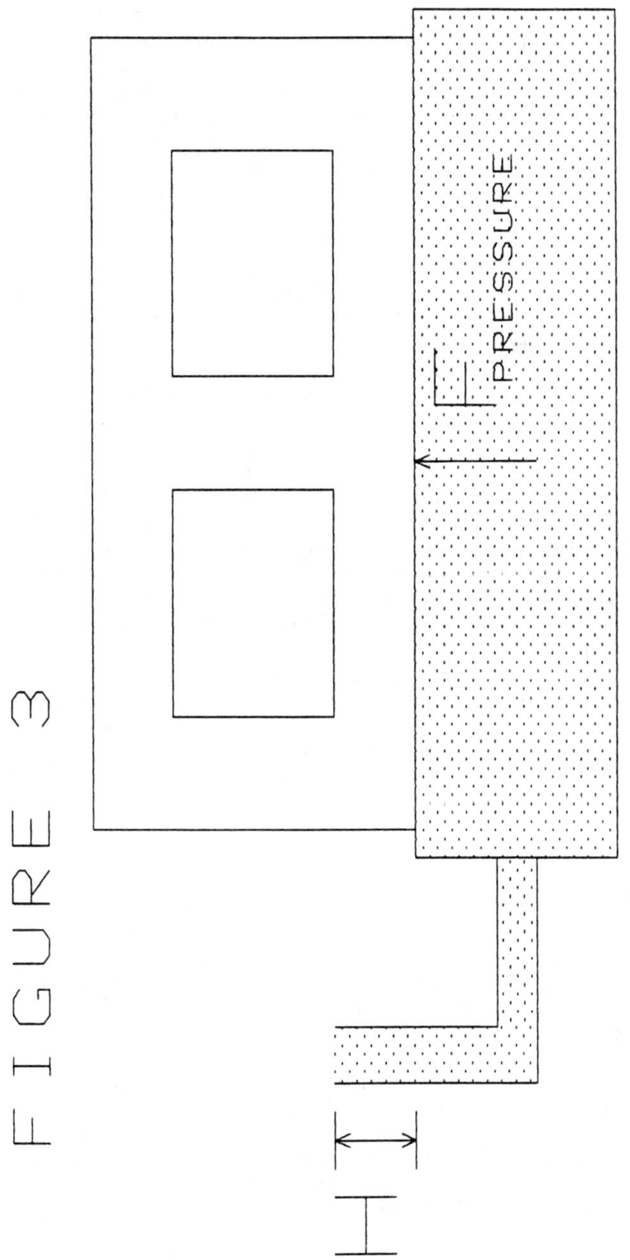

FIGURE 3

Although counter-intuitive, it graphically demonstrates and easily
quantifies the resistance to water flow offered by the material's
permeability and hydrophobicity. For example, an untreated masonry block
can easily wick water 5 cm (two inches) above the water height in the
tube. That means that in untreated masonry, capillary action can be as
forceful as the conventional 100 km/h (62 mph) wind often approximated as
a 5 cm water head. From this type of test we are developing quantitative
models for water migration in masonry materials.

CONCLUSION

The object of this paper has been to describe and encourage a way
of thinking about moisture migration tests in terms of the fundamental
forces of the environment and the material's properties that resist
penetration. Each of these methods (necessarily) attempt to approximate
the highly dynamic, complex, and varying forces of a rainstorm with
selected simplifications.

Comparing the true water resistance of one system to another
requires that rigorous test methods be developed and selectively
implemented based on a fundamental analysis of their procedures and
conditions. Consideration should be given to the type, severity, and
variability of the driving forces to which a sample is exposed.
Insightful analysis of observations and data, designed to measure basic
material properties (such as permeability and hydrophobicity), is
critical to the development and evaluation of products that will continue
to give the long term performance characteristic of masonry structures.
Finally, design, detailing and construction of exterior walls should
reflect our understanding of the fundamental driving forces to which a
building is exposed and the intrinsic properties of the materials used in
their construction. More work is needed to develop test methods upon
which to build a conclusive model for predicting moisture penetration in
masonry structures.

REFERENCES

(1)   Barkofsky, P.R., Driscoll, M.E., "A Theoretical Model of Water
       Penetration into Concrete Masonry Units," _Masonry: Design and
       Construction, Problems and Repair, ASTM STP 1180_, John M. Melander
       and Lynn R. Lauersdorf, Eds., American Society For Testing and
       Materials, Philadelphia, 1993.

(2)   Dullien, F.A.L., _Porous Media: Fluid Transport and Pore Structure_,
       Academic Press, Inc., San Diego, CA 1992, pp 8-11.

(3)   "Measurement of Water Absorption Under Low Pressure," RILEM,
       International Union of Testing and Research Laboratories for
       Materials and Structures, Paris, France.

(4)   "Field Check of Metal Curtain Walls For Water Leakage (AAMA
       501.2-83)," _Methods of Test for Metal Curtain Walls_, American
       Architectural Manufacturers Association, Des Plaines, IL.

(<u>5</u>)   Suprenant, B.A., "Analyzing Wet Walls, "<u>Masonry Construction</u>,"
        July 1991.

(<u>6</u>)   Chin, D., Gates, R.E., "Evaluation of ASTM E514 Water Penetration
        and Leakage Test to Assess Performance of Integral Water Repellent
        Admixture," <u>Masonry: Design and Construction, Problems and Repair,
        ASTM STP 1180</u>, John M. Melander and Lynn R. Lauersdorf , Eds.,
        American Society for Testing and Materials, Philadelphia, 1993.

Michael A. Vickers[1]

COMPARISON OF LABORATORY FREEZE-THAW PROCEDURES

REFERENCE: Vickers, M. A., "Comparison of Laboratory Freeze-Thaw Procedures," <u>Masonry: Design and Construction, Problems and Repair, ASTM STP 1180,</u> John M. Melander and Lynn R. Lauersdorf, Eds., American Society for Testing and Materials, Philadelphia, 1993.

ABSTRACT: The failure modes of three laboratory freeze-thaw procedures are compared with field failures. Sets of fifteen plant-fired brick were selected. Five brick from each set were subjected to each procedure.

Two omni-directional procedures were used: The Standard Method as set forth in ASTM C67, Standard Methods of Sampling and Testing Brick and Structural Clay Tile (with a 4-h cold water saturation), and a modified ASTM (with a 5-h boil saturation).

One uni-directional method was used: The Face Freeze-Thaw Procedure. This procedure was developed in the Acme Brick Company Research Laboratory, Denton, Texas, and utilizes the 5-h boil saturation. After each freeze cycle, only the face of the brick is thawed while the body remains frozen. The test specimens are withdrawn after 100 cycles, or when a specimen has failed. A sample is considered to have failed when a laminar crack develops parallel to the brick face. The crack usually develops 10-15 cycles before the face comes off.

KEYWORDS: physical properties, frost resistance, water absorption, compressive strength, omni-directional, uni-directional, laminar spalling, process defect, face freeze-thaw

---

[1]Director of Research and Production Services, Acme Brick Company, Denton, Texas.

Numerous attempts have been made to correlate physical properties of fired clay units with frost resistance. Butterworth [1] reports that examination of 160 varieties of brick from the UK did not show a satisfactory correlation. Robinson, Holman, and Edwards [2] after testing over 5,000 bricks from 34 manufacturing plants in the USA found that 31.5% of durable bricks were rejected while 22.8% of non-durable bricks were accepted. Van der Klugt[3] reports similar results. These tests were in accordance with ASTM C67 or a modified procedure (48 h water immersion before freezing), and prompted the conclusions that resistance to freezing is related to neither water absorption nor compressive strength [4]. The only reliable test is actual exposure to the environment for five years [2]. The lack of reliable test method is probably the chief reason for the slow progress made over the years in eliminating freeze-thaw failures [5], and, clearly, a better test method and criterion are required [6]. ASTM C67 is an omni-directional test procedure and specifies that the brick be tested with a 4-h cold water saturation.

In 1988, eight years of in-house physical property and freeze-thaw results were compared. These data included over 11,000 brick from 16 Acme Brick Company plants. The data were averaged by body type (red burning, buff burning, shale or clay, etc.) and by all pass or all fail freeze-thaw. These brick were all tested according to ASTM C67 with one modification. The brick were subjected to freeze-thaw while containing the 5-h boil saturation instead of the 4-h cold water saturation. The boil saturation was used to increase the severity of the test and hopefully produce results that would duplicate field failures. Typical results are shown in Table 1.

TABLE 1

	No. of Brick	IRA	Cold Absp(%)	Boil Absp(%)	C/B	Compressive Strength (psi)
All Pass	68	20.2	8.2	10.5	0.78	6,424
All Fail	43	22.4	8.7	11.3	0.77	6,362

Note: 145 psi = 1 MPa

All of the results were similar in that there was very little difference between the physical properties of brick that passed freeze-thaw and those that failed.

DISCUSSION

The results detailed above show that omni-directional freeze-thaw procedures (where all faces of the brick are exposed to the same conditions) do not produce a correlation between physical properties and freeze-thaw failures. The failures that do occur are identifiable as a process defect such as a dryer or cooling crack instead of laminar spalling as seen in the field. These failure types are also reported by Van der Klugt [3].

In response to this universal problem British Ceramic Research Limited has developed a Panel Freezing Test [4] that subjects the samples to uni-directional freeze-thaw conditions. With a uni-directional procedure heat transfer is restricted to one stretcher face of a brick just as it is in a wall exposed to natural conditions.

The BCRL procedure consists of constructing a panel 10 brick high and 3 brick wide. After curing for 28 days the panel is soaked in water for 7 days. After soaking the panel is installed in the cabinet and exposed to 100 freeze-thaw cycles. Excellent results have been achieved when the panel test results are compared against natural exposure of the same type brick in test walls [7].

One shortcoming of the panel test is that it does not allow for direct correlation of physical property measurements with freeze-thaw performance. The face freeze-thaw procedure was developed to enable a direct correlation of measured properties with freeze-thaw performance. This will enable the laboratory to furnish the manufacturer with the parameters needed to insure a trouble free product.

TEST PROCEDURE

The face freeze-thaw procedure is similar to C67 with two major differences:

1. When placed into the freezing tray the moisture content is the 5-h boil saturation.
2. During the thawing cycle, only the face is thawed while the body remains frozen.

After each freeze cycle, the face can be thawed by either of two different methods:

1. In a sink with hot water:
    a. The sample pan is placed on supports inside a second, larger pan;
    b. Hot water is run into the larger pan for approximately three minutes [Figure 1].
2. On a hot plate or stove:
    a. A larger pan is placed on a hot plate and partially filled with water;
    b. The water is brought to a slow boil;

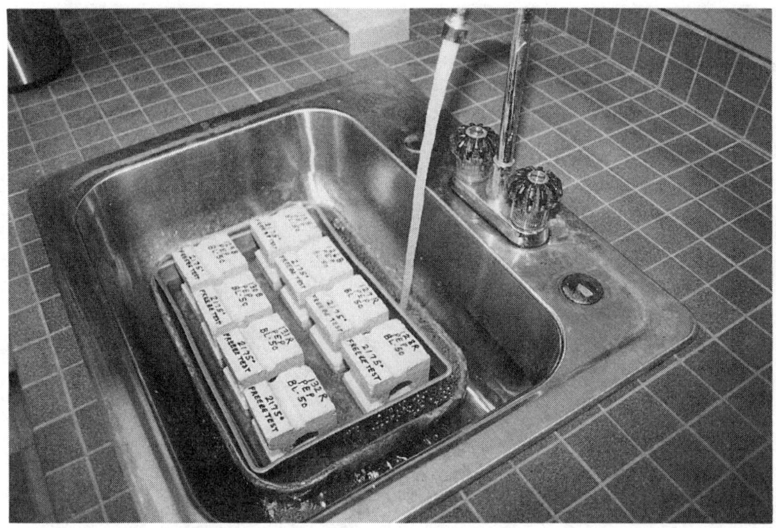

Figure 1--Face procedure thaw cycle using hot water

Figure 2--Face procedure thaw cycle using gas hot plate

Figure 4--Face procedure thawed sample

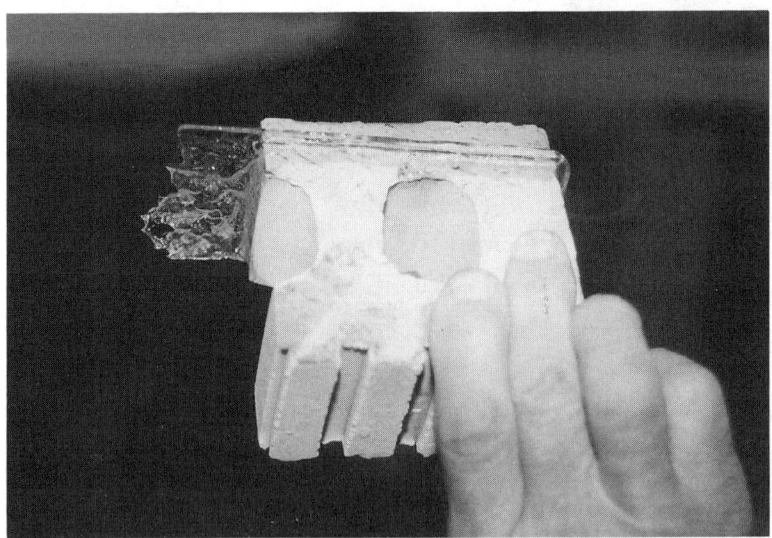

Figure 3--Face procedure thawed sample

c. The sample pan with the frozen samples is placed on supports in the larger pan [Figure 2].
d. The sample pan is removed and returned to the freezer just as the water thaws but the brick body is still frozen and covered with frost.

Thawing time will vary with the mass of the samples and the relative size of the two pans. The thawed samples shown in Figure 3 and Figure 4 have a body temperature of -7.8°C and a face temperature of 30.4°C.

To compare the freeze-thaw results of the three test methods a minimum of 15 plant fired brick from 10 different plants were selected. These samples were not collected at random but were consecutively extruded and fired at the same kiln location. This was done so that the primary variable would be the freeze-thaw procedure instead of normal product variability.

Five brick from each group were tested according to ASTM C67 with a 4-h cold water saturation (Table 2), a modified procedure with a 5-h boil saturation (Table 3), and the face procedure (Table 4).

TABLE 2--Freeze-thaw
with 4-h Cold Water Saturation

Plant	24 h Cold Absorption(%)	C/B	Compressive Strength (psi)	Cycles Min	Avg	Max	No Pass 50 Cycles	Failure Mode
A	4.8	0.63	11,853	91	98	100	5	Dryer
B	4.5	0.65	8,004	76	95	100	5	Dryer
C	2.6	0.65	12,088	100	100	100	5	None
D	1.4	0.77	21,573	100	100	100	5	None
E	6.6	0.69	8,817	100	100	100	5	None
F	8.0	0.78	5,232	31	74	100	4	Lamination
G-1	8.5	0.70	9,513	32	54	100	1	Dryer
G-2	11.2	0.65	3,837	46	78	99	4	Dryer
K-1	4.5	0.65	9,520	100	100	100	5	None
K-2	5.1	0.92	12,675	100	100	100	5	None
L	7.9	0.64	12,074	100	100	100	5	None
M	5.1	0.65	12,088	100	100	100	5	None

Note: 145 psi = 1 MPa

TABLE 3--Freeze-thaw
with 5-h Boil Saturation

Plant	24 h Cold Absorption(%)	C/B	Compressive Strength (psi)	Cycles Min	Avg	Max	No Pass 50 Cycles	Failure Mode
A	5.2	0.65	13,650	12	28	36	0	Dryer
B	4.0	0.61	9,185	11	46	83	2	Dryer
C	3.3	0.71	13,198	100	100	100	5	None
D	0.8	0.70	26,328	100	100	100	5	None
E	6.4	0.67	9,382	2	10	25	0	Bridge
F	8.0	0.77	5,952	14	27	56	1	Lamination
G-1	7.2	0.68	8,268	3	7	16	0	Dryer
G-2	11.3	0.65	3,259	2	11	14	0	Dryer
K-1	4.9	0.67	9,612	100	100	100	5	None
K-2	5.8	0.79	9,005	35	81	100	4	Lamination
L	7.8	0.77	8,617	13	34	53	1	Bridge
M	4.9	0.64	12,098	18	23	26	0	Dryer

Note: 145 psi = 1 MPa

TABLE 4--<u>Face Freeze Thaw</u>

Plant	24 h Cold Absorption(%)	C/B	Compressive Strength (psi)	Cycles Min	Avg	Max	No Pass 50 Cycles	Failure Mode
A	5.2	0.65	10,639	75	95	100	5	Spall
B	4.7	0.66	7,395	100	100	100	5	None
C	3.0	0.69	12,875	100	100	100	5	None
D	1.3	0.69	15,889	100	100	100	5	None
E	6.6	0.68	9,334	14	52	100	3	Spall
F	8.1	0.78	5,904	37	69	100	4	Spall
G-1	8.3	0.68	8,984	100	100	100	5	None
G-2	11.0	0.64	4,305	36	78	100	4	Spall
K-1	5.2	0.68	8,175	100	100	100	5	None
K-2	4.9	0.95	15,184	100	100	100	5	None
L	7.7	0.79	7,968	100	100	100	5	None
M	4.9	0.64	12,297	100	100	100	5	None

Note: 145 psi = 1 MPa

RESULTS

Very few samples (10%) tested according to ASTM failed freeze-thaw [Figure 5].

Numerous samples failed (62%) when tested according to the modified procedure, but 53% of the failures were on a process defect (bridge crack, dryer crack, or cooling crack) and did not resemble field failures [Figure 6].

The face procedure appears to duplicate the laminar spalling seen with field failures [Figures 7 and 8]. Five of the groups that failed the modified procedure passed the face procedure.

CONCLUSIONS

The face freeze-thaw procedure duplicates the laminar spalling of field failures. This will provide the laboratory testing tool needed to determine the physical properties required to eliminate field failures.

NEXT

A series of plant extruded and dried brick will be fired in lab kilns. These firings will be controlled to produce under fired, normal, and over fired samples. Each brick will have a Bell Fire-Chek Key in the center of the face to determine the heat work for each sample. The physical properties will be measured according to ASTM C67 and one-half of each sample will be subjected to the face procedure. Key readings, physical properties, and freeze-thaw results will be correlated.

A second series of lab fired samples will be sent to the BCRL to compare the panel test results with those of the face procedure.

Figure 5--Sample after 100 freeze-thaw cycles using the
ASTM procedure

Figure 6--Sample after 18 freeze-thaw cycles using the
modified procedure

Figure 7--Sample after 46 freeze-thaw cycles using the
          face procedure

Figure 8--Sample after 46 freeze-thaw cycles using the
          face freeze-thaw procedure

REFERENCES

[1]   Butterworth, B., "Frost Resistance of Bricks and Tiles, A Review," _Journal of British Ceramic Society_, Vol. 1, No. 2, 1964, pp 203-223.

[2]   Robinson, G.C., Holman, J.R., and Edwards, J.F., "Relation Between Physical Properties and Durability of Commercially Marketed Brick," _Ceramic Bulletin_, Vol. 56, No. 12, 1977, pp 1017-1079.

[3]   Van der Klugt, J., "Frost Testing by Uni-directional Freezing", _Ziegelindustrie Industrial_, No. 2, 1989, pp 92-98.

[4]   Peake, F. and Ford, R.W., "A Panel Test For Brickwork", _British Ceramic Research Association_, Technical Note 358, 1984, pp 1-7.

[5]   Arnott, M.R. and Litvan, G.G., "Quality-Control Test for Clay Brick Based on Air Permeability," _Ceramic Bulletin_, Vol. 67, No. 8, 1988, pp 1412-1417.

[6]   Winslow, D.N., Kilgour, C.L., and Crooks, R.W., "Predicting the Durability of Bricks", _Journal of Testing and Evaluation_, Vol. 16, No. 6, 1988, pp 527-531.

[7]   Peake, F. and Ford, R. W., "Calibration of the British Ceramic Research Limited (BCRL) Panel Freezing Test Against Exposure Site Results", _British Ceramic Research Limited_, Research Paper 762, 1988, pp 1-7.

John E. Lovatt[1]

PREDICTING DURABILITY OF BRICK VENEER WALLS IN COLD CLIMATES

REFERENCE: Lovatt, J. E., **"Predicting Durability of Brick Veneer Walls in Cold Climates,"** Masonry: Design and Construction, Problems and Repair, ASTM STP 1180, John M. Melander and Lynn R. Lauersdorf, Eds., American Society for Testing and Materials, Philadelphia, 1993.

ABSTRACT: Bricks in brick veneer walls of modern buildings in cold climates are exposed to frequent cycles of freezing and thawing, and may have high moisture content during these cycles. Poor air barriers, humidified buildings, higher insulation levels, and less emphasis on details to guide rainwater away from brick surfaces all contribute to this severe environment. Freezing while wet is a major contributor to spalling and premature failure of bricks. A model has been developed to provide a quantitative estimate of the time to freezing-induced failure of bricks in a veneer wall, given the local climate, the brick characteristics, and building design and operating conditions.

KEYWORDS: brick veneer, durability, freeze-thaw cycles, condensation, air leakage, capillary action, wetting of bricks, modelling

---

Brick masonry has traditionally been considered a low-maintenance, high durability material for use in building walls. This reputation is based on centuries of experience with brick in buildings, under a wide range of climate conditions. In large buildings exposed to very cold winters, solid masonry walls are heavy, expensive, and provide poor thermal resistance. In these types of applications, brick is now used almost exclusively as an exterior cladding, or veneer, attached with masonry ties to some interior construction which transmits loads due to wind and the dead weight of bricks to the building frame. Although it is expensive, it is popular as a result of its reputation for durability and attractive appearance.

Brick veneer walls of modern buildings in Northern climates are exposed to a very severe environment. In older, uninsulated buildings, heat escaping from the building warmed the bricks in winter, so that they would freeze very slowly, infrequently, and to a limited depth. Poor air and vapor barriers allowed moisture to move to the inside surface where it would dry in the warm interior air. Buildings were leaky and not humidified, so that condensation was never a problem. Exterior facades had ample protrusions and drips to direct rain and meltwater away from the brick surface.

---

[1] Senior Project Engineer, MORRISON HERSHFIELD LTD., 1980 Merivale Road, Nepean, Ontario, Canada K2G 1G4

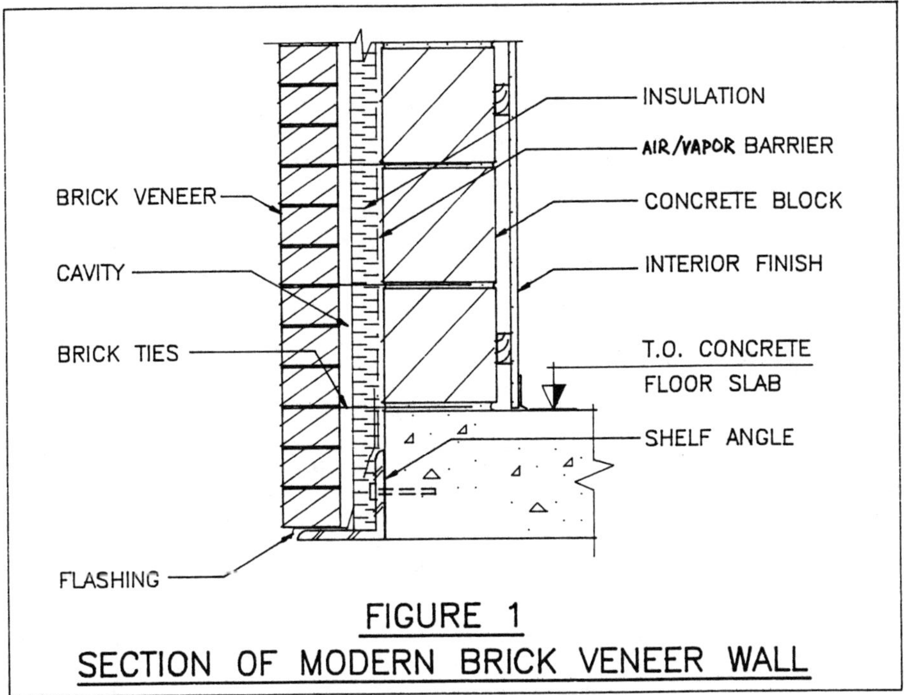

FIGURE 1
SECTION OF MODERN BRICK VENEER WALL

Figure 1 shows a section of a modern brick veneer wall. The
insulation is located inboard of the cavity behind the brick veneer, so
that the brick temperature is much closer to the outdoor temperature.
The first cold surface encountered by indoor air leaking through a poor
air barrier is the backside of the brick, and if indoor air is moist
enough, condensate and frost will be deposited on this surface. In
general buildings are less leaky, and mechanical humidification and
ventilation systems are able to provide higher indoor humidities and to
force indoor air out through air barrier flaws to prevent drafts. The
vapour barrier prevents moisture diffusion through the inner wall in
both directions, so that once the brick is wet it cannot dry to the
interior. The new architectural style provides vast expanses of flat
facade unmarked by corbels and protrusions, and unable to direct rain
and meltwater away from the brick surfaces.

The result is that bricks in veneer walls of modern buildings
freeze more often, faster, to a greater depth, and are generally closer
to being saturated with water when they freeze. Due to the physical
properties of brick, saturated freezing is uniquely damaging to it, and
there are an ample number of cases of buildings in Canada where the
brick veneer has required replacement within ten years due to widespread
spalling and cracking. Replacement of brick veneer walls is a very
expensive and time-consuming problem and has a large impact on the
useability and marketability of the building; owners and operators of
buildings having brick veneer walls require some means of determining
whether walls in their buildings are in danger, what the remaining
lifetime is, and what steps can be taken to extend that lifetime.

PHYSICAL PROCESSES OF FREEZING DAMAGE IN BRICK VENEER WALLS

<u>Moisture Uptake</u>

Masonry materials contain voids in their structure which can absorb water. If water is available on the surface of the material, it will be absorbed at a rate dependent on the relative size of individual pores and the total pore volume. This characteristic is called suction, and is often measured since it determines the nature of the bond between brick and mortar. Water enters the pores of the material under the forces of:

- vapor diffusion,
- capillary action, and
- gravity

In general, the material may be dried via vapor diffusion and gravity, but not via capillarity. It has been found that bricks which are more favourable to capillary action (a function of pore size and connections) are much more susceptible to saturated freezing damage. Those masonry materials which have a large proportion of total pore volume between 0.1 and 1 micron in diameter are damaged in fewer cycles. Pores of this size provide fast water absorption through capillary action, but are slow to dry via diffusion and surface evaporation. Smaller pores are generally not well-connected to each other and so require longer time to become saturated. Larger pores are filled less quickly and completely by capillary action. Many large pores cannot be completely filled by capillary action; thus there is air trapped in the pore which compresses to take up the expansion of the water as it freezes.

Applying this knowledge to the situation of a brick veneer wall, six mechanisms can be identified for moving water or moisture into and out of the bricks in the wall:

- Wind-driven rain hitting the outside face of the wall. This is most likely to result in freeze-thaw damage in Maritime climates, where the temperature hovers around freezing and winter rain is more common.
- Meltwater drainage and leaks. Whenever melting of snow cover takes place and the drainage details of the wall allow, meltwater from surfaces that gather snow (roofs, parapets, window sills, projections) will run down the outside surface of the brick wall, and through leaky flashing details at the top of the wall. Melting of frost buildup on the inside face of bricks is difficult to account for. If the frost melts from the brick side, some will be absorbed as water but much of the frost buildup may fall off as ice to the bottom of the cavity. If the frost melts from the cavity side, much of it will drain down the still-frozen inner surface of the brick.
- Condensation on the inside face of the brick. When moist indoor air leaks out into the cavity and is cooled, some of the moisture condenses out on the cold inside surface of the brick. If this surface is above freezing, the moisture will be absorbed into the

brick; if it is below freezing it will build up on the brick surface as frost.

As previously discussed, frost is less of a source of wetting than liquid condensation; excessive frost buildup may in fact plug the cavity and stop the air leakage. Frost is a major source of wetting only when the drainage of the cavity to the outside is blocked; then when the frost melts, it pools at the bottom of the cavity and provides a source for wetting of the lower bricks via capillary action.

Air at 0°C can hold just under 4 grams moisture per kilogram dry air. This corresponds to a Relative Humidity of 23 % at typical indoor temperature of 22°C. The water available for liquid condensation is a function of the indoor RH over 23 % and the volume of indoor air leaking out. The amount that actually does condense out on the brick is dependent on the brick inside surface temperature, the cavity air temperature, and the length of the air flow path through the wall. The worst case occurs when the indoor RH is high, the wall is leaky and the building is pressurized, the wall insulation is significant, and a large part of the leakage path is in the cavity.

- Hygroscopic or capillary action. Wherever there is a point source of liquid water on the surface of a brick, it will move into the brick in all directions under the action of capillarity. In the pore sizes commonly found in bricks, a capillary water column easily the height of the wall can be supported if there is enough liquid water. Capillary action fills pores of smaller sizes at a faster rate, and larger pores at a slower rate. Bricks which have poor connections between pores are not as affected, since only pores which are directly connected to the source of liquid water can be filled by capillarity.

- Drying due to gravity drainage. Bricks which have been wetted from above or the side, and contain a large proportion of pore area in pores and cracks over 1 mm will lose this water through gravity drainage down and out of the wall. This will occur immediately upon the end of saturation at the top of the wall, and as the water level drops in the lower courses of bricks in the wall. Usually this accounts for a small proportion of the total water content of the brick because pores and cracks of this size are not typical of brick.

- Drying due to diffusion and evaporation. This is the main mode of drying of bricks. On the outside face it proceeds at a significant rate whenever the temperature is above freezing and the RH is below 100 %. This rate is dramatically increased by solar heating of the brick, which increases the moisture capacity of the air and induces a convection current which quickly replaces saturated air with dry air against the face of the brick. Water is brought to the brick surface by diffusion and surface tension.

<u>Freezing of Wet Bricks</u>

Many studies have investigated the effects of the geometry of freezing action and the rate of freezing [1],[2],[3]. There is general consensus that freeze fronts which advance into the masonry unit from more than one direction are more damaging than unidirectional freeze fronts. There is also evidence that a faster rate of freezing results in more damage to the masonry unit. The generally-accepted reason for this behaviour is that water is extruded from the brick by the advancing freeze front; when the rate of freezing is quick there is less time to extrude water, while there is less area available to extrude it into if there are two or more freeze fronts. The size of pores also governs the temperature at which the water in them freezes. Freezing begins at 0ºC in the largest pores and cracks, but water in the smallest pores will freeze at much lower temperature.

Water expands by 10% during freezing within the material; if there is not enough room within the pores to allow for this expansion, it is taken up in the material itself. Masonry materials are notoriously inelastic and brittle in tension; the stress imposed by the ice expansion causes initiation and propagation of cracks until the ice expansion has been accommodated. When the material next thaws, this area is added to the existing void area which can be completely saturated with water. After a certain number of cycles of freezing, thawing, and saturation of the new volume, a crack propagates completely across a section of the material. A common mode of cracking is parallel and close to the outside face of the brick; this is known as spalling. The spalled piece falls to the ground below, potentially damaging property and injuring people.

The deterioration process continues, possibly at a faster rate due to exposure of more surface area within the brick, and the unit may quickly become so badly damaged that it no longer carries any loads within the wall. Cracking, buckling, and failure under moderate wind loads soon follows.

Ambient temperatures well below freezing are required to initiate freezing within some section of a brick wall, since water in the brick does not start to freeze until about -3ºC. Heat escaping from the building is one of the major factors determining the temperature profile of the brick. The air temperature within the cavity is a dominant factor in determining the brick temperature profile.

The cavity air temperature is affected by the following characteristics of the wall construction and the cavity condition:

- Size and location of ventilation and drainage openings to outside. If the wind is able to establish flow through the cavity, the temperature will be close to that outside. Similarly if ventilation openings exist at top and bottom of the cavity, a convective flow will be established which will increase heat loss and bring the temperature closer to that outside.
- Quality of air barrier and pressure profile in building. If the air barrier is sound, no flow will occur within the cavity caused by inside-outside pressure difference. If it is not sound, positive pressure within the building will cause inside air to flow into the cavity and bring its temperature closer to that inside. Negative pressure will cause outside air to flow into the cavity and bring its temperature closer to that outside. A poorly-located air barrier can allow convective flow within the wall to warm the cavity without any leakage to outside occurring.
- Insulation level of wall system. In an uninsulated wall, the cavity temperature will be around halfway between inside and

outside temperature. In a wall insulated to standard levels of about RSI 1 to 1.6, the cavity temperature will be outside plus about 10 to 20 % of the difference. Air leakage to and from the building has much more effect on the air cavity temperature of an insulated building than an uninsulated building.

- Openness of the cavity. In some cases, the cavity may be nearly or completely blocked by mortar droppings, fallen pieces of brick or insulation, or built-up frost, ice, or water. Typical cavities are between 25 and 50 mm wide, and are laced with brick ties which provide surface area to accumulate frost and debris. A closed cavity reduces airflow, but also impedes drainage. This is difficult to detect without making openings in the wall.

Three types of freezing events can be defined based on the change in environment which caused the freezing front to expand, and the source of heat being removed from the brick wall. They are:

- Type 1: Ambient air temperature drop, building heat sourced
- Type 2: Sun shading or setting; solar heat sourced
- Type 3: Change in building operation, building heat sourced

In any brick veneer wall, there is a range of outside temperatures within which some section of the brick is freezing at $-3^{\circ}C$. This range is large for uninsulated or leaky walls where there is high heat flow from the building interior; it is small for tight, well-insulated walls. When the outside temperature drops within this range, the $-3^{\circ}C$ temperature front moves, and a new section of brick freezes. The size of the section frozen also depends on the amount of heat escaping from the building into the brick; in an uninsulated or leaky building, only a small thickness will freeze for each degree drop in outside temperature; with a well-insulated, tight wall the whole brick may freeze within a $3^{\circ}C$ temperature drop. These events will be referred to as Type 1 freeze events.

On walls that face the sun, the outside face of the brick will thaw to a significant depth on a sunny day where the outside air temperature never goes above $-3^{\circ}C$. Solar absorption can increase the temperature on the outside face of the brick by $20^{\circ}C$ on a sunny day. Often, snow buildup will melt, run down the brick faces, and be absorbed into the thawed portion of the brick. When the sun goes down or moves off the brick face, it quickly cools down to ambient air temperature. Two freezing fronts move in opposite directions, one quickly in from the outer face of the brick, and one more slowly from the still-frozen inside of the brick. Since the absorbed water is trapped in the brick between these two fronts, this type of event, referred to as Type 2, is potentially more severe than Type 1, where there is only one front.

In a Type 3 event, the ambient air temperature may remain constant, but a change in indoor conditions drops the cavity air temperature below freezing. This is most commonly caused by a change in building pressure affecting airflow through a leaky envelope. Shutting down ventilation fans which pressurize the building when in operation can cause a pressure reversal on some walls, which may result in a dramatic drop of up to $15^{\circ}C$ in cavity temperature. This can initiate a fast-moving freeze front on the inside face of the brick, which will meet a slow-moving front advancing inwards. During the operational period, the inside face of the brick may be above freezing and absorb significant condensation from the air leaking out. This water will be trapped between the two fronts, resulting in potentially more severe freeze events than Type 1. Under the appropriate conditions, this event may occur once every winter day and be the dominant form of brick deterioration. However, the inside face is affected, and the damage is therefore not visible unless the wall is opened for inspection. It does not involve an ongoing hazard to people and property below until very serious damage has been done to the bricks.

Brick Properties Affecting Freeze-Thaw Damage Resistance

Bricks are made by forming a mixture of clay and/or ground shale with water into the desired shape, and firing the bricks in a kiln at temperatures around 1000ºC. The materials contain some mixture of refractory materials such as limestone, silicon, aluminum and iron oxides, and fluxes such as oxides of sodium and potassium.

During firing, the water is driven off and some of the brick materials are partially melted and fused together. The total pore area is reduced and the average pore size is increased. The total heat work of firing is therefore a key factor in determining the freeze resistance of the brick; longer firing at higher temperatures produces larger, poorly connected pores with lower total pore volume. The proportion of refractory materials to fluxing materials is also important; a higher proportion of fluxes will require less heat work to increase pore size and reduce total pore volume. A well-fired brick will end up with more than 50 % of its pore volume in pores larger than 1 mm in diameter, above the critical size range for frost damage.

The porosity of bricks can be measured in several ways. The standard test for cold absorption involves first drying and weighing the brick, then immersing it in room temperature water for a period of 24 hours, after which the original weight is subtracted from the new weight resulting in the weight of water absorbed. This is expressed as a percentage of the dry weight of the brick; it is typically in the range of 7 % to 25 %. Since water is only half the density of dry brick, at saturation it takes up from 15% to over 50% of the total volume of the brick.

Another common measurement is the boiling absorption test. The brick, which has just undergone a 24 hour cold immersion, is then immersed in boiling water for five hours. At the end it is weighed again, and the difference in weight is expressed as a percentage of the dry weight of the brick. This value is higher than the cold absorption number, because the higher water vapor pressure is more able to drive out and compress air remaining in the pores, and expansion of the brick provides more space in the voids.

Dividing the cold absorption by the boiling absorption gives the Saturation Coefficient, which is theorized to represent the proportion of total pore volume which is easily filled with water. The remaining space is thought to be available to accommodate freezing expansion of water in the pores by compressing the air; thus a brick having a low saturation coefficient is less likely to be damaged by freezing when wet.

The above tests are incorporated into the Canadian Standard used to grade bricks in terms of durability, CSA A82.1-M87. A brick is considered Grade SW frost-resistant if:

- the compressive strength is above 20.68 MPa and total water absorption via boiling test is less than 17% and the Saturation Coefficient is less than 0.78;   or
- if the Saturation Coefficient is greater than 0.78 but the Cold Absorption does not exceed 8% on average, and the compressive strength is acceptable; or
- the unit survives 50 cycles of omni-directional freezing and thawing without breakage or loss of more than 0.5 % mass.

This standard is identical to ASTM C62-87 Standard Specification for Building Brick (Solid Masonry Units Made From Clay or Shale) in this specification.

There is currently debate about how well the above tests correlate with freezing durability in field situations. The compressive strength is generally not well correlated with in-service failures; the Saturation Coefficient is well-correlated but it is clear that damage can occur at values below 78 %; and the omni-directional freezing test is not representative of field conditions, where the freeze front usually moves through the brick from one side to the other. The freezing cycle test is considered too lengthy by manufacturers and too short by others.

A research program has been underway at the National Research Council of Canada since 1978, to investigate various aspects of clay brick manufacture, testing, and in-service durability. It is expected that one result will be proposals for new tests and standards which better correlate with resistance to damage due to freezing while wet.

An earlier phase of this work was reported in reference [4]. The "One-hour Cold Absorption" test is described, and several correlations between it and the Saturation Coefficient, level of firing, and raw material variation are presented in graphical form. The author recommends the use of a combination of the absorption and the saturation coefficient as a measure of durability.

<u>Local Climate Factors Affecting Brick Durability</u>

The number of times a brick or section of brick in a brick veneer wall is subject to freezing is primarily a function of the local climate. It is not the coldest climates that are worst, but those that have the most variability during the winter. Hourly weather data such as the outside air temperature is available for many years for most locations in Canada. Typical annual numbers of freeze events are easily developed by counting the number of temperature drops from this data.

Sunshine and night sky radiation effects on a brick wall result in Type 2 freeze events. Locations which have high frequencies of cold, clear, sunny winter days will impose the greatest number of Type 2 freezing events on walls exposed to the winter sun. In many locations, data concerning the equivalent temperature of a surface including radiation effects are available in the form of a Sol-Air Temperature. This data can be used to determine the increased number of freeze-thaw cycles which bricks in South-facing walls are exposed to.

The likelihood that bricks are wet when a freezing event occurs is related partly to the frequency and duration of wind-driven rain, snowfall, and thawing cycles. This data can also be developed from existing weather data for most locations.

MODEL OF SPALLING FAILURE IN BRICK VENEER WALLS

The purpose of this model is to assess the expected number of years of survival of a given brick veneer wall system against saturated freezing under a given set of operating conditions, climate factors, and brick characteristics. The model can also be used to assess particular local conditions within a wall, such as bricks located under a faulty drainage detail or bricks located at the bottom of a poorly drained cavity wall.

Model Input

The following information describing the wall system, the building, the climate of the location, and the bricks themselves, is used in the model:

Direction: This is in 45° increments, and determines the walls solar exposure and exposure to driving rain.

Cavity Temperature Index (CTI): This defines the temperature profile of the brick given the inside and outside temperatures. It represents the proportion of the total equivalent insulation value of the wall which is contributed by the brick. The equivalent value includes the effect of airflow through the cavity.

The value of this parameter can be determined by measuring the temperature of the cavity under appropriate winter climate and operating conditions, ensuring that stack effect and ventilation system pressurization is accounted for. The measurement should not be taken under unusually high wind speeds or unusual directions. The CTI is calculated as

$$T_o = (T_s + T_a) / 2 \qquad (1)$$

where

T_o = effective outdoor temperature, $^\circ C$,
T_s = Sol-Air temperature at brick surface, $^\circ C$,
T_a = ambient air temperature, $^\circ C$

and

$$CTI = (T_c - T_o) / (T_i - T_o) \qquad (2)$$

where

T_c = cavity air temperature, $^\circ C$
T_i = indoor air temperature, $^\circ C$

If the cavity temperature cannot be measured, CTI can be estimated based on the wall constructions in Table 1.

If the building ventilation system pressurizes the interior of the wall being analyzed when operating, and is shut down during daily unoccupied periods so that pressure becomes equal or less than outside, two values of CTI will be required; CTI_o for when the building is occupied, and CTI_u for when it is unoccupied.

TABLE 1-- <u>Cavity Temperature Index CTI</u>

Insulation	Air Barrier	Inside-Outside Pressure Difference	CTI
None	Poor	Positive	0.6
RSI 1	Poor	Positive	0.45
None	Good	Not Relevant	0.4
RSI 1	Good	Not Relevant	0.2
None	Poor	Negative	0.2
RSI 1	Poor	Negative	0.1

Insulation Value (RSI): The insulation value is combined with the measured CTI to determine the contribution of airflow to the temperature in the cavity behind the brick veneer. It can be found from on-site inspection of the wall construction or from the building drawings.

Several other parameters relating to the geometry of the wall may be required for analysis of specific problem areas. The following parameters relate to conditions inside the building.

Inside Temperature Ti: The inside temperature is easily measured and is assumed constant in the model equations. Most commercial buildings have adequate mass and internal heat generated to maintain temperature within a few degrees range even during unoccupied periods.

Relative Humidity RH: This is a key parameter governing the amount of moisture deposited as condensate on the inner brick face of walls through which air is exfiltrating. The RH is easily measured using a psychrometer and is assumed constant in the model equations.

Typical Number of Occurrences of Temperature Drops within Freezing Ranges: A typical year of data for hourly Sol-Air Temperature and Ambient Air Temperature must be analyzed to identify and sort drops in temperature into bins representing 3°C ranges of both Sol-Air temperature and the difference between Sol-Air temperature and Ambient air temperature. Sol-Air temperature depends on the direction the wall faces, but in winter the sun generally affects only walls facing within 60° of South, so there are four tables required for each weather location, as follows:

- Southeast Facing Walls
- South Facing Walls
- Southwest Facing Walls
- All other walls

Table 2 is an example of a Temperature Drop Table.

A similar table is required for each location, which contains numbers of occurences of daily minimum temperatures falling within the 3°C temperature ranges above.

Cold/Boil Saturation Coefficient: This can be estimated from the original manufacturer's data for the brick type used; however this is not as reliable as carrying out the ASTM C67 test procedure.

One-Hour Cold Absorption: This can be measured in the course of the C/B test procedure, by removing and weighing the brick after one hour immersion in cold water.

TABLE 2--<u>Temperature drop table</u>

Calgary, Alberta - South East facing wall: Number of occurrences in a typical year								
Range of Sol-Air Temperature Drop	\(Ta-Ts\) at end of drop, $^{\circ}$ C							
	+3	0	-3	-6	-9	-12	-15	-18
-3oC to -6oC	9	26	11	7	4	1	1	0
-6oC to -9oC	8	24	12	8	2	1	0	0
-9oC to -12oC	5	21	9	5	2	0	0	0
-12oC to -15oC	4	17	8	3	1	1	0	0
-15oC to -18oC	5	19	7	4	2	1	0	0
-18oC to -21oC	2	13	5	5	3	1	0	0
-21oC to -24oC	2	12	3	2	2	0	0	0
-24oC to -27oC	2	8	3	1	0	0	0	0
-27oC to -30oC	1	4	1	0	0	0	0	0

<u>Derivation of Model Equations</u>

The equations and methodology of the model are too complex and extensive to present in detail in this paper. The following is a brief outline of how the model works:

- Choose the number of sections that the brick thickness is divided into. Each section is modelled separately.
- Assign each bin of the Temperature Drop Table for the wall location and facing direction to a section. This represents the number of Type 1 freeze events imposed on that section in a typical year.
- Assign Type 2 freeze events to sections. The first line of the Temperature Drop Table is used, without double-counting events which are already assigned to the section.
- Assign Type 3 freeze events to brick sections using the Daily Minimum Table.
- For each brick section, the sum of all bins of each type assigned to that section represents the annual number of times that section was frozen.
- For each of these times, the degree of saturation of the brick is estimated using statistical distributions of weather and known building conditions to determine wetting and drying of the brick veneer wall. This involves determining how long the brick section was thawed prior to it being refrozen, what volume of water, if any, was available for absorption, and how much water left the brick via drying processes. A calculation is included for each of the six potential wetting and drying processes.

- For each event, the freezing expansion is calculated as a function of the estimated degree of saturation and the durability index of the brick.
- Another equation is used to calculate what proportion of the freezing expansion is irreversible.
- The cumulative annual irreversible damage to the brick is calculated as a multiple of the irreversible damage from each freeze event.
- One of the brick sections will have the highest annual damage; this is the section having the shortest expected lifetime. If it is the outside or inside face section, failure by spalling can be expected. The number of years to failure is the number of times the annual cumulative damage would have to be multiplied by itself to reach the critical damage level at which cracking or spalling occurs.

The logic used in developing the relationships used in the model is described as follows.

The Durability Index DI was developed by inspection of Figure 28, page 47 of reference [1]. DI reflects the material characteristics of bricks that are most correlated with long-term survival of freezing events: low total porosity, slow rate of water uptake, and low proportion of total porosity which can be easily filled with water. Figure 28 of reference [1] indicates how the brick composition and the effect of firing heat work relate to the values of the two parameters Cold Absorption (CA) and Saturation Coefficient (SC). A vector representing lines of equal durability was estimated to have a defining equation of 13.5 SC + CA, where typical values for SC are 0.6 < SC < 0.95 and for CA are 5 < CA < 13.

The Saturation Rate Coefficient (SRC) was developed from the following estimate of wetting times due to diffusion and capillary action, related to the DI. A relationship of the form:

$$DS(x,t) = 1 - \exp\ (-SRC * (t/x)) \tag{3}$$

was used, where x is the distance in cm into the brick from the source of water on the surface and t is the time in hours that the surface is saturated. The estimates of wetting time were, for a brick of DI = 18, DS(4.5,1) = 0.5, and for a brick of DI = 22, DS(4.5,1) = 0.7. A linear relationship was assumed between SRC and DI, resulting in

$$SRC = 9.932 - .3785 * DI \tag{4}$$

These estimates are based on one hour cold absorption data compared to total 24 hour cold absorption data, and they need further refinement.

The optimum number of brick sections occurs when the distance the freeze front moves for a Sol-Air temperature drop of one bin range (set as $3^{\circ}C$) is one section. An approximate relationship for this is

$$NSt \sim CTI\ (Ti - To)\ /\ 3 \tag{5}$$

where NSt is the number of sections through the brick thickness which are modelled separately. Since the size of the movement of the freeze front is not linear with drop in outside temperature, this approximation does not give perfect results; some bins may not be assigned a section because they affect an area significantly smaller than one section.

The cavity air temperature Tc is affected by the ambient air temperature and by the Sol-Air temperature. Air which is drawn into the cavity from outside begins at ambient temperature but must pass through the outer brick section affected by the Sol-Air temperature. The calculation assumes that the actual temperature of outside air reaching the cavity is the average of Ta and Ts.

The assumptions and reasoning for assigning potential Type 2 freeze events is as follows. The brick is thawed from the outside by solar radiation, which is suddenly removed as the sun sets or the wall is shaded. The rate of temperature drop is high because the brick section is nearly surrounded by materials below freezing temperature. All thawed sections of brick refreeze very soon after the Sol-Air temperature drops below freezing (some of the inner sections refreeze before this point). Thus all the potential events for all sections are represented in the first line of the Temperature Drop Table. Those sections which would have thawed due to the ambient temperature are eliminated from Type 2 to avoid doublecounting. The maximum Sol-Air temperature is selected as part of the moisture calculation for each event, and its value determines whether the section was thawed previous to the potential freeze event.

Similar logic applies to the assignation of Type 3 freeze events. Only those sections of brick which would not have been subject to a Type 1 freeze event can be subject to a Type 3 freeze event. Only those sections having their centrelines between the "operating" freeze front location and the "unoccupied" freeze front location are affected by a Type 3 freeze event. Assumptions are that the Daily Minimum Ambient air temperature is the same as the Minimum Sol-Air temperature, and that the minimum occurs during the unoccupied period at night.

Determining the moisture content of brick when it freezes requires a knowledge of the history of moisture movement into and out of the brick. Brick moisture content cannot be tracked continuously; we assume that what happened in the previous 24 hours will be the most significant part of the history. If no water entered and no water left the brick section in 24 hours, we assume that the brick section is dry. This may be too short a period for bricks which have coatings which inhibit drying. Further investigation is required.

The probability distributions of Peak Sol-Air and Peak Ambient Air temperature are estimated here, but could be developed from weather data for the location.

The rate at which the temperatures rise to and fall from the peak determine the total heat work available to thaw bricks, which is quantified as Degree-Hours above zero (DHO). It also determines the time above freezing during which liquid water can enter and exit the brick. They are also selected from a probability distribution.

The amount of moisture in a section due to driving rain on the outer surface is calculated as a simple capillary-diffusive movement of moisture into the brick from the outside surface while it is raining. A lower DI for the brick forces water to penetrate at a slower rate. This assumes that the volume of rain falling on the surface is easily in excess of the maximum rate of absorption; the excess simply drains down the outside of the wall.

TABLE 3--_Estimate of Condensation Rate_

INSIDE AIR RH	CAVITY TEMPERATURE	CWSM
23 %	0°C	0
33 %	0°C	1
33 %	5°C	0
42 %	5°C	1
48 %	10°C	0
56 %	10°C	1

Wetting from meltwater drainage is calculated by determining whether the outside surface of the brick is wetted by meltwater, and applying the same capillary-diffusive movement as above if it is. The length of time the brick surface is saturated can be no more than the time required to melt all the remaining snow.

In calculating saturation due to condensation, the Leakage Modifying Factor LMF approximates how much leakage of indoor air into the cavity contributes to the Cavity Temperature Index. If the CTI is much higher than it would be if only conduction was supplying heat, then indoor air leakage is assumed to provide the rest and LMF is set high. The Condensate Water Supply Multiplier (CWSM) equation uses the following assumptions. The key rate of air leakage is set at 1 L/sec/M^2 wall area, and Table 3 shows the condensation conditions within the wall.

At CWSM = 0, water is just beginning to condense on the brick face; at CWSM = 1, the rate of condensation is high enough to saturate the surface of the brick. The brick can only be wetted while the cavity temperature is above 0°C. The movement of water into the brick from the inside face is governed by the capillary-diffusive equation during the period the brick section is not frozen.

Calculation of hygroscopic water intake is straightforward - the capillary-diffusive equation is applied with the distance from the point source and the time the section was unfrozen.

Calculation of drying due to evaporation on the outside and the inside faces of the brick, requires an estimate of diffusive moisture transfer rates within the brick. The following assumptions and estimate are used to simplify the problem of determining the water vapor pressure gradient. No drying takes place at temperatures below freezing nor when the surface is saturated (by rain, meltwater, or condensation). The drying rate is 0.5 % per degree-hour of ambient air temperature above 0°C and 1 % per degree-hour of Sol-Air effect when the temperature is above 0°C. Sol-Air effect is estimated as twice that of external air because the external air is heated when near the brick face, and because convection of the heated air up the brick surface quickly removes moist air from the face.

An inverse exponential function is used to describe the diffusion rate through the brick to the surface. The diffusion is faster for bricks having a lower durability index; the larger pore passages allow better air circulation and greater surface area in contact with air. The equation is calibrated so that 240 DHO will completely dry a saturated brick of 92 mm thickness with DI = 15, and 480 DHO will do the same with a brick of DI = 20. This calibration is estimated; proper calibration is required based on experimental evidence.

The equation for freezing expansion is based on the data from Figure 18 of Reference [4]. The actual values of CA and SC for the bricks in this graph are not explicitly reported in Reference [4]. A linear relationship between Degree of Saturation (DS), DI, and Freezing Expansion (FE) is assumed; in reality DS and DI are not independent and the slope of the line should be affected by the value of DI.

The equation for irreversible damage is based on the data presented in Figure 17 of Reference [4]. Some calibration of the coefficients is required to get the best fit to the data; the coefficients "a" and "b" are currently estimated as 0.05 and 1.25 respectively.

The cumulative damage to each section caused by all freeze events in a typical year is the product of multiplying the irreversible damage from all freeze events in all temperature bins assigned to that section. The years to failure is the number of times the annual cumulative damage would have to be multiplied by itself to reach the critical damage point at which cracking or spalling would occur.

SUMMARY

Premature deterioration of bricks in brick veneer walls is a serious problem in cold climates when the moisture exposure of the bricks is high. The model described in this paper provides a means of combining local weather conditions, building design and operating conditions, and brick characteristics in a logical way to assess the expected lifetime of brick veneer walls exposed to freeze-thaw cycles.

REFERENCES

[1]   Litvan, G.G., "Freeze-Thaw Durability of Porous Building Materials," _Durability of Building Materials and Components, ASTM STP 691_, P.J. Sereda and G.G. Litvan, Eds., American Society for Testing and Materials, Philadelphia, 1980, pp 455-463.

[2]   Brand, R.G., "High Humidity Buildings in Cold Climates - A Case History," _Durability of Building Materials and Components, ASTM STP 691_, P.J. Sereda and G.G. Litvan, Eds., American Society for Testing and Materials, Philadelphia, 1980, pp 231-238.

[3]   Ritchie, T., "Factors Affecting Frost Damage to Clay Bricks," Building Research Note 62, Division of Building Research, National Research Council Canada, Ottawa, 1968.

[4]   Kung, J.H., "Frost Durability of Clay Bricks - Evaluation Criteria and Quality Control," _Proceedings of the CBAC/DBR Manufacturer's Symposium, March 7-8, 1984_, Division of Building Research, National Research Council Canada, Ottawa, 1984.

Norbert V. Krogstad[1] and Richard A. Weber[2]

USING MODIFIED ASTM E 1105 TO EVALUATE THE RESISTANCE OF MASONRY BARRIER, MASS AND SKIN WALLS TO RAIN

REFERENCE: Krogstad, Norbert V., Weber, Richard A. "Using Modified ASTM E 1105 to Evaluate the Resistance of Masonry Barrier, Mass and Skin Walls to Rain" **Masonry: Design and Construction, Problems and Repair, ASTM STP 1180,** John M. Melander and Lynn R. Lauersdorf, Eds., American Society for Testing and Materials, Philadelphia, 1993.

ABSTRACT: Test procedures are proposed to evaluate masonry barrier, mass and skin wall systems using ASTM E 1105, *Test Method for Field Determination of Water Penetration of Installed Exterior Windows, Curtain Walls, and Doors by Uniform or Cyclic Static Air Pressure Difference*. Variables involved in these tests, such as water flow rates, size and positioning of spray grids and duration of testing are also reviewed. Studies comparing the penetration rates at different flow rates and at different air pressures are reviewed. Systematic procedures are presented for masking off portions of walls in order to determine significant sources of leakage. Further procedures are outlined to isolate sources of leakage after areas of leakage are found. Several case studies are reviewed on different types of wall systems. Each of the case studies includes a discussion of the problems found and how ASTM E 1105 tests were used to determine the leakage sources.

KEY WORDS: masonry, barrier walls, drainage walls, skin walls, mass walls, water testing, spray rack, water penetration, flashing, interior leakage

---

Many new and existing masonry walls contribute to interior leakage as a result of problems in design and construction. Sometimes, however leakage problems are wrongly attributed to the masonry when, in fact, other components within the wall systems are to blame. Windows, vents, sealant joints, coping and roof base flashing membranes can all contribute to interior leakage. For this reason, it is necessary to test the masonry walls in addition to curtain walls, roof flashings and other components to determine sources of interior leakage and to prevent unnecessary repairs. This paper presents a testing program that can be used to systematically isolate and rule out possible sources by

---

[1]Consultant, Wiss, Janney, Elstner Associates, Inc., 330 Pfingsten Road, Northbrook, IL 60062

[2]Architect/Engineer II, Wiss, Janney, Elstner Associates, Inc.

simulating wind-driven rain conditions as closely as possible. The test program can also be used as part of a quality control program for new construction. The test program is particularly suited for barrier, mass and skin walls as defined in the following section.

Masonry Wall Types

Water will penetrate the outer wythes of most masonry walls. The majority of the water penetrates the masonry through cracks and bond separations between the mortar and the masonry units. Water penetrating through a masonry wall system can damage property or interior finishes. For these reasons, masonry walls are designed with considerations for water penetration. These considerations result in four main masonry wall types: 1) Barrier Wall, 2) Drainage Wall, 3) Mass Wall and 4) Skin Wall. Any one of these four systems, if properly designed, constructed and maintained, will prevent interior leakage.

A **Barrier Wall** is a masonry wall that contains a nearly impervious barrier in a plane within the system parallel to and behind the exterior surface. A typical wall section for a barrier wall is shown in Fig. 1. The barrier often consists of a solid grouted collar joint immediately behind the veneer. The back-up wall is commonly masonry or other material capable of withstanding the hydrostatic pressures generated during the filling of the collar joint. The barrier is intended to prevent water from penetrating through the wall system. Water stopped at the plane of the barrier travels down the wall system on the exterior face of the barrier where it drains to the exterior at through-wall flashing and weepholes.

A **Drainage Wall** is a masonry wall that contains a continuous cavity in the plane of the wall at the back face of the exterior wythe known as a drainage cavity. A typical drainage wall is shown in Fig. 2. The back-up wall can be masonry or another wall system. The cavity is designed to allow water to run down the back of the exterior wythe of masonry. The flashing system at the base of the wall segment will direct water back to the exterior. Cavity and veneer walls are common types of drainage walls.

A **Mass Wall** contains several wythes of masonry, but no impervious barrier or open cavity. A typical detail of a mass wall is shown in Fig. 3. In a Mass Wall, water penetrating the exterior wythe is absorbed by the masonry until it can evaporate or travel down the wall to any flashing that the system may contain.

A **Skin Wall** is a masonry wall that contains a nearly impervious barrier on the exterior surface of the wall or a nearly impervious exterior wythe itself. A typical skin wall is shown in Fig. 4. Flashing may be installed at the base of skin walls but only as a secondary line of defense if the impervious barrier fails. Skin walls are commonly single wythe walls, either field laid or panelized.

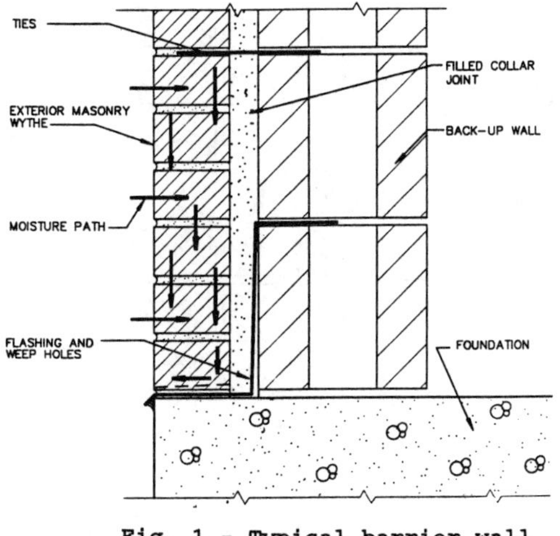

Fig. 1 - Typical barrier wall

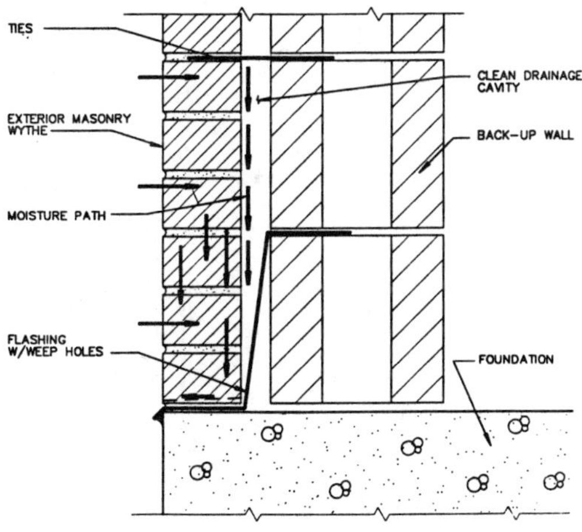

Fig. 2 - Typical drainage wall

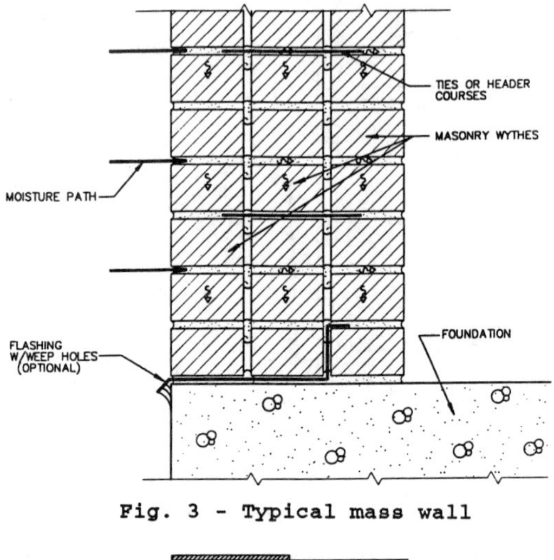

Fig. 3 - Typical mass wall

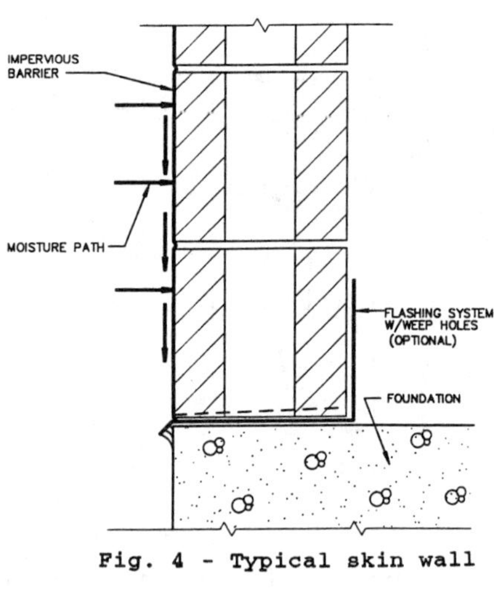

Fig. 4 - Typical skin wall

TESTS CURRENTLY AVAILABLE

Because the principle behind the resistance to water penetration of each wall type is different, test methods that are well suited for one wall system may not be appropriate for another. Cavity walls can be evaluated based on testing of the drainage system. Several tests are currently available to do this. Such tests, however, would not be appropriate for barrier, mass and skin walls because they do not contain a drainage space.

Current test methods for evaluating the performance of masonry wall systems that we are aware of are as follows:

1.   ASTM E 514 entitled *Standard Test Method for Water Penetration and Leakage Through Masonry*. This test, as accepted by American Society for Testing and Materials (ASTM), is only to be used as a laboratory test to compare penetration rates through masonry wall systems. Currently, there is no field version of this test accepted by ASTM to evaluate in-place walls. A field modified version of this test is described in a paper by Monk[1] appearing in ASTM STP 778 entitled **"Adaptations and Additions to ASTM Test Method E 514 (Water Permeance of Masonry) for Field Conditions.** In field versions of this test, the rate of water penetration into the exterior surface of the wall is measured when a 1.1 m^2 (12 sq ft) area is subjected to wind-driven rain conditions. This test exposes only a small percentage of a masonry wall to the test conditions. For this reason, it would not be effective in testing large wall areas, barrier, mass and skin walls. The quantities of water entering walls may be too small to recreate leakage. Increasing the size of the chamber is possible, however, such a chamber would be extreme, cumbersome and, in our opinion, would not be practical.

2.   Wall Drainage Test described in a paper appearing in ASTM STP 1063 entitled *Masonry Wall Drainage Test - A Proposed Method for Field Evaluation of Masonry Cavity Walls for Resistance to Water Leakage* by Krogstad[2]. This test can be used to evaluate the performance of drainage systems. Water is introduced into masonry cavities at the back face of the veneer. The rate of water entry is determined from ASTM E 514 tests or other methods. Large lengths of walls can be tested in a short period of time. This test simulates water penetration that occurs during rainstorms. This test cannot be used, however, to test the effectiveness of a·barrier wall, mass or skin wall because they do not contain a free drainage cavity.

3.   AAMA 501.2-83 entitled *Check of Metal Curtain Walls for Water Leakage*[3]. This test is intended to check joints and gaskets in metal curtain walls. The test uses a special nozzle with a pressure gage attached to a 1.91cm (3/4 in.) hose. The test is performed by spraying water at a specified pressure to the joint in question. Each joint in the system is tested for 1 minute per foot of length. The test is used to locate sources of water leakage in a wall system such as cracks or sealant joints. However it does not test a large area of the wall at one time and does not simulate the effect of a long rainstorm on walls.

4.   Fire Hose Testing. As in AAMA 501.2, the water is applied via a hose to the wall surfaces. A 6.35cm (2 1/2 in.) fire hose is commonly used for this purpose. Advantages of this test are that it is relatively easy to perform without advanced equipment and it can be used to test a relatively large area at a single test. However, it is not precise in its application of the water since the pressure at any hydrant can vary and it does not apply a uniform spray to the entire test area. This is a means of quickly identifying areas of leakage but not sources of leakage. If the pressure is too great, however, it may create leaks that may not otherwise occur.

5.   Rilem Tube: This test is performed by sealing a 2.54 cm (1 in.) diameter clear plastic cylinder to the wall. This cylinder is open on the wall surface and contains a .84 cm (1/3 in.) diameter head pipe oriented vertically attached to the top side of the cylinder. This pipe and cylinder are filled to a specific level to create an exposure to rain and using water head pressure, an exposure to pressure differential. During the test, the water level is monitored to determine the amount of loss to calculate the penetrate rate. This test only tests a 1 in. diameter circle of wall at a time. It cannot be used to evaluate the performance of a wall system. The leakage rates determined by this test vary considerably depending on the placement of the tube.

6.   ASTM E 1105 entitled *Standard Test Method for Field Determination of Water Penetration of Installed Exterior Windows, Curtain Walls and Doors by Uniform or Cyclic Static Air Pressure Difference*. The test consists of sealing a chamber to the interior and exterior face of the assembly to be tested, supplying air to a chamber mounted on the exterior or exhausting air from a chamber mounted on the interior, at a rate required to maintain the desired air pressure difference across the assembly. Water is applied to the exterior surface of the wall by using a spray rack. This rack consists of nozzles spaced on a uniform grid based on calibration rates to deliver a uniform spray against the exterior surface at a rate of 3.4 L/sq.m/minute (5 U.S. gallons/sq ft/hour). Continuous observations are made of the interior surfaces during the test to record any leakage. A typical test setup is shown in Fig. 5. The advantage of this test method is that large areas of walls can be tested at one time in a relatively controlled fashion. The test would be appropriate for barrier, mass and skin walls because it does not require that a cavity be present. If a pressure difference is applied to the walls, however, the test can be very cumbersome.

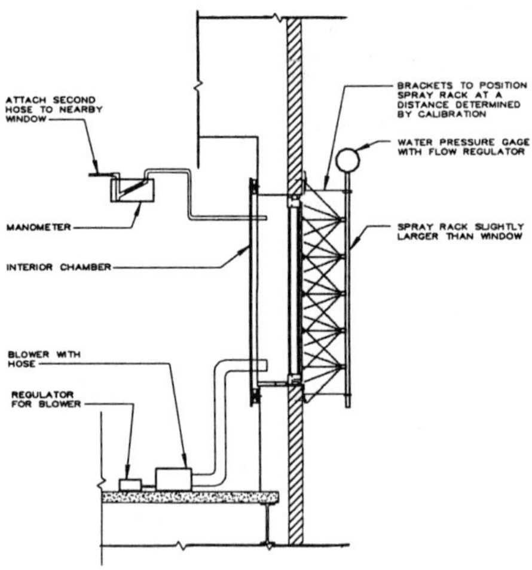

Fig. 5 - Typical ASTM E 1105 setup

PROPOSED TEST PROCEDURE

We recommend testing masonry barrier, mass and skin walls by using ASTM E 1105 tests without an air pressure difference alone or in combination with AAMA 501.2 tests to identify particular leakage sources. The proposed procedure that has been used with success is as follows:

1.   Prior to performing any test, a thorough understanding of the existing conditions should be obtained. This would consist of a survey of areas of reported leakage to examine trends or patterns and to perform a condition survey of the exterior at the leakage areas to identify any potential sources or particular problem areas.

2.   Select the area(s) to be tested. If leakage is common at several areas of the building, the test areas should be at typical conditions. If large cracks or failed sections of sealant exist at leakage locations, AAMA 501.2 testing would be a logical first test. However, if the leakage source is not apparent, ASTM E 1105 testing is practical.

3.   If the ASTM E 1105 testing is needed to determine the source of leakage, all areas above and at the sides of test area are to be masked as necessary to isolate the test area as shown in Fig. 6. The masking can be performed with polyethylene sheets taped in place or held in place with wood furring. The masonry wall can be divided into several components testing from the bottom to the top in order to isolate contributions to leakage.

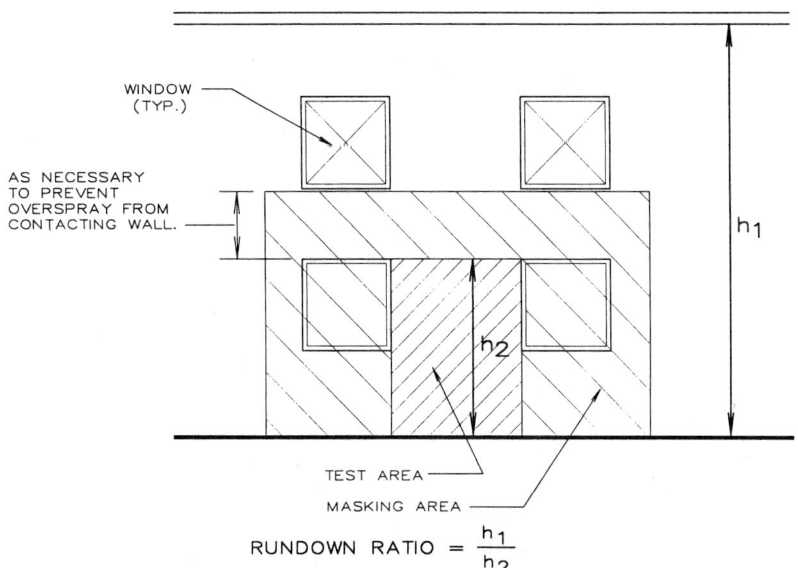

Fig. 6 - Typical masking system

4.   Perform ASTM E 1105 tests by installing a spray rack over the area to be tested. The spray rack applies water in a uniformly distributed spray to the entire test area. The water supply to the spray rack is monitored with a pressure gage and globe valve. The rate of water application and other test parameters will be discussed in the following section. During testing, the interior of the test area is monitored for leakage.

5.   If leakage is observed, the test area can be masked off further to isolate specific sources or areas can be tested with the AAMA 501.2 nozzle test. Spot testing in areas of suspected penetration sources with the nozzle test described in AAMA 501.2 can determine the influence of cracks and/or sealant failures on the leakage.

6.   After testing, observe the drying patterns of the masonry to determine which areas are the last to dry out. This may indicate voids in the wall system where water can accumulate. Masonry openings could be made to verify the presence of these voids.

TEST PARAMETERS

ASTM 1105 test for masonry walls shall be based on the following parameters:

Water Application Rate
<u>   </u>

The water application rate to be used in the test shall be based either upon the actual weather data for the particular region or the current default values listed in ASTM E 1105. Both of these approaches are reasonable.

When using actual weather data, we recommend beginning with ten-year, 30-minute rainfall intensities as obtained from **Technical Paper No. 40 Rainfall Frequency Atlas of the United States** published by the U. S. Department of Commerce[4]. The data from this atlas represents rainfall intensities on the ground surface. If rain strikes the wall at a 45 degree angle, the rainfall intensity on the wall will be identical to that on the ground. The rainfall intensity on the wall, however, is further increased by rundown of rain water from the wall surfaces above. Depending on the test location, we recommend increasing the water spray intensity by a factor ranging from 1 to 4 or more. The actual factor can be calculated by dividing the height of wall above the test area by the height of the test area. This should be reduced by 5 percent to account for water that either is absorbed by or penetrates through the masonry wall above. This percentage has been determined from Field Modified ASTM E 514 tests that have been performed on walls in the past.

If actual weather data is not readily available, the default value of 20.3cm (8 in.) per hour can also be used. Tests that we have performed using Field Modified ASTM E 514 on several different masonry walls has indicated very little change in water penetration values when the application rate is changed. By varying rates from 6.35cm (2 1/2 in.) per hour to 19.05cm (7 1/2 in.) per hour, leakage rates varied by approximately 20 percent. Approximately 2/3 of the United States has 10-year, 30-minute rainfall intensities exceeding 6.35cm (2 1/2 in.) per hour. Therefore, for the majority of the country, the current default rate listed in ASTM E 1105 is reasonable even if the test is positioned near the top of the wall where rundown is not a factor.

Test Duration

The test duration should be based on how fast leakage is experienced in the building during an average rainstorm. If leaks occur early in the rainstorm, one-hour test duration should be used. If reported leaks occur only after sustained rainstorms lasting several hours or several days, the test duration should be increased. A test duration of 3 hours is not unreasonable in these instances. If leakage occurs during testing, the test can be terminated to prevent damage.

Air Pressure

Wind pressure will have a sizable effect on the rate of rain penetration into masonry walls. The ratio between water penetration rates with an air pressure differential of 500 Pa (10 psf) and with no air pressure differential has been found to vary from 2.9 to 5.9.[5]. However, the duration of high winds is only a small percentage of the duration of the rain. For this reason, it is reasonable to neglect air pressure when performing these tests.

CASE STUDIES

The following case studies illustrate how we have used ASTM E 1105 to determine the source of observed leakage in masonry walls.

Case Study No. 1 - Dormitory Building

The building is three stories tall with a structural steel frame. The wall system consists of a multi-wythe barrier wall. It is composed of an exterior wythe of brick masonry with an interior wythe of concrete masonry. The collar joint is 5.1cm (2 in.) wide. This joint is filled with grout and reinforcing in both directions.

The leakage observed by residents consisted of ponding of water on the floor slabs during and after heavy rains of long duration. Mold and mildew were observed in several areas where the interior finish was removed. Water was observed on the interior surface of the concrete block in the stairwells. These areas do not contain any interior finish.

A detailed field investigation was performed on the building consisting of a condition survey of the exterior of the building, water testing and construction openings. The investigation included both the masonry and the windows. The overall condition of the mortar was observed to be fair to poor. The joints contained many voids and bond separations. Efflorescence was also observed at several areas. The locations for testing were determined to achieve representative data for the entire building. Since the exterior condition of the walls and the leakage patterns were typical, the test areas included areas where bond separations and other defects were present.

The masonry spray grid test consisted of mounting three 3m (10 ft) x 3m (10 ft) spray grids, one above the other, to test a 3m (10 ft) wide x 9.1m (30 ft) high area. No windows were in the test area. Prior to testing, the windows on each side of the test area were masked with butyl tape and polyethylene to eliminate them from the test. The spray racks were calibrated to apply water at a rate of 3.4 liters/sq meter/per minute (5 gallons/sq ft/hour). The tests were performed for three hours or until interior leakage was observed. This was to account for absorption of the concrete block, insulation and the interior finishes which would delay the observation of leakage on the floor.

Interior leakage was defined as the ponding of water on the floor.

All tests exhibited interior leakage prior to three hours. After leakage was observed, the test was terminated and the interior finishes were removed to determine moisture patterns on the block. In all cases, water was observed coming through the interior surface of the block. After the test had been shut off for quite some time, moisture patterns were observed in the exterior brick during dryout. Openings were made through the areas of brick masonry that were the last to dry out and at the areas corresponding to the observed leakage. These openings revealed the cold joint in the grout where, in some cases, mortar had accumulated due to masonry placement. In other cases, large voids were observed that allowed water to accumulate in the collar joint and easily bridge to the concrete block. Leakage was attributed to these voids and to the mortar bridges through the collar joint.

The windows were also tested in accordance with ASTM E 1105. All windows tested also exhibited leakage. However, this testing of windows is not within the scope of this paper.

The use of ASTM E 1105 without air pressure on the masonry incorporated with systematic masking permitted determination of the leakage sources. This case study also shows that the test can be used to test as large an area as the available water pressure will allow. In this case study, water pressure was achieved by tapping into a fire hydrant and two building spigots. Booster pumps were not required.

Case Study No. 2 - High-Rise Condominium Building

This building consists of a 26-story, reinforced concrete frame including deep concrete spandrel beams. The exterior walls consist of an exterior wythe of brick masonry with an interior wythe of concrete masonry. The collar joint between wythes was 1cm (0.4 in.) wide and partially filled with mortar. The interior finish was wood furring and plaster. The masonry walls were mainly in horizontal bands alternating with horizontal bands of metal frame windows. Flashing was installed at shelf angles at each floor line. The wall system is a hybrid between a barrier wall and a drainage wall. The leakage observed in this building consisted of dampness in the plaster wall and ceiling finishes, and water accumulation on the floors and window sills after certain storms of short duration. Repairs had been performed to the building with limited success. The repairs consisted of tuckpointing selected areas of masonry and resealing the perimeter sealant joints. These repairs were based on a condition survey and repair recommendations by others.

The field investigation consisted of a brief condition survey of the exterior and a leakage survey of the residents, water testing of the exterior and construction openings. The investigation included both the masonry and the windows. The overall condition of the masonry was good with some vertical cracking observed at corners. The condition of the sealant at the windows was observed to be poor. The locations for testing were based mainly in the leakage surveys. A scaffolding drop was set up at the area of the building where leakage was the most common. This occurred on the east elevation.

The spray grid consisted of two 1.8m (6 ft) high x 2.4m (8 ft) long racks installed horizontally to test a 1.8m (6 ft) high x 4.9m (16 ft) long area. The first test was always performed at the lowest possible leakage location with masking installed immediately above and on the sides of the test area. For example, when investigating reported moisture at a ceiling, the first test would be the masonry at that level with masking over the window above. If no leakage was observed during testing, the test would be performed at the windows above and mask the masonry above the windows. This would proceed until the leakage was

located. In this case study, the limestone sills were included in the
masonry testing. The test was run at 3.4 liters/sq meter/minute
(5.0 gallons/sq ft/hour) which is the rate specified in ASTM E 1105.
The random factor for the test location was 3 or more. The test was run
without air pressure at the masonry and the windows.
0.5
        During window testing, several areas of leakage were observed.
This leakage was in the form of water accumulating on the sill and
floor. This was attributed to the window cap beads and perimeter
sealant. During masonry testing, leakage was also observed. This was
in the form of water accumulating on the floor and dripping from the
ceiling.   After leakage was observed during masonry testing, the test
was terminated, and suspected areas of water penetration were tested
with the nozzle specified in AAMA 501.2. This was performed at cracks
in the masonry from the bottom to the top and on the limestone sills.
When the joints in the sills were sprayed, the leakage would recur. A
construction opening at this location exhibited discontinuities in the
sill flashing near these locations allowing water to bypass the flashing
and penetrate the interior wythe of masonry.

        In this case study, the reported leakage would be solved with a
repair addressing the windows and the window sills. No repairs were
required for the masonry walls to solve the immediate leakage problems.
The water source for the grid was the building fire standpipe in the
roof penthouse. This case study shows that the proposed test can be
combined with other tests to form a detailed masonry investigation.

Case Study No. 3 - Warehouse/Office Building

        This building consists of a two-story tall steel frame structure.
The exterior walls were single wythe 30.5cm (12 in.) thick concrete
masonry in horizontal bands alternating with aluminum ribbon windows.
Flashing was installed above the windows to direct any water in the
cores of the masonry to the exterior. A water repellent was added to
the concrete block during fabrication. Admixtures were also added to
the mortar to increase bond to the block to reduce water penetration.
The wall system was a combination of a skin wall and a cavity wall. In
this case, the cores of the concrete masonry units worked as the wall
cavity.

        The water leakage investigation consisted of water testing and
performing general construction openings through the existing
construction. The locations of the tests were determined based on
leakage reports of the building occupants. One test was performed at
each elevation. The majority of the leakage reportedly occurred at the
window heads and accumulated on the sills during moderate storms.

        The spray grid used was 3m (10 ft) x 3m (10 ft). The tests were
performed at the lowest possible leakage location with masking above as
described in Cast Study No. 2. The rate used in this test was
3.4 liters/sq meter/minute (5.0 U. S. gallons/sq ft/hour). The rundown
factor was from 1 to 3. The test was run at the window levels for 1 1/2
hours without any interior leakage. The grid was raised to test the
masonry above the window. Leakage did occur during this testing within
one hour. Bond separations were apparent within the test area. These
were tested with the nozzle specified in AAMA 501.2. The leakage
recurred within two minutes.

        This testing program indicated the mechanism of leakage for the
building. The spray grid testing indicated that the leakage was due to
the masonry wall system. The nozzle testing confirmed that the water
entered the wall system via bond separations between the mortar and the
block. Openings made in the masonry walls indicated several problems
with the masonry flashings.

CONCLUSIONS

Currently, there is little published literature concerning procedures for testing barrier, mass and skin masonry wall systems. A test is required in order to avoid making unnecessary repairs in masonry walls. ASTM E 1105 test has been used with success in determining leakage sources in these types of walls. This test can achieve a significant amount of data in a relatively short period of time without a great deal of cost.

REFERENCES

[1]   Monk, C. B., Jr., Adaptations and Additions to ASTM Test Method E 514 (Water Permeance of Masonry) for Field Conditions," *Masonry: Materials, Properties, and Performance ASTM STP 778*, J. G. Brochelt, Ed., American Society for Testing and Materials, Philadelphia, 1982, pp. 237-244.

[2]   Krogstad, N. V., "Masonry Wall Drainage Test - A Proposed Method for Field Evaluation of Masonry Cavity Walls for Resistance to Water Leakage," *Masonry: Components to Assemblages, ASTM STP 1063*, John H. Matthys, Editor, American Society for Testing and Materials, Philadelphia, 1990.

[3]   Architectural Aluminum Manufacturers Association, AAMA 501.2 Methods of Test for Metal Curtain Walls.

[4]   Hershfield, D. M., National Technical Information Service, Technical Paper No. 40, *Rainfall Frequency Atlas of the United States for Durations from 30 Minutes to 24 Hours and Return Periods from 1 to 100 Years.*

[5]   Krogstad, N. V.; Monk, Jr., C. B.; White, W. E., "Field Evaluation of Masonry Water Permeance.," *Fifth North American Masonry Conference*, Volume IV, June 3-6, 1990.

David Chin[1] and Robert E. Gates[2]

EVALUATION OF ASTM E 514-90 WATER PENETRATION AND LEAKAGE TEST TO ASSESS
PERFORMANCE OF INTEGRAL WATER REPELLENT ADMIXTURES

REFERENCE: Chin, D., Gates, R.E., "Evaluation of ASTM E 514-90 Water
Penetration and Leakage Test to Assess Performance of Integral Water
Repellent Admixtures," Masonry: Design and Construction, Problems and
Repair, ASTM STP 1180, John M. Melander and Lynn R. Lauersdorf, Eds.,
American Society for Testing and Materials, Philadelphia, 1993.

ABSTRACT: Integral water repellent admixtures have played an important
role in improving performance of architectural concrete block by
minimizing penetration of water into the concrete block wall. The ASTM
E 514-90 Test Method for Water Penetration and Leakage through Masonry
is commonly used to measure the performance of unit masonry subjected to
a 4 hour simulated wind-driven rain. A laboratory test program was
conducted to evaluate the effectiveness of the ASTM E 514-90 test method
in determining performance of walls constructed with materials which
contain integral water repellent admixtures, to determine the effect of
extending the test period from 4 hours up to 72 hours and to compare the
performance of four different integral water repellent admixtures.

Suggestions are made regarding potential modifications that can be made
to the E 514-90 test in order to effectively evaluate the performance of
walls constructed with concrete masonry units and mortar which contain
an integral water repellent admixture.

KEYWORDS: water penetration, leakage, leaks, permeance, permeability,
integral water repellent admixture, concrete masonry.

---

[1] Technical Supervisor, W. R. Grace & Co. - Conn., Construction Products
Division, 62 Whittemore Ave., Cambridge, MA 02140.

[2] Operations Manager, Forrer Unit of W. R. Grace & Co., - Conn., 7221 W.
Parkland Court, Milwaukee, WI 53223.

The use of architectural concrete block has become increasingly popular.
An important consideration in the use of architectural block is to
maintain a clean, attractive appearance and provide protection against
water and weather. Integral water repellent (IWR) admixtures have
played an important role in improving the performance of architectural
block by minimizing penetration of water into the concrete block wall,
which reduces building maintenance costs and controls efflorescence.

Integral water repellent admixtures are added to the concrete and mortar
as they are being mixed and are homogeneously distributed throughout the
concrete or mortar. The IWRs prevent the intrusion of water, such as
that produced by rain, into the concrete block by coating the surfaces
of the concrete and the internal pores of the concrete matrix with a
layer of molecules which produces a hydrophobic surface that repels
water [1,2]. A number of test methods have been used to measure the
effectiveness of integral water repellents and include tests for
absorption, initial rate of absorption, initial surface absorption,
permeability, and water permeance. However, actual performance in an
exterior wall depends on many other factors, such as workmanship, the
ability of the masonry unit and mortar to minimize water permeance, good
bond between mortar and masonry unit, microcracks, shrinkage, proper
design and weathering.

Of the test methods listed above, the ASTM E 514-90 Test Method for
Water Penetration and Leakage Through Masonry, which measures the
resistance to water penetration and leakage through unit masonry
subjected to simulated wind-driven rain (simulating 140 mm (5.5 in.) of
rain/hour accompanied by a 100.6 km/h (62.5 mph) wind), is the principal
test used to assess performance of integral water repellent admixtures
[3, 4, 5, 6].

Some major modifications to the original ASTM E 514-74 test method were
made in 1986/90 and a comparison of the 1974 and 1986/90 test methods is
given in Table 1. The major changes included (1) reducing the curing
time from 28-60 days to 14 days, (2) eliminating preconditioning the
wall for 24 hours prior to the start of the test, (3) reducing the test
period from 72 hours to 4 hours, and (4) eliminating the rating given to
a wall at the end of the test. These modifications were made to
simplify the test method, but it is possible that the present 4 hour
test may be insufficient to effectively determine the performance of
walls constructed with materials which contain integral water repellent
admixtures.

The objectives of this study were to:

1.   Evaluate the effectiveness of the ASTM E 514-90 test method in
     determining performance of walls constructed with materials which
     contain integral water repellent admixtures.

2.   To determine the effect of extending the test period from 4 hours
     up to 72 hours.

3.   To compare the performance of different integral water repellent admixtures.

TABLE 1 − − Comparison of ASTM E 514-74 and E 514-86/90 test conditions and observations

	E 514-74	E 514-86/90
Curing of walls	28-60 days in lab air	7 days in plastic 7 days in lab air
Preconditioning	expose 24 hrs to test condition, then allow to dry	not required.
Water Penetration & Leakage Test		
Observations Required	9	5
Time of First Dampness	X	X
Time of First Visible Water	X	X
% Dampness	24, 72 hrs.	4 hrs.
Back of Wall (top trough)		
Time for Leakage to Begin to Flow from Flashing	X	−
Maximum Rate of Leakage & Time Maximum Leakage Observed	X	−
Rate of Leakage at 24 hrs.	X	−
Total Water Collected	−	4 hrs.
Leakage from Interior Cavity (bottom trough)		
Time for Leakage to Begin to Flow from Flashing	X	−
Maximum Rate of Leakage & Time Maximum Leakage Observed	X	−
Rate of Leakage at 24 hrs.	X	−
Total Water Collected	−	4 hrs.
Rate Water Permeance of Wall	Rate as E, G, F, P or L	Not required

SCOPE OF INVESTIGATION

A laboratory test program was conducted to evaluate the effect of varying the E 514-90 test period from between 4 hours to 72 hours, and to compare the performance of four different integral water repellent admixtures.

EXPERIMENTAL DETAILS

Integral Water Repellent Admixtures

Four commercially available integral water repellent admixtures were evaluated and are identified as S1, S2, S3 and S4. Admixture S1 is similar to S2 and S3 is similar to S4, but S1/S2 are different from S3/S4. S2, S3 and S4 were added to both the concrete masonry units and to the mortar, but S1 consisted of two products, one which was formulated for use with concrete blocks and another for use with mortars (S1-M).

Concrete Masonry Unit and Mortar Properties

The concrete masonry units were a normal weight, split face 203 x 203 x 406 mm (8 x 8 x 16 in.) architectural block which contained two cavities. The concrete blocks were manufactured at a concrete block plant and produced with and without integral water repellent admixtures. The mix design of the concrete and gradation of the aggregates are summarized in Table 2 and physical properties of the architectural concrete masonry units are summarized in Table 3. The concrete blocks were manufactured using S1, S2, S3 and S4. S1 was tested at 296, 444, 592, 739 and 887 mL/45.36 kg cement (10, 15, 20, 25 and 30 oz/cwt cement), S2 at 296, 592 and 887 mL/45.36 kg (10, 20 and 30 oz/cwt), S3 at 444 mL/45.36 kg (15 oz/cwt) and S4 at 444 mL/45.36 kg (15 oz/cwt). The recommended addition rates for S1 and S2 are 739 ± 148 mL/45.36 kg cement (25 ± 5 oz/cwt), and for S3 and S4, about 444 ml/45.36 kg cement (15 oz/cwt).

The walls in this test series were constructed using architectural concrete blocks because integral water repellent admixtures are widely used in blocks of this type. Standard 203 x 203 x 406 mm (8 x 8 x 16 in.) common concrete blocks with two cavities were also used to construct one test wall and their physical properties are also included in Table 1. This wall was constructed to compare the performance of the architectural blocks with standard concrete blocks.

TABLE 2 - - Mix design of concrete and aggregate gradation.

Mix Design

	kg	lb
Type III cement	158.8	350
Coarse Aggregate	861.8	1,900
Fine Aggregate	771.1	1,700
Total	1791.7	3,950

Aggregate Gradation

% Retained on Each Sieve

Sieve No.	Fine Aggregate	Coarse Aggregate
3/8		1.2
4	0.1	52.9
8	8.5	33.3
16	17.1	8.4
30	32.2	2.3
50	27.6	-
100	11.9	-
Pan	2.6	1.9
Total	100.0	100.0

TABLE 3 - - Physical properties of concrete masonry units.

Mix No.	Admix	mL/ 45.36 kg Cement	oz/cwt	Compressive Strength Gross Area (MPa)	Unit Weight (kg/m³)	(pcf)	Absorp. (%)
1	Ach Control	0	0	31.01	2,085.1	130.3	7.6
2	S1	296	10	20.13	2,085.5	130.3	7.6
6	S1	444	15	39.88	2,145.1	134.0	5.6
3	S1	592	20	29.11	2,116.9	132.3	6.8
5	S1	739	25	33.46	2,189.2	136.8	6.3
4	S1	887	30	40.73	2,166.0	135.4	5.8
8	S2	296	10	44.58	2,109.2	131.8	7.1
7	S2	592	20	32.83	2,126.2	132.9	6.6
9	S2	887	30	40.51	2,153.4	134.6	6.1
10	S3	444	15	27.67	2,102.2	131.4	6.4
14	S4	444	15	27.41	2,119.8	132.5	6.2
-	Std Control	0	0	6.50	2,258.4	141.1	4.7

Each result is the average of three specimens

Type S portland cement-lime mortars were used in constructing the walls and proportioned according to ASTM C 270-89 Specification for Mortar for Unit Masonry and tested according to ASTM C 780--91 Test Method for Preconstruction and Construction Evaluation of Mortars for Plain and Reinforced Unit Masonry. The mix design for the mortar is given in Table 4 and the physical properties of the mortars are summarized in Table 5. All of the mortars contained an integral water repellent admixture, except for the mortar used to construct the two Control walls. The blocks and mortar used to construct the Control walls contained no IWR admixtures. The mortar admixture that was used for each particular wall was the one recommended by the producer of the IWR admixture contained in the block that was used in that wall, and was used at the recommended addition rate for a Type S portland cement-lime mortar - 828 mL/45.36 kg (PC + lime) (28 oz/cwt) for the S1-M/S2 admixtures and 192 mL/45.36 kg (PC + lime) (6.5 oz/cwt) for the S3/S4 admixtures.

TABLE 4 -- -- Mix design for Type S portland cement-lime mortar.

Mix Design		kg	lb
Type I cement		23.68	52.2
Type S Hydrated Lime		5.03	11.1
Masonry Sand		94.35	208.0
	Total	123.06	271.3
S1-M, S2	828 mL/45.36 kg (PC + lime)		(28 oz/cwt)
S3, S4	192 mL/45.36 kg (PC + lime)		(6.5 oz/cwt)
Mortar Penetration		45-55 mm	

TABLE 5 - - Physical Properties of Mortar

WALL NO.	BLOCK ADMIX	mL/ 45.3 kg Cement	oz/cwt	BATCH SIZE	Mortar Admix mL/ 45.3 kg	AIR (%)	INITIAL PENET. (mm)	COMPRESSIVE STRENGTH		
								7-DAY (MPa)	28-DAY (MPa)	56-DAY (MPa)
1	Achit'l Control	0	0	1x	0	5.1	53	22.950	27.209	36.017
3	S1	296	10	1x	828	9.2	45	22.605	28.405	34.576
9	S1	444	15	1x	828	—	50	—	—	—
16	S1	444	15	--	828	—	—	—	—	—
13	S1	592	20	1x	828	9.8	52	—	—	—
14	S1	592	20	1x	828	—	56	21.695	26.147	33.784
10	S1	739	25	1x	828	10.0	46	21.916	28.229	29.538
22	S1	739	25	1x	828	11.0	52	14.349	17.395	—
15	S1	887	30	2x	828	8.2	53	20.572	25.007	31.137
2	S2	296	10	1x	828	10.0	57	20.496	27.271	33.425
11	S2	592	20	1x	828	10.5	47	22.591	27.009	31.606
12	S2	592	20	1x	828	11.2	48	20.510	26.602	31.985
4	S2	887	30	1x	828	9.6	49	21.819	27.884	34.114
17	S3	444	15	2x	192	16.0	51	16.382	19.914	23.184
18	S3	444	15	--	192	—	—	—	—	—
19	S4	444	15	2x	192	17.0	48	15.320	17.967	22.295
20	S4	444	15	--	192	—	50	—	—	—
26	Std Control	0	0	2x	0	6.5	52	19.359	24.831	31.840

A double batch of mortar was used to construct Walls (15, 16), (17, 18) and (19, 20).

Mortar admixture dosage rate is based on 45.3 kg (100 lb) portland cement + lime.

Each compressive strength result is the average of:
2 cylinders at 7 days, 6 cylinders at 28 days, and 2 cylinders at 56 days.
Cylinder = 76.2 mm x 152.4 mm (3 in. x 6 in.).

Construction of Walls

Eighteen test walls were constructed according to ASTM E 514-90. The walls were single wythe wall panels 1422 mm wide by 1829 mm high (56 in. x 72 in.), or (3.5 block wide x 8 block high). The 1422 mm (56 in.) wide panel was used in order to avoid having a head joint of a half unit included in the 914 mm (36 in.) wide test area. The walls were constructed during a one week period by two masons whose work would be rated good.

Each wall was constructed by one mason and required approximately 1.5 hours to complete, with the masonry work being done over a period of 60 min. The wall was constructed on an inverted steel channel and the bottom course was laid on a bed of mortar which covered the face shells and webs, but not the cavities. Full bedded mortar joints were used and the walls were constructed one course at a time by applying mortar the full length of the bed joint (3.5 blocks), then buttering the ends of a block one at a time before setting on the bed joint. The joints were initially struck and tooled with a concave jointer after the top course was laid and a final tooling was done approximately 30 - 60 min. later.

The S1 and S2 walls were constructed in a random order, based on concrete block admixture addition rate, to avoid potential systematic errors which might have occurred if the walls had been built in order from low-to-high admixture addition rate, or vice versa. The walls were cured according to ASTM E 514-90 which requires curing for 7 days enclosed in plastic and for a minimum of 7 days in laboratory air. The total curing time for the walls ranged from 35-42 days.

Flashing was built into the wall such that water which leaked through the exposed face and passed through to the back of the wall was collected in the top trough, which was located between the first and second course. Water which leaked into the interior cavities of the wall and did not pass through to the back was collected in the bottom trough. The interior cavity water collected in the bottom of the wall and slowly leaked through mortar joints or blocks and was collected in the bottom trough located at the back of the wall.

Water Permeance

The water permeance tests were conducted according to the ASTM E 514-90 procedure in which the wall is exposed to the simulated wind-driven rain test for 4 hours, except that the test period was continued to 72 hours. The 72 hour test is similar to the ASTM E 514-74 procedure except that the walls were not preconditioned. The preconditioning step in the E 514-74 procedure calls for the walls to be subjected to the simulated wind driven rain for 24 hours, then allowed to dry before the start of the test.

Architectural split face block contain a smooth face, which is normally the back of the wall, and an uneven face which is the architectural face and is normally the exposed face. In these tests, the test frame was attached to the smooth face making this the exposed face, and the architectural face became the back of the wall. This was necessary so that the test frame could be sealed to the exposed wall surface. The back of the wall was white washed according to the ASTM E 514-90 procedure to facilitate observations regarding leakage and dampness.

The test conditions involve subjecting 1.08 m² (12 ft²) of the test wall to a simulated wind driven rain (simulating 100.6 km/h (62.5 mph) wind and 140 mm (5.5 in.) rain/hour).

The simulated 100.6 km/h (62.5 mph) wind is obtained by pressurizing the 914 mm wide x 1219 mm high (36 in. x 48 in.) chamber to 500 Pa (10 lbf/ft²) which equals the pressure produced by a 48.8 mm (1.92 in.) head of water. The simulated 140 mm (5.5 in.) rain/hour is obtained by spraying water down the face of the wall at a flow rate of 12.9 L/0.093 m²/h (3.4 gal/ft²/h or 154.4 L/h (40.8 gal/h) for 1.15 m² (12 ft²).

RESULTS

Mortar

An IRW admixture was added to mortars used with concrete blocks that contained an IWR admixture and was the companion mortar admixture recommended by each manufacturer of the specific IWR admixture. In order to eliminate the mortar as a variable, the mortar admixture addition rate was kept constant for the various walls and added at the recommended addition rate for a Type S portland cement-lime morter, i.e. 828 mL/45.36 kg (PC + lime) (28 oz/cwt) for S1-M/S2 and 192 mL/45.3 kg (PC + lime) (6.5 oz/cwt) for S3/S4.

The mix design for the mortar is given in Table 4 and the physical properties of the mortars are summarized in Table 5. The mortar was used at the workability level requested by the masons and had morter cone penetrations of approximately 45-55 mm as determined by ASTM C 780-91. Compressive strengths were determined using 76.5 x 152.4 mm (3 x 6 in.) cylinders and plastic air contents were determined using a 0.007 m³ (0.25 ft³) pressure meter. A batch of mortar was mixed for each wall, although a limited number of double batches were made if two walls which used the same mortar were being constructed at the same time. Mortar was used within 1.5 hours after mixing and generally did not require retempering, except for several mixes which had low initial mortar penetrations.

The S1-M/S2 mortar admixtures entrained approximately 5-6% more air than the Control mortars, and the S3/S4 mortar admixtures produced 10-12% more air than the Control. The compressive strengths for the S1-M/S2 admixtures were approximately equal to those for the Control mortars (96% of the Control mortar at 28 days) even though the air contents for the S1-M/S2 admixture mortars were higher, a condition that would generally result in lower strengths.

The mortar strengths for the S3/S4 products were significantly lower than the Control mixes (70% of the Control at 28 days), probably due to the higher air contents of the mortars. The higher air-lower strength characteristics of these mortars may be one of the factors which contributed to the poorer performance of the walls which were constructed using the S3/S4 admixtures.

ASTM E 514-90 Test Results (4 Hour Results)

The ASTM E 514-90 test requires that five observations be made - time of appearance of first dampness and first visible water on the back of the wall, % dampness on back of wall at 4 hours, and the total water collected from each trough at 4 hours. The results for the 4 hour test are summarized in Table 6 and Figure 1. No water leaked through to the back of any wall and only five of the eighteen walls had leakage from the interior of the wall into the bottom trough. The leakage from the interior was highest for the two Control walls, 6.1 L and 12.1 L, (1.6 and 3.2 gal.) and low for three of the four walls which contained S3/S4, 1.4-2.1 L (0.37 - 0.55 gal). No leakage occurred for the walls which contained S1/S2 admixture.

The first dampness results were variable and there was poor agreement between most of the six duplicate walls. However, the results indicate that dampness for the S3/S4 admixture walls appeared significantly earlier than for the S1/S2 admixture walls. At one hour all of the walls which contained S3/S4 admixtures were damp, while only three of the twelve S1/S2 admixture walls showed any dampness. The average first dampness and % dampness for the S3/S4 admixture walls was 42 minutes and 6.6%, respectively, and for the S1/S2 admixtures, 155 minutes and 2.1%, respectively.

The short period of time for which the test walls are exposed to the simulated "wind-driven rain" make it difficult to draw conclusions regarding the performance of the various walls. But the results do indicate that the S1/S2 admixture walls performed better than the S3/S4 admixture walls, and that the walls which contain an IWR admixture perform significantly better than the Control walls.

The four hour test was not able to provide information regarding the effect of concrete block admixture addition rate upon performance. No significant differences in performance were observed at four hours for the walls constructed using blocks which contained different addition rates of S1/S2 admixtures.

ASTM E 514 24 Hour Results

The 24 hour results are included in Table 7 and summarized in Figure 2. Very little water leaked though to the back of the wall, but significantly more water was collected at the bottom trough which collects the water which leaked through the face shell into the interior wall cavity. Averaging the results for the various admixtures indicates that the S1/S2 admixtures have reduced the amount of water leakage into the wall by an order of magnitude, 31.61 L (8.35 gal) leakage for

Table 6--ASTM E 514-90 4 hr results.

Wall No.	Admix	mL/ 45.36 kg Cement	oz/cwt	First Damp. (hr:min)	Condition at 4 hrs.		
					Damp. (%)	Total Leakage Back Wall (L)	Int. Cavity (L)
1	Ach Control	0	0	0:05	9	0	11.95
3	S1	296	10	0:30	6	0	0.00
9	S1	444	15	5:00	0	0	0.00
16	S1	444	15	2:40	4	0	0.00
13	S1	592	20	2:00	2	0	0.00
14	S1	592	20	0:20	5	0	0.00
10	S1	739	25	3:00	1	0	0.00
22	S1	739	25	0:10	5	0	0.00
15	S1	887	30	6:30	0	0	0.00
2	S2	296	10	1:06	1	0	0.00
11	S2	592	20	1:30	0	0	0.00
12	S2	592	20	2:00	0	0	0.00
4	S2	887	30	6:00	0	0	0.00
17	S3	444	15	0:19	10	0	1.41
18	S3	444	15	0:12	12	0	1.82
19	S4	444	15	1:00	3	0	0.00
20	S4	444	15	0:27	11	0	2.09
26	Std Control	0	0	5:00	0	0	6.09

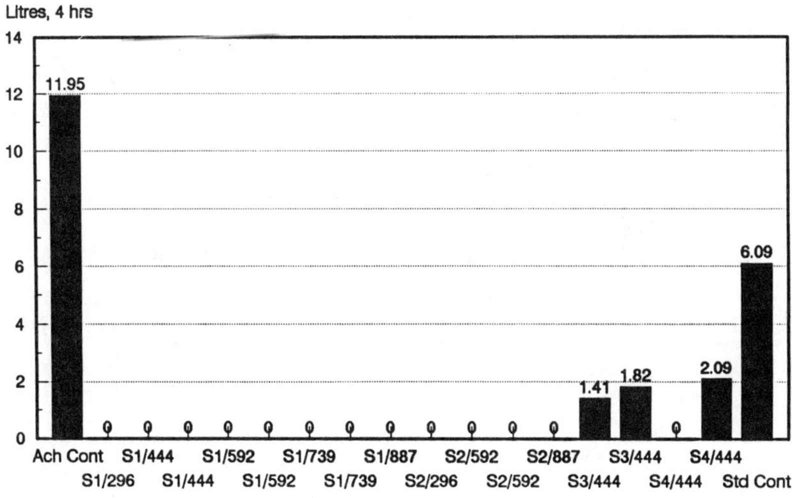

Fig. 1--ASTM E 514-90 4 h leakage from cavity.

TABLE 7 – – ASTM E 514 Water Permeance and Leakage Results

Wall No.	Admix	mL/ 45.36 kg Cement	oz/ cwt	First Damp	Dampness 4 h (%)	24 h (%)	48 h (%)	72 h (%)	Total Leakage Back of Wall (Top Trough) 4 hr (L)	24 hr (L)	48 hr (L)	72 hr (L)	Total Leakage Interior cavity (Bottom Trough) 4 hr (L)	24 hr (L)	48 hr (L)	72 hr (L)	Wat. inside cavity 72 hr (L)	Total Wat.= leak+ cavity 72 hr (L)
1	Ach Control	0	0	0:05	9	16	20	20	0.00	0.00	0.00	0.00	11.95	31.62	74.00	92.13	22.85	114.97
3	S1	296	10	0:30	6	21	23	23	0.00	0.70	1.64	2.17	0.00	12.90	37.84	52.38	9.49	61.87
9	S1	444	15	5:00	0	5	19	20	0.00	0.03	0.63	1.81	0.00	1.89	7.80	20.70	38.75	59.45
16	S1	444	15	2:40	4	18	22	28	0.00	0.17	2.35	4.99	0.00	0.82	8.77	28.40	50.51	78.91
13	S1	592	20	2:00	2	12	12	13	0.00	0.00	0.27	0.32	0.00	4.46	20.13	33.49	12.08	45.57
14	S1	592	20	0:20	5	17	22	23	0.00	0.37	0.79	1.31	0.00	2.04	6.86	10.99	6.04	17.03
10	S1	739	25	3:00	1	7	17	18	0.00	0.18	0.87	1.69	0.00	0.12	3.14	7.46	21.31	28.76
22	S1	739	25	0:10	5	14	15	17	0.00	0.19	0.46	0.70	0.00	1.48	4.66	9.06	14.04	23.10
15	S1	887	30	6:30	0	14	16	17	0.00	0.11	0.34	0.60	0.00	0.00	1.68	6.22	19.08	25.30
2	S2	296	10	1:06	1	8	16	20	0.00	0.00	3.23	3.54	0.00	0.18	18.35	26.33	13.49	39.82
11	S2	592	20	1:30	0	25	27	27	0.00	1.16	2.29	3.56	0.00	1.95	23.20	58.09	19.94	78.03
12	S2	592	20	2:00	0	6	18	18	0.00	0.07	0.60	1.19	0.00	0.26	4.97	13.19	13.13	26.32
4	S2	887	30	6:00	0	7	13	13	0.00	0.00	0.05	0.12	0.00	1.88	18.69	24.14	15.94	40.08
17	S3	444	15	0:19	10	18	21	21	0.00	0.46	0.79	1.30	1.41	33.57	64.41	79.95	5.86	85.81
18	S3	444	15	0:12	12	18	19	19	0.00	0.57	0.68	0.68	1.82	37.52	70.68	98.44	11.36	109.80
19	S4	444	15	1:00	3	14	18	18	0.00	0.02	0.35	0.93	0.00	28.26	49.20	64.05	10.95	75.00
20	S4	444	15	0:27	11	22	22	25	0.00	0.94	1.20	1.46	2.09	28.71	52.69	68.50	12.72	81.22
26	Std. Control	0	0	5:00	0	14	11	11	0.00	0.00	0.00	0.00	6.09	41.29	70.91	95.62	3.82	99.44
	AVERAGE RESULTS																	
	S1				3	14	18	20	0.00	0.22	0.92	1.70	0.00	2.96	11.36	21.09	21.41	42.50
	S2				0	12	19	20	0.00	0.31	1.54	2.10	0.00	1.07	16.30	30.44	15.63	46.06
	S-1&2				2	13	18	20	0.00	0.25	1.13	1.83	0.00	2.33	13.01	24.20	19.48	43.69
	S3				11	18	20	20	0.00	0.51	0.74	0.99	1.61	35.55	67.55	89.20	8.61	97.80
	S4				7	18	20	22	0.00	0.48	0.78	1.19	1.04	28.48	50.95	66.28	11.83	78.11
	S-3&4				9	18	20	21	0.00	0.50	0.76	1.09	1.33	32.01	59.25	77.74	10.22	87.96

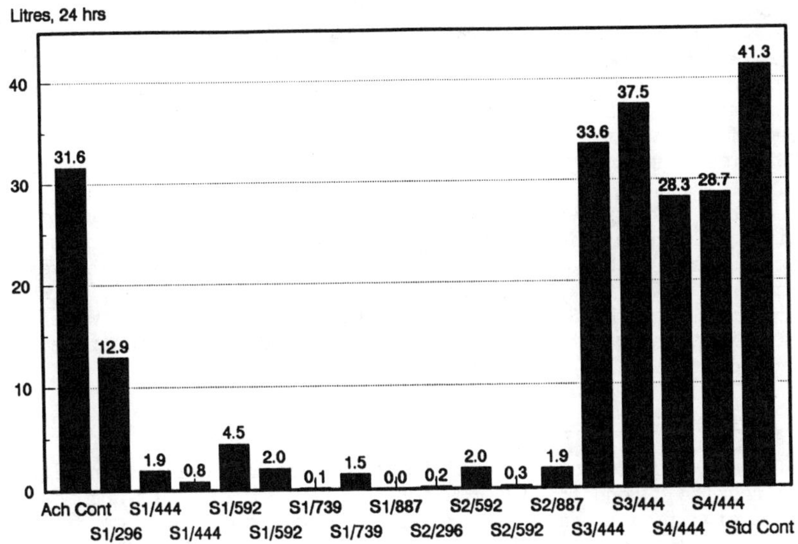

Fig. 2--24 h leakage from cavity into bottom trough.

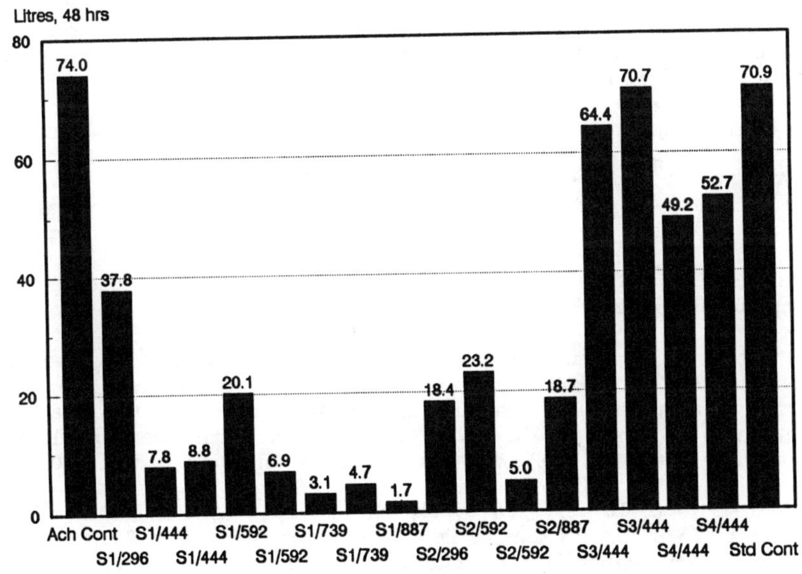

Fig. 3--48 h leakage from cavity into bottom trough.

Architectural Control vs. an average of 2.35 L (0.62 gal) for
S1/S2 admixtures. This compares with the S3/S4 admixtures which reduced
leakage only by about half, 18.77 L (4.96 gal) average leakage. The
percent of the back of the wall that was damp at 24 hours was slightly
higher for the S3/S4 admixtures than for the S1/S2 admixtures.
Agreement between the duplicate walls for water leakage and % dampness
at 24 hours was good.

The 24 hour results, especially water leakage into the cavity, provide
significantly more information regarding performance differences between
admixtures and within admixtures at varying dosage rates. The results
indicate that (1) the S1/S2 admixtures provide significantly better
performance than the S3/S4 admixtures and (2) the amount of water
leaking into the cavity of the wall was reduced as addition rate was
increased for S1/S2.

ASTM E 514 48 Hour Test Results

The 48 hour results are included in Table 7 and summarized in Figure 3.
These results extend the information obtained at 24 hours and again
provide significantly more information regarding performance differences
between admixtures and within admixtures at varying dosage rates.

ASTM E 514 72 Hour Test Results

The test was terminated at 72 hours and the results are included in
Table 7 and summarized in Figure 4. The water that was present inside
the interior cavity at the end of the test was collected by drilling
drainage holes in the bottom mortar joint and these results are
summarized in Figure 5. A summary of the water permeance test results
at 24, 48 and 72 hours is given in Figure 6.

Very little water leaked through to the back of the wall, generally less
than one gallon of water over a period of 72 hours. Most of the water
that penetrated the exposed face traveled down the interior cavities of
the wall and was collected at the bottom trough. For this reason, the
leakage into the cavity provides the best information regarding
performance. The average leakage into the cavity at 24 - 72 hours is
summarized in Figure 7 and indicate that the S1/S2 admixtures perform
significantly better than the S3/S4 admixtures.

The effect of admixture addition rate upon performance is summarized in
Figures 8 - 12 and it can be seen that the rate of leakage into the
cavity as a function of time decreases as addition rate for the S1/S2
admixtures increases. For S1, tests were done with concrete block
containing 296, 444, 592, 739 and 887 mL S1/45.36 kg cement (10, 15, 20,
25 and 30 oz S1/cwt). There is a large reduction in leakage between 0
to 296 mL (0 to 10 oz), 296 to 444 mL (10 to 15 oz), and 444 to 739/887
mL (15 to 25/30 oz); the minimum leakage occurs between 739 to 887 mL
(25 to 30 oz), which is the recommended addition rate.

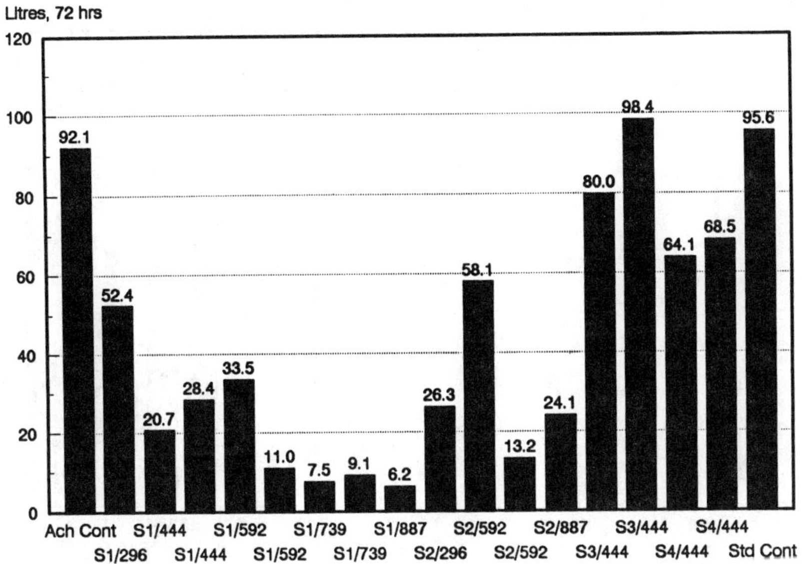

Fig. 4--72 h leakage from cavity into bottom trough.

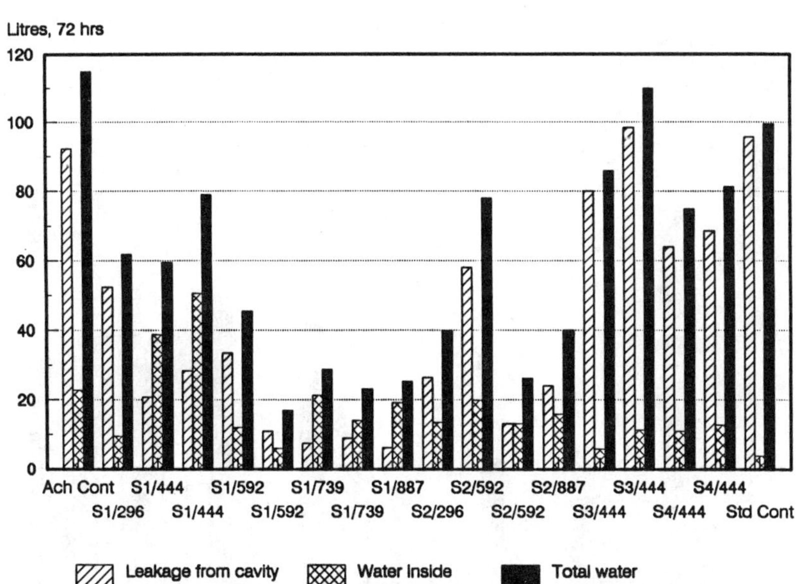

Fig. 5--Total water (leakage + drainage) collected
from cavity into bottom trough at 72 h.

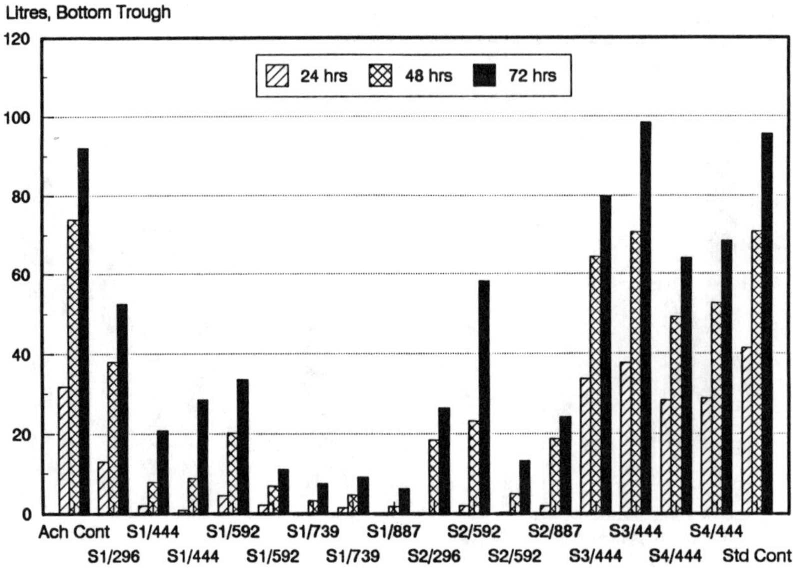

Fig. 6--Comparison of 24, 48 and 72 h leakage
from cavity into bottom trough.

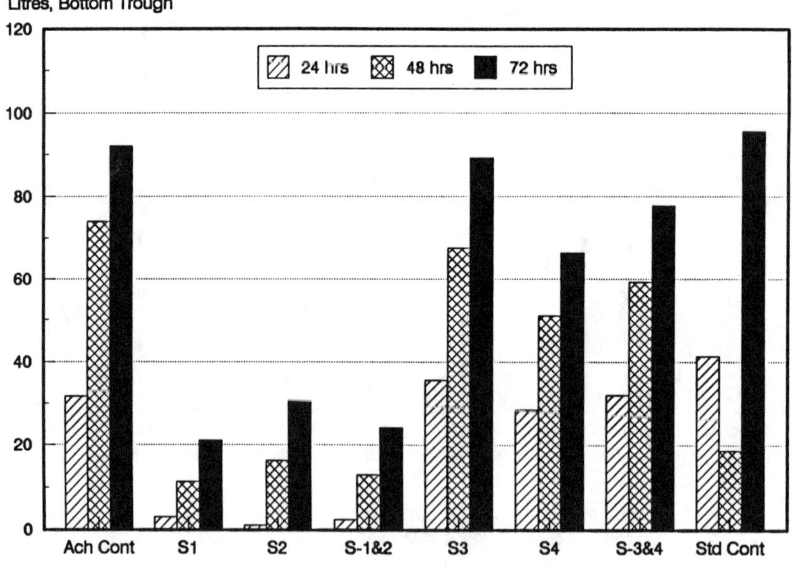

Fig. 7--Average leakage from cavity for different IWRs.

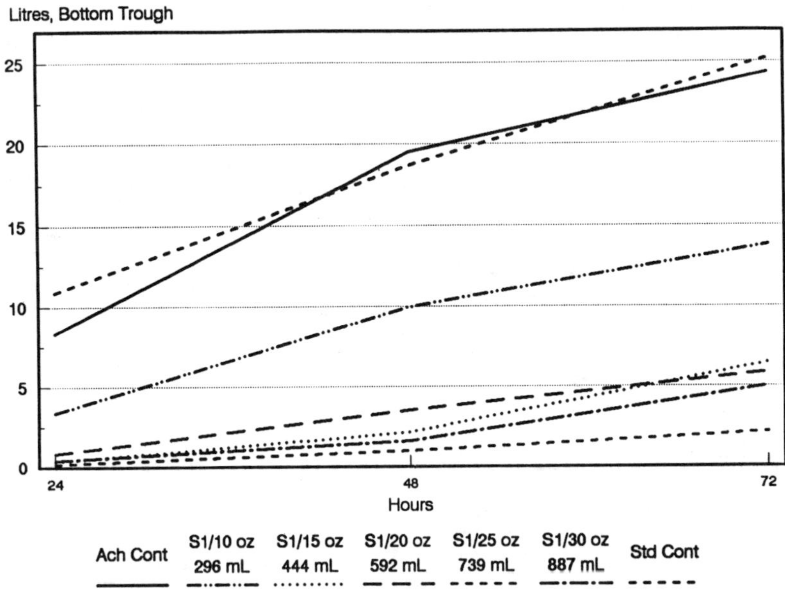

Fig. 8--Effect of S1 dosage upon leakage from cavity.

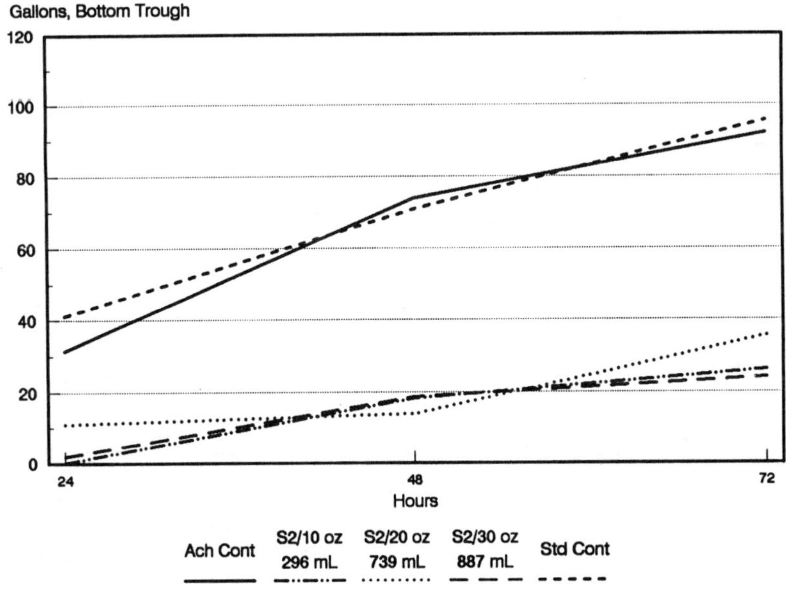

Fig. 9--Effect of S2 dosage upon leakage from cavity.

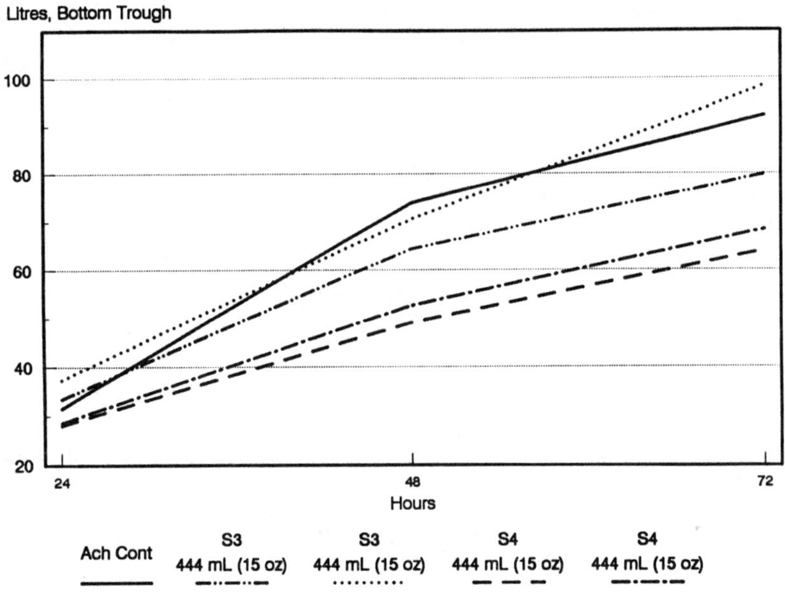

Fig.10--Effect of S3 and S4 upon leakage from cavity.

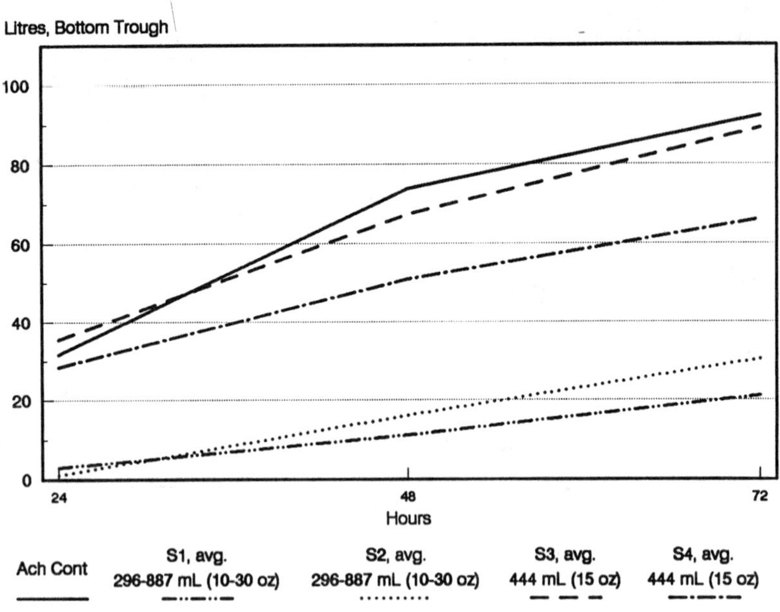

Fig. 11--Comparison of S1/S2 and S3/S4 upon
leakage from cavity.

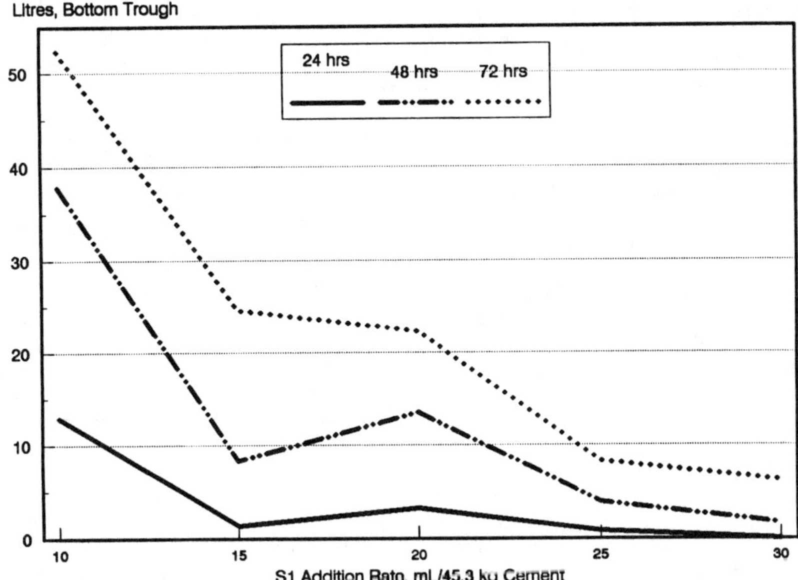

Fig. 12--Effect of S1 dosage and time upon
leakage from cavity.

S2 was tested at 296, 592 and 887 mL/45.36 kg (10, 20 and 30 oz/cwt) but
the results are not as clear cut as those for S1, mainly because only
four walls were tested rather than eight, and because one of the
duplicate walls had a higher rate of leakage than the others. The
overall performance of S2 was not quite as good as that for S1.

The performance of S3 and S4 at 444 mL/45.36 kg (15 oz/cwt) were similar
and indicate that the S3/S4 products did not perform as well as the
S1/S2 admixtures at their recommended addition rates.

DISCUSSION

Leaks

The water that penetrated through the face wall was either absorbed by
the concrete blocks, trickled to the bottom of the wall inside the
cavity, or migrated to the back face shell where it appeared either in
the form of dampness or as a leak.

Most of the leaks through the face wall occurred at the mortar joints, which could be observed at the top of the wall by looking down the cavities of the blocks. Looking inside the cavity of the wall provides an excellent vantage point from which to evaluate the ability of the wall to resist water penetration or leakage. You can quickly determine the source of water penetration, if water is being absorbed by the concrete blocks, or if water is leaking through the blocks or mortar joints. Unfortunately, the observations are somewhat limited because observations can only be made near the top 3-4 courses of the wall, even with the use of mirrors and lights.

The types of leaks that occurred depended upon whether or not the blocks and mortar contained an IWR admixture. For the Control walls, the water tended to be absorbed by the concrete blocks and mortar and appeared more as dampness than as visible water. Most of this absorbed water was conveyed by gravity to the interior base of the wall and was collected at the bottom trough. No water was collected in the top trough and the only water that reached the back of the wall appeared as dampness for both Control walls (architectural and standard gray block).

For blocks and mortar which contain an IWR admixture, the majority of the leaks occurred through the mortar joints with little or no water being absorbed by the blocks and mortar. Leakage through the concrete blocks was observed (looking down the cavities of the blocks) only for the lower addition rate S1/S2 admixtures and for the S3/S4 admixtures, and occurred mainly near the top of the wall near the area where the water spray was impinging upon the face of the block.

Each wall contained one or more leaks through the mortar joints, although the leaks tended to be very small in size. An occasional larger leak was observed which was large enough to allow air to flow through the crack. Potential reasons for the leaks are microcracks, mortar shrinkage which pulled the mortar away from the block substrate, poor bond between the mortar and the block, the inability of the mortar admixture to minimize water permeance, or workmanship.

In general, it would be expected that more leaks would tend to occur at head joints than bed joints, presumably because compression of the mortar in the bed joint by the weight of the block would tend to improve the bond in the bed joint over that of the head joint. Although some of the leaks in the bed joints near the top of the wall were obvious, it was generally difficult to determine the location of any leaks in the head joints. It is possible that any leaks which did occur at head joints may have flowed slowly down to the bed joint and appeared to be a bed joint leak.

By looking down the interior of the wall, it could be seen that the mortar joints with the S3/S4 admixtures had a significantly higher rate of leakage than the S1/S2 admixtures. These observations are in agreement with the total amount of water that leaked from or was collected from the interior of the wall. The exact reasons for this high rate of leakage through the mortar joints for the S3/S4 admixtures are not known, but may be due to the higher air content and lower strength of the mortar, or to a poor bond between the mortar and the block, or because of the inability of the S3/S4 admixed mortar to minimize water permeance.

Two of the S1/S2 admixed walls at 592 mL/45.36 kg (20 oz/cwt) had a
significantly higher rate of leakage than expected based on IWR
admixture addition rate. The reasons for this higher rate of leakage
could easily be seen by looking down the inside of the cavity and
observing that these walls contained several large mortar joint leaks.
Duplicate walls had been constructed for 592 mL/45.36 kg (20 oz/cwt) S1
and S2 and these walls performed well, which was not in agreement with
the walls with the large leaks. The reasons for these larger leaks
could not be determined, but the presence of these larger leaks helps to
explain the discrepancy in the results for these two walls.

Effect of Test Period

The water penetration and leakage test results were reported at 4, 24,
48 and 72 hours. Under the particular conditions tested, i.e. using
this particular normal weight, architectural split face concrete block
with cavities, the results indicate that the 4 hour results provide very
little information regarding performance, especially for walls that
contain an IWR admixture. Significantly more information was obtained
as the length of the test was increased to 24, 48 and 72 hours.

Of the five observations made - time of appearance of dampness on back
face, time of appearance of first visible water on back face, % dampness
at the back of the wall, leakage through the back of the wall (top
trough), and leakage into the interior cavity (bottom trough) - leakage
through the face of the wall into the interior cavity provides the best
information for differentiating performance. The amount of water
collected at the bottom trough is a direct measure of the integrity of
the wall and its resistance to water penetration and leakage. The other
four observations, although important, provide only an indirect
indication because they depend upon the chance occurrence that the water
finds its way to the back of the wall. In this series of tests, the
results for % dampness and leakage to the back of the wall were low,
even at 72 hours, and did not provide a good method for differentiating
performance.

The results for water leakage through the face wall into the interior
cavity for 24 - 72 hours are plotted for various admixtures in Figures 8
- 12. Linear regression analyses were done and the results are
summarized in Table 8. The adjusted coefficient of determination (r^2
adjusted) indicates that there is a good fit to these lines, except for
592 mL/45.3 kg (20 oz/cwt) S1 and S2, both of which had one wall which
had a significantly higher rate of leakage than the second duplicate
wall.

As expected, the amount of water that leaks through the face into the
interior cavity increases with time, with a minimum of 24 hours needed
to provide a good indication regarding performance in terms of the
ability of a wall to resist water penetration and leakage. At 24 hours,
performance differences can be seen between the Control, S3/S4
admixtures and S1/S2 admixtures, with performance increasing in that
order.

Performance differences between different addition rates of the S1/S2
admixtures can also be observed at 24 hours, but are made clearer at 48
and 72 hours, as indicated in Figures 8, 9 and 12. The 72 hour results
provide definite information that the effectiveness of the S1/S2
admixtures increases as addition rate is increased from 296 to 887
mL/45.3 kg (10 to 30 oz/cwt).

Table 8 - - Linear regression analysis of ASTM E 514 results

Equation: $Y = a + bX$

Admixture	Addition Rate mL/ 45.3 kg Cement (oz/cwt)		Constants a	b	R^2 adjusted
Ach Control	0	0	1.72	0.33	96.3
S1	296	10	0.99	0.21	98.2
S1	444	15	1.25	0.95	81.8
S1	592	20	0.74	0.09	56.5
S1	739	25	0.36	0.03	89.3
S1	887	30	0.35	0.23	74.5
S2	296	10	1.25	0.11	87.5
S2	592	20	1.85	0.14	44.6
S2	887	30	0.90	0.10	89.3
S2	444	15	0.24	0.34	95.4
S4	444	15	0.40	0.25	96.5
Std Control	0	0	1.40	0.34	97.9

Y = litres of water leakage from interior cavity into bottom trough

X = hours

The results of these tests suggest that the standard ASTM E 514-90 test
of 4 hours may not be effective in providing realistic information
regarding the performance of walls constructed with materials that
contain IWR admixtures. The test should be conducted for a minimum of
24 hours to determine the effectiveness of an IWR admixture. Extension
of the test to 48 or 72 hours is helpful in determining the effect of
IWR admixture addition rate in the concrete block in order to determine
the optimum addition rate.

The walls in this test series were constructed without weep holes which
led to a build-up of water in the interior of the walls. The amount of
water inside the wall at the end of the 72 hour test varied from 6.06 -
50.35 L (1.6 - 13.3 gal) and ordinarily would not be included in the
results. This information is important and provisions should be made to
obtain this information, either by drilling drainage holes at the end of
the test, or by constructing walls with weep holes. The latter

suggestion is preferred since weep holes will facilitate collection of water from the interior of the wall without affecting the test procedure. Weep holes are important if the test is conducted for longer than the required 4 hours, especially since the height of the water inside the wall at the end of 72 hours ranged from 0.25 -- 2.5 courses high.

CONCLUSIONS

The results of this study indicate that:

1.   The ASTM E 514-90 Test Method for Water Penetration and Leakage Through Masonry is only moderately effective in evaluating walls in which the materials contain no integral water repellent agents, due to the relatively short test period of 4 hours.

2.   The ASTM E 514-90 test method is not effective in evaluating the performance of walls which contain integral water repellent admixtures. The test should be extended to 24 hours in order to effectively evaluate the performance of walls constructed using concrete masonry units and mortar which contain an integral water repellent. A further extension to 48 or 72 hours may be required if tests are being done to determine the effect of admixture addition rate upon performance.

3.   Of the five observations required by ASTM E 514-90, the amount of water that leaks through the face wall into the cavity and is collected in the bottom trough provides the best information regarding the performance of the wall. This information provides a direct measure of the integrity of the wall and its resistance to water penetration and leakage.

4.   The walls should be constructed with weep holes to facilitate collection of water from the interior of the wall.

5.   Some extremely useful observations may be made at the top of the wall by looking inside the cavity of the wall. This provides an excellent vantage point from which to evaluate the ability of the face wall to resist water penetration and leakage and to determine if leaks are occurring through mortar joints or concrete masonry units.

6.   A comparison of four different integral water repellent admixtures indicates that S1 and S2 performed significantly better than the S3 and S4 admixtures. The performance of the S1/S2 admixtures improved as the addition rate of the admixture contained in the concrete block was increased from 296 to 887 mL/45.36 kg (10 oz to 30 oz/cwt) cement.

REFERENCES

[1] Hewlett, P. C., Edmeades, R. W., Holdsworth, R.L., "Integral
     Waterproofers for Concrete," in Concrete Admixtures: Use and
     Applications, edited by Rixom, M. R., The Construction Press, New
     York, 1977, pp. 55-66.

[2] Rixom, M. R., "Concrete Waterproofers," in Chemical Admixtures for
     Concrete, John Wiley & Sons, New York, 1978, pp. 134-144.

[3] "Laboratory Research on Water Permeance of Masonry," The Masonry
     Cement Research & Education Group, 1980.

[4] Birkeland, O. and Svendsen, S.D., "Norwegian Test Methods for Rain
     Penetration Through Masonry Walls," Symposium on Masonry Testing,
     ASTM STP 320, American Society for Testing and Materials,
     Philadelphia, 1963, pp. 3-15.

[5] Krogstad, N. V., Monk, C. B., Jr., and White, W. E.,
     "Field Evaluation of Masonry Water Permeance," Proceedings, The
     Masonry Society, Fifth North American Masonry Conference, Vol, IV,
     June, 1990, pp. 1463-1471.

[6] Krogstad, N. V., "Masonry Wall Drainage Test - A Proposed Method
     for Field Evaluation of Masonry Cavity Walls for Resistance to
     Water Leakage," Masonry: Components to Assemblages, ASTM STP 1063,
     J. H. Matthys, editor, America Society for Testing and Materials,
     Philadelphia, PA 1990.

Paul R. Barkofsky[1] and Mary E. Driscoll[1]

A THEORETICAL MODEL OF WATER PENETRATION INTO CONCRETE MASONRY UNITS

REFERENCE: Barkofsky, P. R., Driscoll, M. E., "A Theoretical Model of Water Penetration into Concrete Masonry Units," Masonry: Design and Construction, Problems and Repair, ASTM STP 1180, John M. Melander and Lynn R. Lauersdorf, Eds., American Society for Testing and Materials, Philadelphia, 1993.

ABSTRACT: The effect of water damage on masonry walls is well known, and a variety of test methods have been developed to measure the susceptibility of masonry to the forces that drive water within wall systems. Several test methods and their correlation to these forces are discussed in a separate paper.

This paper offers a qualitative model of water penetration that relates the inherent properties of masonry and the external conditions to which it is exposed to its ability to resist water leakage.

Darcy's equation of flow through porous media describes the resistance to flow offered by the tortuous networks of channels in a permeable material and quantifies the force necessary to overcome this resistance. Considering the environmental conditions that a wall is exposed to, we can modify Darcy's equation and use it to characterize masonry's propensity for water transmission driven by the forces of rain velocity, capillary pressure, differential air pressure, and gravity based on masonry material properties. In addition, with data typical of environmental exposure characteristics, the singular and additive effects of these forces can be computed and compared.

KEYWORDS: water penetration, water infiltration, theoretical model, masonry, Darcy's law, permeability, wind-driven rain, capillary pressure, air pressure, gravity.

---

[1] Research Engineer and Technical Supervisor, respectively, W. R. Grace & Co. - Conn., Construction Products Division, 62 Whittemore Ave., Cambridge, MA 02140.

SCOPE OF THE PAPER

The effects of water penetration into masonry units are well known and
much work has been done to measure and predict water infiltration under
the very dynamic conditions of our environment. The forces of kinetics
(such as wind-driven rain), capillary pressure, differential pressures,
and gravity that act to drive water into a masonry structure can be
calculated or measured. But it may be impossible to rigorously compute
the net effect of these forces, given the dynamic and variable
conditions of a typical installation.

As is often the case in the construction industry, we propose a
simplified model of water infiltration to begin to unravel these
complicated issues. We will use Darcy's equation to describe masonry's
resistance to water penetration against the forces to which it is
exposed.

To develop this model, we will consider a highly idealized situation for
which the following statements are true.

1.   The masonry material is isotropic, that is, there is no spatial
      variability in the pore morphology throughout the medium. This
      implies that the material's properties, such as porosity (the
      fraction of bulk sample volume occupied by pore space) and
      permeability (the material's resistance to penetration by a
      fluid), are constant.

2.   The masonry material is also homogeneous throughout the unit.

3.   Infiltration occurs only in one dimension.

4.   The sample is sufficiently large so that non-uniformities caused
      by the edges of the sample can be neglected.

Of course, such a system is artificial. In a real masonry unit, the
material properties vary from surface to interior mainly due to
orientations in the concrete matrix that are introduced during the
manufacturing process. Also, the material properties of a masonry wall
vary because of joints and surface imperfections (both of which may
offer relatively little resistance to water penetration). Infiltration
may be complicated by water running down the hollow cores of the unit.
Finally, the edge effects of each masonry unit and of an entire wall may
contribute substantially to the overall infiltration of water.

Again, our purpose here is to highlight the forces acting on a masonry
unit that tend to drive water into the material. By doing so, we hope
to gain an insight into better design and construction ideas as well as
improvements in standard tests and methodologies for measuring masonry's
resistance to water penetration.

DARCY'S LAW

Darcy's law is a relatively simple equation for modeling slow, steady-state, unidirectional flow of a single fluid through a saturated porous medium. This law can be expressed as

$$Q = (kA/\mu)(\Delta P/L) \tag{1}$$

where Q is the volumetric flow rate, A is the cross-sectional area normal to the flow, μ is the viscosity of the fluid, ΔP is the hydrostatic pressure drop across the sample, L is the length of the sample in the direction of flow, and k is the "specific permeability" of the porous medium. Darcy's law assumes that there are no physical or chemical changes to the medium due to the flowing fluid. This equation merely states that the flowrate (Q) through a porous medium is related to the driving force (ΔP) by a proportionality constant, the specific permeability.

The specific permeability, or simply "permeability," is a measure of the porous medium's resistance to laminar flow of a Newtonian fluid through its pore structure. Its value is solely determined by the medium's pore structure. As a result, permeability is independent of flow mechanisms or fluid properties. In an ideally uniform medium, the permeability is assumed to be independent of direction. The value of the permeability for a specific medium is determined by developing either mathematical models of the system or empirical relationships.

The unit of permeability is the "darcy." A porous material has a permeability of 1.0 darcy (D) if a fluid whose viscosity is 0.001 Pa·s (1.0 cP) will flow at 1.0 cm^3/s through a cube whose sides are 1.0 cm long under a pressure of 101.325 kPa (1.0 atm). A unit analysis of this definition shows that 1 D = 0.9869 μm^2.

Dullien [1] states that "good" concrete has a permeability of less than 0.1 mD and porosities ranging from 6% to 10%. The permeability of a typical masonry unit may be somewhat higher; its value would depend upon its raw materials, special additives, and processing conditions (such as density, compaction, hydration, and curing). While surface imperfections and cracks could increase the overall permeability of a masonry unit, mortar joints may increase the overall permeability of a masonry wall.

The form of Darcy's law is similar to that of other linear transport laws such as Ohm's law of electricity, Fick's law of diffusion, and Fourier's law of heat conduction. For example, Fourier's law can be expressed as

$$q = k_T A(\Delta T/L), \tag{2}$$

where q is the heat-transfer rate, A is the cross-sectional area normal to the direction of heat flow, ΔT is the temperature difference across the conductive medium, L is the length of the sample in the direction of heat flow, and k_T is the thermal conductivity. Although thermal conductivity is a function of temperature, it is similar to permeability in that it is (a) a property of the conducting medium and (b) assumed to be independent of direction. So Fourier's law relates the flow of heat (q) to the driving force (ΔT) through a proportionality constant related to the material (k_T).

GENERALIZED DARCY EQUATION

In developing a more realistic model for the flow of fluid through a system, we should consider the effect of gravity on the flow. Darcy's law can be modified to account for these gravitational effects. This is accomplished most readily by expressing the pressures as heights of liquids (or pressure "heads"). Thus, the Darcy equation becomes

$$Q = (kA/\mu)(\Delta P/L) \qquad (3)$$

where P (the total pressure) is defined as

$$P = P + \rho gz. \qquad (4)$$

Here, P is the fluid's hydrostatic pressure, ρ is the fluid's density, g is the acceleration due to gravity, and z is the vertical height above an arbitrary, horizontal datum level. Figure 1 illustrates a device, called a piezometer, which can be used to determine the value of P. Often, P is indicated through a parameter called the "piezometric head" (ϕ) whose dimension is length:

$$\phi = P/\rho g = (P/\rho g) + z \qquad (5)$$

Here the quantities $P/\rho g$ and z are called the "pressure head" and "elevation head," respectively. Figure 1 shows that P is the difference in pressure due to the flow of the fluid through the medium; without flow, P is constant throughout the system.

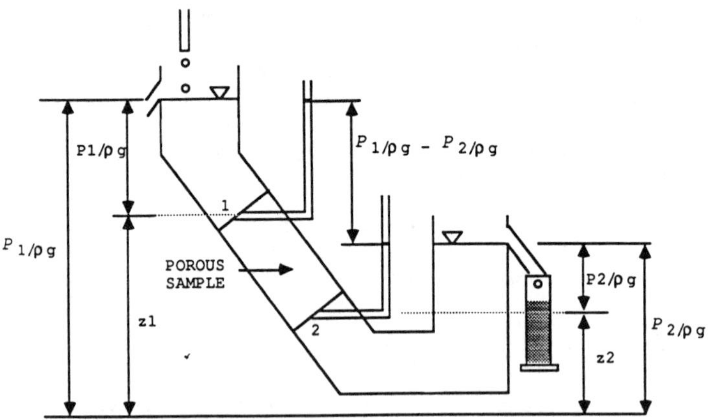

Figure 1. A piezometer showing piezometric head, elevation head, and pressure head (adapted from [2]).

FORCES CONTRIBUTING TO WATER PENETRATION

A masonry unit exposed to the environment is subject to several forces which conspire to drive water into and through the medium. In describing these forces and their contribution to water infiltration, we will focus upon a "control volume" of the porous medium. So that we may later apply the Darcy equation, we must define this control volume as that section of material which is already saturated with infiltrated water (as shown in Figure 2).

The forces acting upon this control volume are:

   1.   wind and rain velocity,
   2.   capillarity,
   3.   differential pressure, and
   4.   gravity.

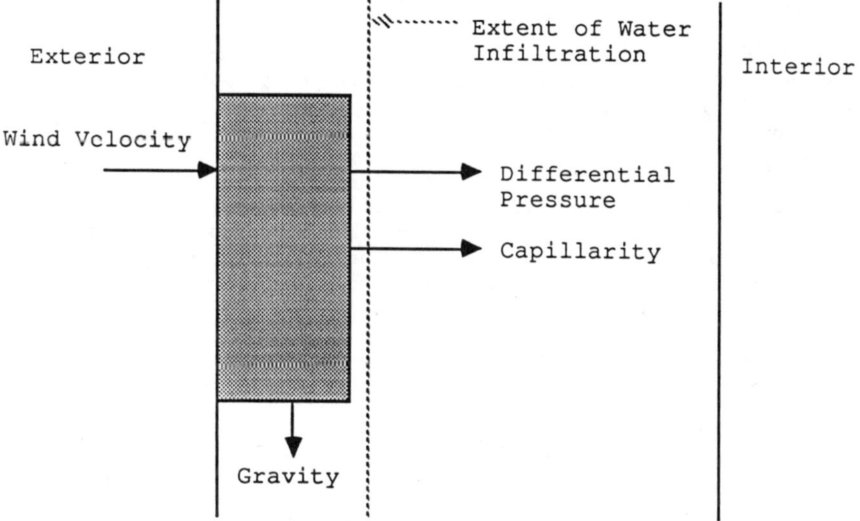

Figure 2. Control volume of saturated masonry material.

The velocity at which rain and wind strike the masonry structure serves to drive some of the moisture into the structure. Capillarity (or capillary pressure) acts to wick some portion of the water into the interior of the masonry's pores. Differential pressure between the interior and exterior of the structure also acts to promote water flow (Darcy's equation). Gravity tends to force a downward flow of the infiltrating water.

Each of these forces is described in the following sections.

Wind and Rain Velocity

The impact of water and wind against the exterior face of a masonry
structure tends to force water into the medium. The force of this
impact is directly related to the velocity of the wind and rain as it
strikes the wall.

In reality, this force is extremely difficult to quantify with any
accuracy. The actual wind velocity against an exposed surface varies
temporally and spatially. During a storm, the wind velocity often
varies instantaneously and erratically. Wind velocity at an exposed
surface also varies with the surface's position relative to other
objects or its location within the larger structure -- for example, some
masonry units might be shielded from the wind by other objects, or they
might be on the leeward side of the building [3]. Finally, the absolute
contribution of the rain droplets to this force would depend on the
"concentration" of the droplet impacts (mass of rain/area of surface)
and the angle of impact at any given moment.

Fortunately, we can calculate an approximate value for this force by
assuming that (a) the wind velocity is constant and perpendicular to the
wall and (b) the rain droplets do not contribute to the force. For this
approximation, we will find the static pressure that results due to the
wind velocity.

Figure 3 shows pressure-measuring device, called a pitot tube, in an air
stream. At point 1, the air velocity is v_1, but at point 2 (the opening
of the pitot tube), the velocity is zero. The impact of the air stream
at the opening of the pitot tube creates a pressure force which raises
the water in the manometer to a corresponding height.

We can calculate the static pressure using the generalized Bernoulli
equation for steady-state, incompressible, non-viscous flow:

$$P_1/(\rho g) + v_1{}^2/(2g) + z_1 \ = \ P_2/(\rho g) + v_2{}^2/(2g) + z_2 \tag{6}$$

where P is the hydrostatic pressure, ρ is the air density, g is the
acceleration due to gravity, v is the air velocity, z is the height
above some datum plane, and the subscripts represent points 1 and 2.
Here we will also assume that the density of the air is constant in our
system. If points 1 and 2 are at the same height, then equation (6)
rearranges to become

$$v_1{}^2/(2g) \ = \ P_2/(\rho g) - P_1/(\rho g). \tag{7}$$

But, using the equation for a manometer, the pressure at point 2 is

$$P_2 \ = \ P_1 + H\rho_w g - H\rho g. \tag{8}$$

where H is the head of the water in the manometer and ρ_w is the density
of water. Substituting equation 8 into equation 7, simplifying, and
solving for H yields

$$H \ = \ [v_1{}^2/(2g)]/[(\rho_w/\rho) - 1] \tag{9}$$

If ρ_w is 1000 kg/m^3 (62.4 lb/ft^3) and ρ is 1.2 kg/m^3 (0.075 lb/ft^3), then H is about 5.1 cm (2.0 in.) of water for a wind velocity of 105 km/h (65 mph). This head is equivalent to about 0.51 kPa (0.074 psi).

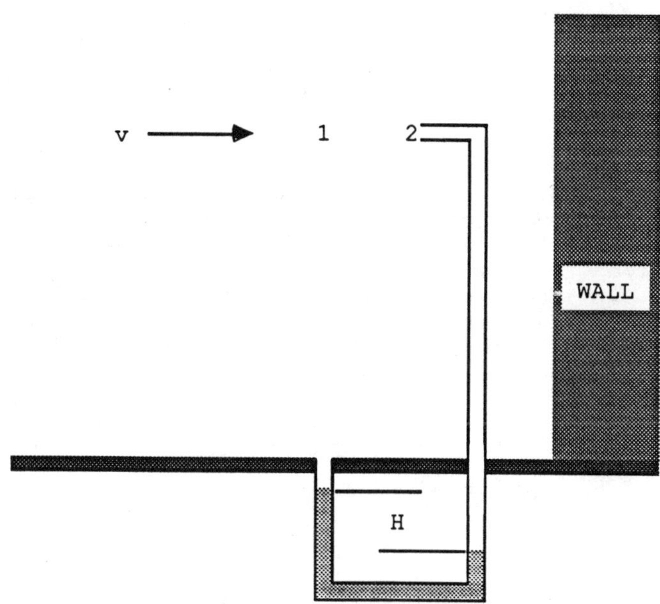

Figure 3.   Pitot tube estimation of static pressure due to wind velocity.

Capillarity (Capillary Pressure)

The rise of liquid in a thin tube is a phenomenon called "capillary pressure" or "capillarity." Capillarity occurs, for example, when spills wick upward into a paper towel or sponge. Likewise, it can occur in a masonry unit as the porous material absorbs water into its internal structure. The source for the water can be either water at the exposed surface or water within the masonry unit itself.

Capillarity is closely related to the forces of surface tension. Surface tension results from molecular forces acting at an interface between two different materials (such as water in contact with air). In the interior of the water, each molecule is surrounded on all sides by like molecules. But at the surface, air molecules lie above the surface water molecules. Raising a surface molecule slightly stretches the cohesive bonds between that water molecule and its neighboring water molecules, thereby creating a resistive force.

The resistive force behaves like a tensile force (hence, surface tension) as it seeks to minimize the surface area of the surface. The force required to overcome these cohesive bonds and break the surface is quantified by the coefficient of surface tension (σ). The value of σ for pure water is about 0.073 N/m (73 dyn/cm).

If we consider a static system of water and air inside a thin (glass) tube, there are cohesive forces between each of the water molecules, but there are also adhesive forces between the water and glass (and corresponding forces for the air as well).

Figure 4 shows the forces acting upon a molecule in close proximity to the tube wall, where vectors F_c and F_a represent the cohesive and adhesive forces, respectively. F_c acts downward and to the right because there are no water molecules above or left of the molecule; F_a acts perpendicular to the wall. If the adhesive force is stronger than the cohesive force, as shown, the net force (F_{net}) is directed down and left. Because the system is in equilibrium, the net force must act perpendicular to the surface. So, the surface curves upward to the tube at a contact angle (θ) and forms a meniscus characteristic of the system. The contact angle for water and glass is about 25.5°.

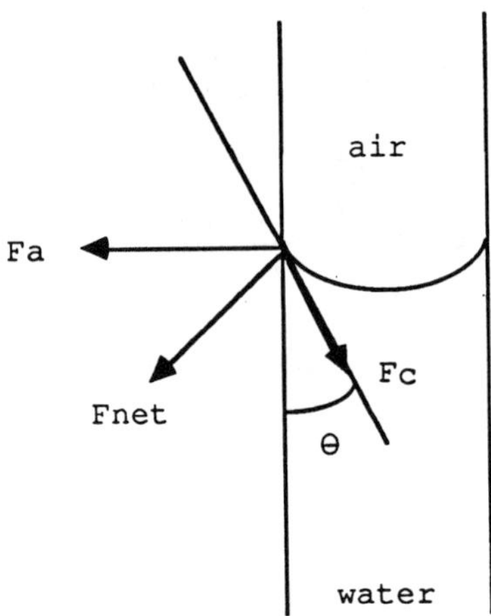

Figure 4.   Forces acting upon a molecule in close proximity to the tube wall.

If a thin tube of varying radius is placed into a liquid, the liquid may spontaneously rise to some equilibrium height (h). This situation is illustrated in Figure 5. One end of the tube is submerged an arbitrary distance into the liquid, the other end is open to the atmosphere. The upward force due to the surface tension supports the weight of the column of liquid. The vertical component of surface tension ($\sigma \cos[\theta+\phi]$) acts along the length of the contact surface ($2\pi R$), so the total vertical force is $\sigma(2\pi R) \cos[\theta+\phi]$. Neglecting the slight curvature of the meniscus, the volume of the liquid in the tube is $\pi R^2 h$, so the weight of liquid is $\rho(\pi R^2 h)g$. Equating these forces yields

$$\sigma(2\pi R) \cos[\theta+\phi] = \rho(\pi R^2 h)g \qquad (10)$$

or

$$h = [2\sigma \cos(\theta+\phi)]/[\rho Rg]. \qquad (11)$$

The tendency for liquids to rise in thin tubes can be thought of as a pressure exerted by the liquid within the tube. This "capillary pressure," measured at point A, can be expressed as

$$P_c = \rho gh = [2\sigma/R] \cos(\theta+\phi). \qquad (12)$$

Capillary pressure can vary, depending on whether the fluid is at rest or flowing. Significant variations between the static and dynamic capillary pressures occur when the fluid is vigorously forced into the porous medium (especially when turbulent flow occurs within the pores) [4]. If we assume that the driving forces are reasonably small and the flow is relatively slow, then the capillary pressure under static conditions is approximately that under dynamic conditions. We will assume these conditions hold throughout the rest of this paper.

By completely contacting the face of an untreated masonry unit with water, we have found that the typical value of static capillary pressure is equivalent to approximately 5 cm (2 in.) of water. This is equivalent to a pressure of about 0.5 kPa (0.07 psi).

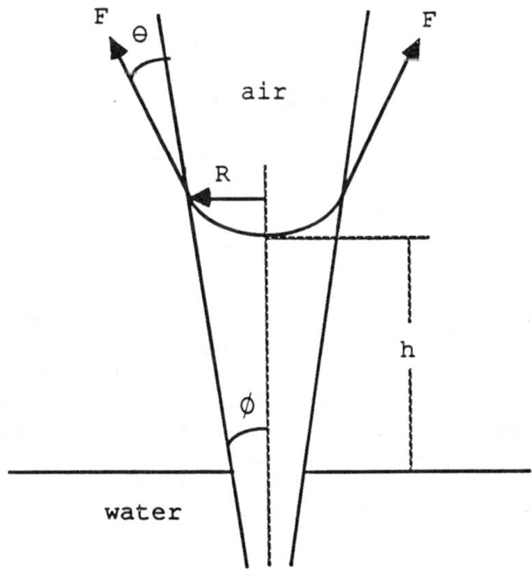

Figure 5. Capillary pressure in a thin tube of varying radius (adapted from [5]).

Differential Pressure Across the Medium

Another driving force for water penetration results from either (a) differences in the static air pressure across the masonry structure or (b) a static head of water within or against the structure.

The air differential pressure can be expressed as

$$\Delta P = P_{ext} - P_{int} \tag{13}$$

where P_{ext} is the exterior pressure and P_{int} the interior pressure. Depending on the sign of ΔP, this force will contribute as a driving or resisting force to water infiltration.

In general, this force occurs from the operation of HVAC systems within the building or from air infiltration and exfiltration due to wind velocities. Consequently, high or low atmospheric pressure can be created within perimeter rooms or even an entire building. In general, the effect of the interior pressure on water penetration is negated by proper building techniques such as adequate vapor barriers.

During high winds, the absolute differential pressure may exceed 125 Pa (0.0181 psi) [6]. Typical values for the differential pressure range from -0.5 cm (-0.2 in.) to 0.5 cm (0.2 in.) of water. This is equivalent to a pressure of -50 Pa (-0.007 psi) to 50 Pa (0.007 psi) [7]. Note that the differential air pressure is only one-tenth that of other two forces.

The pressure from a static head of water is

$$P = \rho g z \tag{14}$$

where z is the height of the water column above the plane of interest. This column of water can result from water pooling against the side or perhaps above a masonry unit. If weep holes become plugged, infiltrated water may also collect within the hollow cores of the units. If the water column is taller than about 5 cm (2 in.) then it can become the greatest driving force in the system.

Gravity

The force of gravity causes infiltrated water to flow downward through the control volume. In a real masonry system exposed to water, one would expect that over a long period of time more infiltrated water would be present within the lower masonry units then the higher ones. But, as we have seen with capillarity, other forces can overcome the force of gravity and cause water to actually flow upward. Consequently, infiltrated water in a porous medium may tend to flow downward, but some portion of it will flow laterally and even upward.

The force due to gravity can be expressed as

$$F_g = \rho g z \tag{15}$$

where ρ is the fluid density, g is the acceleration due to gravity and z
is the height above some datum level.

The force of gravity was included in the generalized Darcy equation
(equation 3) developed earlier.

CUMULATIVE EFFECT OF THESE FORCES

Having defined the forces, we are now in a position to examine the
cumulative effects of these forces. First, we will look at water
infiltration from a qualitative standpoint. Then we will take a
somewhat quantitative approach to the problem.

We will assume that water infiltrating into our control volume (Figure
2) can enter only from the exposed surface (in reality water could enter
from the top due to gravity or from the bottom due to capillarity).
Water at the surface may be either stagnant water or perhaps wind-driven
rain. Given a sufficient wind velocity, water in contact with the
exposed surface may penetrate into the material to some depth.
Capillary pressure will tend to draw the stagnant surface water or
infiltrated, wind-driven rain into the bulk of the material through its
pores. If we assume the interior static pressure is less than that of
the exterior, then this differential pressure across the masonry unit
will also tend to force the water further into the medium. Finally, the
force of gravity will attempt to draw the water downward. If
evaporation is negligible, the infiltrated water can only exit the
control volume (a) at the plane parallel to the surface or (b) at the
bottom plane (due to gravity).

A force balance performed on our control volume shows that the total
force on the system is the sum of the individual forces from wind-driven
rain, capillarity, differential pressure, and gravity. Here we assume
that there is no lost work in the system -- that is, all of the energy
in the system goes into producing the pressure terms we defined above.

We have defined our control volume such that after a period of time,
water has infiltrated to the extent that the control volume is saturated
with water. This definition allows us to apply Darcy's law and modify
it to include all of the driving forces in the system. To simplify the
mathematics and make our system unidirectional, we will first assume
that the effect of gravity on the flow in our control volume is
negligible.

Again, the generalized Darcy equation can be expressed as

$$Q = (kA/\mu)(\Delta P/L) \tag{3}$$

where

$$P = P + \rho gz. \tag{4}$$

In this equation, P is the hydrostatic pressure of the fluid, while P
(the total pressure) accounts for the effect of gravity on the pressure
at the measuring point. If the measuring points are at the same height,
$\Delta P = \Delta P$. Because the forces on our system are cumulative, we can

express the pressure as the sum of the incremental pressures from each force:

$$P = P_{wdr} + P_c + P_{dp} \tag{16}$$

where P_{wdr} is the pressure from wind-driven rain, P_c is the static capillary pressure, P_{dp} is the differential air pressure across the system. Consequently, the Darcy equation can be expressed as

$$Q = [kA/\mu][(P_{wdr} + P_c + P_{dp})/L] \tag{17}$$

In terms of pressure heads, equation 16 becomes

$$Q = [kA/\mu][(H_{wdr} + H_c + H_{dp})/\rho_w g L] \tag{18}$$

where ρ_w is the density of water. As mentioned in the previous section, typical values for H_{wdr}, H_c, and H_{dp} are 5 cm (2 in.), 5 cm (2 in.) and 0.5 cm (0.2 in.), respectively. If a column of water is standing against, above, or within a masonry unit, its pressure head contribution at the measuring point equals the height of the water above it.

It is interesting to note that the pressure forcing wind-driven rain into the control volume is essentially equal to the capillary pressure drawing water further into the medium. Consequently, merely contacting the face of our idealized masonry unit with water is the equivalent of applying 5 cm (2 in.) of water pressure to the surface. By equation 17, pressurizing the water in contact with the unit to 5 cm (2 in.) of water creates an equivalent total pressure of 10 cm (4 in.) on the unit.

CONCLUSIONS AND FUTURE WORK

In developing our simplified model of water infiltration, we identified several intrinsic properties of masonry material which impact the rate of water penetration. The specific permeability of the material, a measure of its resistance to infiltration, includes such material characteristics as the connectivity, "tortuosity," and size distribution of the pores. The capillary pressure of the masonry unit is effected by the pore diameter and contact angle of the entire medium.

We also highlighted several key conditions that need to be monitored and controlled in a testing environment. These variables include:

- ΔP, the differential pressure across the medium;
- L, the length of the sample;
- v, the wind velocity;
- z, the hydrostatic head or height of water column;
- ρ, the density of water; and
- μ, the viscosity of water.

Unfortunately, real masonry materials often contain irregularities, surface flaws, and cracks which tend to "short circuit" the resistance of the bulk material. These shorts can have a dramatic influence on the results of water penetration tests.

Consequently, in testing masonry materials for their resistance to water penetration, one must consider the intrinsic material properties, key test conditions, and material anomalies at the planning stage of the test. In addition, one must be sensitive to these three areas during the analysis of tests results in order to draw proper conclusions about material performance. Another paper [8] considers these issues in reviewing several test methods for water resistance.

As we mentioned at the outset, this paper is simply a beginning to understanding the complex interactions involved with water penetration. Further work along these lines should examine more closely the discontinuities in masonry materials and relax our simplifying assumptions.

REFERENCES

[1]   Dullien, F. A. L. _Porous Media: Fluid Transport and Pore Structure_, Academic Press, Inc., San Diego, 1992, pp. 83-84.

[2]   Dullien, F. A. L. _Porous Media: Fluid Transport and Pore Structure_, Academic Press, Inc., San Diego, 1992, p. 238.

[3]   _ASHRAE Handbook, 1985 Fundamentals_, American Society of Heating, Refrigerating and Air Conditioning Engineers, Inc., Atlanta, 1985, pp. 14.1-14.17.

[4]   Dullien, F. A. L. _Porous Media: Fluid Transport and Pore Structure_, Academic Press, Inc., San Diego, 1992, pp. 415-422.

[5]   Tipler, P. A., _Physics_, Worth Publishers, Inc., New York, 1982, pp. 383-384.

[6]   _ASHRAE Handbook, 1985 Fundamentals_, American Society of Heating, Refrigerating and Air Conditioning Engineers, Inc., Atlanta, 1985, p. 14.10.

[7]   _ASHRAE Handbook, 1985 Fundamentals_, American Society of Heating, Refrigerating and Air Conditioning Engineers, Inc., Atlanta, 1985, p. 22.7.

[8]   Driscoll, M. E., Gates, R. E., "A Comparative Review of Various Test Methods for Evaluating the Water Penetration Resistance of Concrete Masonry Wall Units," _Masonry: Design and Construction, Problems and Repair_, _ASTM STP 1180_, John M. Melander and Lynn R. Lauersdorf, Eds., American Society for Testing and Materials, Philadelphia, 1993.

Kurt R. Hoigard, Robert J. Kudder, and Kenneth M. Lies[1]

INCLUDING ASTM E 514 TESTS IN FIELD EVALUATIONS OF BRICK MASONRY

REFERENCE: Hoigard, K. R., Kudder, R. J., and Lies, K. M., "Including ASTM E 514 Tests in Field Evaluations of Brick Masonry," _Masonry: Design and Construction, Problems and Repair_, _ASTM STP 1180_, John M. Melander and Lynn R. Lauersdorf, Eds., American Society for Testing and Materials, Philadelphia, 1993.

ABSTRACT: Field evaluations of brick masonry are performed for quality control or investigative purposes. One of the many indicators of masonry performance is the permeability of the outer wythe. Field measurements of water permeability can be made using an adaptation of the procedures and equipment of ASTM Standard Test Method for Water Permeance of Masonry (E 514). This paper discusses case studies of field applications of E 514, the techniques used to obtain consistent and valid information, and some concerns about misapplication of the procedure. Modified ASTM E 514 has been found to be a useful field technique for masonry evaluations.

KEYWORDS: brick masonry, evaluation, material compatibility, permeance, testing, workmanship

---

The use of field tests modeled on ASTM Standard Test Method for Water Permeance of Masonry (E 514) in the evaluation of brick masonry can be controversial. Test users report three different approaches to using E 514 in the field: an adaptation using standard test parameters, accompanied by both quantitative and qualitative evaluations of the wall response [1,2]; variations on the standard test parameters to accommodate other test objectives [3,4]; or abandonment of the test because there is no universal or absolute standard against which to judge test results [5]. Differences in the way the E 514 test is used in the field indicates the absence of consensus on how the test should be conducted and how the results should be interpreted. The writers have found that the first approach, using standard test parameters and conducting the test the same way all the time, provides useful information about wall performance. This paper presents the test methodology and interpretation of results which the writers use as one of many elements in the field evaluation of masonry wall

---

[1]Mr. Hoigard is an Associate, Dr. Kudder is a Principal, and Mr. Lies is an Associate, with Raths, Raths & Johnson, Inc., 835 Midway Drive, Willowbrook, Illinois 60521.

performance. In discussing successful and rational applications of E 514, the writers hope to keep a legitimate and valuable evaluation tool from being unjustifiably discredited because of misapplication and inappropriate expectations.

Permeance testing was originally developed for the evaluation of barrier walls [6,7]. The test procedure was standardized for laboratory use in ASTM E 514, and is being used for testing the exterior wythe of cavity walls. Early versions of the E 514 test standard included a rating system based on the measured permeance of water through the wall. The rating system has been dropped from the standard. Like other test standards, E 514 now describes a procedure with acceptance criteria and interpretation of results determined by the user. The procedure is intended for laboratory use. Successful programs in which a modified E 514 procedure was used in the laboratory, a mock-up, and then in the field [8] as a quality control test, clearly indicate that it can be adapted to accommodate the special requirements of field use and the testing of a completed, in-place wall.

Water permeability is a legitimate performance criteria for a masonry wall. Circumstances under which E 514 test procedures are useful in the laboratory or on a mock-up, and in the field, include:

- Evaluating compatibility and prequalifying masonry units, mortar materials and mortar mix proportions.

- Prequalifying masons' skills.

- Construction quality control testing, based on the results of pre-construction tests. It is reasonable to expect larger variations in field work than in laboratory or mock-up work, so acceptance criteria should be adjusted accordingly.

- Comparison of various locations on a building presumed to be of uniform construction.

- Evaluation of performance improvements achieved by repairs, by testing the same area before and after repairs are made.

- Evaluation of repair longevity, by periodic retesting of the same repaired area.

- Introducing controlled and repeatable forced permeation so that water paths can be traced and evaluated.

- Evaluation of permeability by comparison of test results to an empirical body of data from many walls tested in the same manner.

The last item in the list is the root of the controversy in the field use of E 514. Even though ASTM has correctly removed an absolute rating system from the E 514 standard, judgements based on measured permeability performance should not be precluded. Handled properly, an empirical body of test data is useful. Misapplications result from sole reliance on quantitative test results, and the misinterpretation of the test results as a measure of only one of the many variables which affect permeability.

THE TEST

The field water permeability test method presently in use basically follows the requirements of ASTM E 514. A 1.22 m by 0.91 m (4 ft by 3 ft) test chamber is sealed to a brick wall, to enable water to be sprayed on the wall at a rate of 154.4 liters (40.8 gallons) per hour while the air pressure within the chamber is maintained at 0.48 kPa (10 psf) above ambient. The differential pressure is measured with either a water manometer or a bourdon tube pressure gage, and a spherical float rotameter is used to measure the flow rate through the spray bar.

Two modifications distinguish the field test from the standard ASTM E 514 test. The first change is the use of a portion of an in-place wall instead a wallette specifically fabricated for use in the test. The second, and most significant change, involves the method used to quantify the rate of permeance. Instead of basing the permeance rate on the amount of water collected from the bottom of the specimen after having passed through it, the modified method measures the amount of water consumed from a closed pumping circuit. Water is pumped from a calibrated tank to the spray bar inside the chamber, and the water that does not return to the tank is assumed to have permeated the face of the wall [1]. Figure 1 shows a typical test set-up.

While plots of measured leakage rate versus time for the standard ASTM E 514 test and the field modified version have different appearances (Figure 2), the curves both stabilize at approximately the same final rate. The primary cause for the different plot shapes can be attributed to water absorption by the masonry. During the early stages of the test, absorption slows the flow rate at which water reaches the back and bottom of the test wall to be collected by the standard E 514 test apparatus, thus producing a permeance curve that starts low and increases asymptotically. Conversely, absorption increases the initial rate at which water is lost from the closed system used for the field modified version of the test, resulting in a permeance curve that starts high and decreases asymptotically to approximately the same rate (for a given test wall) as the standard method. The presence of sealers has been observed to affect the shape of the permeance rate plot, sometimes producing plots with permeance rates that increase rather than decrease with time.

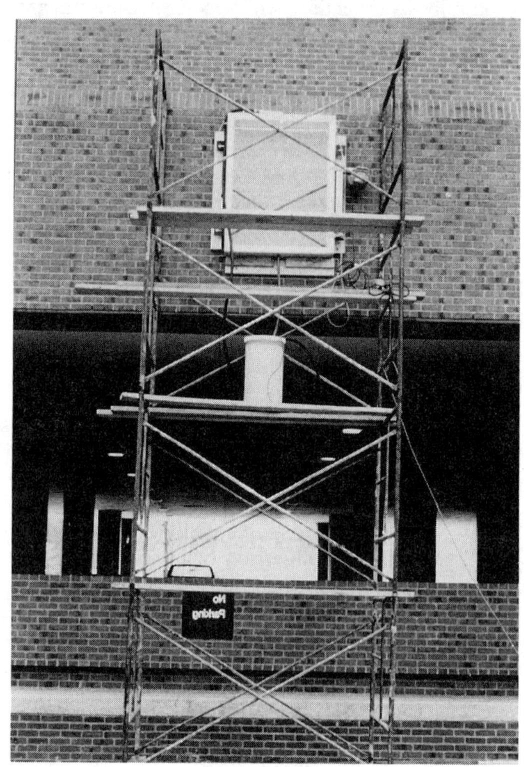

FIG. 1 --   Typical set-up for modified E 514 test.

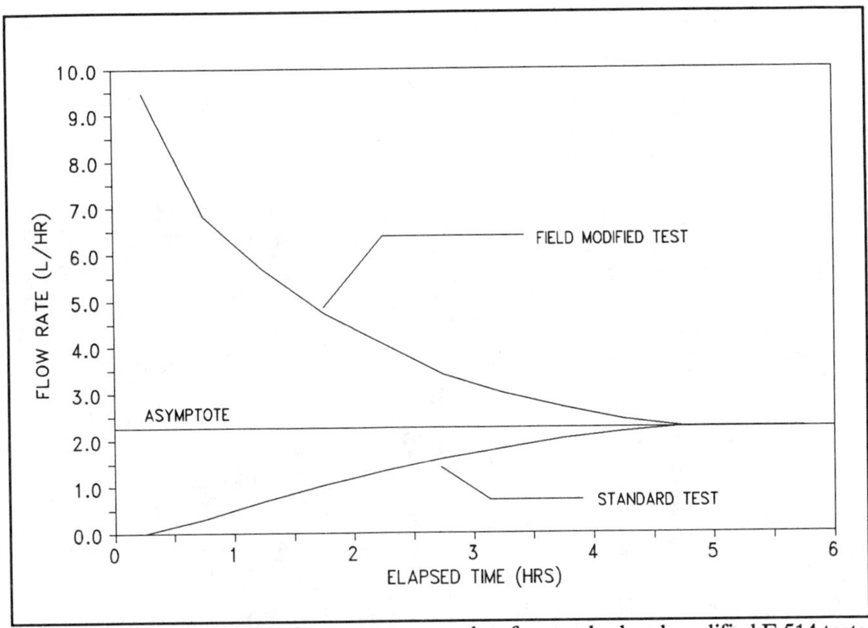

FIG. 2 --   Comparison of permeance rate plots for standard and modified E 514 tests.

It is the responsibility of the engineer conducting an evaluation including the E 514 test to consider all of the conditions present that may be contributing to the indicated permeance. Masonry walls are hand crafted using variable materials, and not all walls will behave the same. Therefore, it is not reasonable to expect identical test results from different areas, even though the test methods used are standardized and are capable of producing repeatable results at a given location.

Prior to initiating a test, it is important to inspect the exterior face of the test area to determine if it is typical of general wall conditions. Structural cracks, damaged bricks and other localized conditions will affect the test results. It is also advantageous to make inspection openings in the wall behind and below the test area to allow the back side of the outer wythe of masonry to be viewed. In situations where the back side of the outer wythe cannot be accessed through openings from the interior, other means such as a mirror or a fiber-optic borescope can be used. Examining the back of a wall is instructive. Fullness of joints, excessive squeezings, mortar droppings, tie installation, etc., can be evaluated by looking at the back side. On occasion, daylight might be seen through the wall.

During the test, important observations can then be made regarding:

- The elapsed time for water to permeate the outer wythe.

- Whether or not water bridges over a cavity or collar joint space to an inner wythe or back-up (Figure 3).

- Whether or not water which permeates the outer wythe is controlled by water barriers and flashing systems.

- Distinguishing the amount of water that actually permeates to the back of the outer wythe, moves vertically through cores, or migrates laterally within the wythe (Figure 4).

One phenomenon that can adversely affect the apparent permeance rate is excessive lateral migration of the test water within the outer wythe of masonry. The occurrence of lateral migration is a useful observation by itself. It can indicate poor mortar bond, bed joint furrowing, or other potential problems worth investigating. Pressurization of the chamber drives water through the outer wythe as intended, but can also force water laterally, increasing the apparent rate of water loss from the closed conduit. Large areas of wet mortar and masonry units immediately adjacent to the test area are symptoms of this phenomenon (Figure 5). Water can travel laterally within the masonry outside the chamber area, then exit and run down the exterior face of the masonry.

One way to control lateral migration of water from a test area is to isolate it by cutting through the outer masonry wythe around the perimeter of the test chamber and sealing the cut surfaces. The result closely resembles the laboratory E 514 specimen. This method is both costly and time consuming, and results in undesirable damage to the wall being tested.

FIG. 3 --   Interior side of CMU back-up is saturated during field E 514 test due to water bridging cavity space.

An alternate method has been developed to control lateral water migration using a double chamber test apparatus. In this method, a second pressurized chamber without a water spray bar is created around the primary test chamber. When the secondary chamber pressure is increased to a level equal to that of the primary chamber, the pressure differential driving the lateral migration is equalized, and the migration of water from the test area is limited to gravity and capillary flows. Figure 6 shows a double chamber configuration in use. Field testing case studies involving specimens which exhibited inordinate lateral migration were tested using the double chamber first without outer chamber pressurization, and then with the outer chamber pressurized (Figure 7). In all cases, a noticeable drop in apparent permeance rate was observed after the outer chamber was pressurized.

FIG. 4 --   Water from test presoak has bridged cavity and is dripping through cores of CMU back-up.

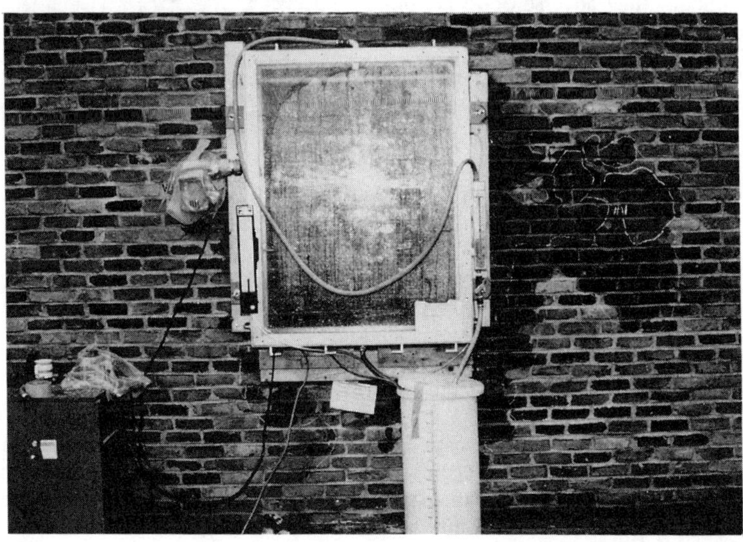

FIG. 5 --   Wet areas adjacent to test chamber caused by lateral migration of water.

EMPIRICAL STANDARD OF PERFORMANCE

One of the benefits of a standardized test method is the accumulation, over time, of historical data. By performing a test in the same manner, the experimenter removes experimental technique from the list of variables. The permeance performance of masonry depends on many variables. The E 514 test measures the performance resulting from the combined effects of the variables, and can be used to determine how the test wall compares with other walls. Based on the writers' experience conducting and analyzing more than 200 E 514 tests modified for field use as discussed above, the following guidelines for interpreting quantitative results have been developed for brick masonry performing essentially as constructed without the adverse effects of deterioration or damage:

FIG. 6 --   Typical set-up for field E 514 test with double chamber.

- Walls with a permeance rate below 1.90 liters per hour per 1.11 square meters of wall area (0.5 gallons per hour per 12 square feet) can be achieved when industry recommendations for workmanship are strictly followed and compatible materials are used. (Note that permeance rates reported in the paper are for the full 1.11 square meter [12 square feet] test area.)

- Walls with a permeance rate between 1.90 and 3.79 liters per hour (0.5 and 1.0 gallons per hour) can be achieved in standard production masonry when industry recognized workmanship recommendations are generally followed and compatible materials are used. Although performance could be better, this range is about what should be expected for ordinary brick masonry construction.

- Walls with a permeance rate between 3.79 and 7.57 liters per hour (1.0 and 2.0 gallons per hour) should be considered suspect. Ordinary production brick masonry construction in reasonable compliance with industry recommendations can be expected to provide better performance than this. Walls with rates of this magnitude, in situations where permeance is critical to the serviceability of the wall, should be investigated in more detail.

- Walls with a permeance rate greater than 7.57 liters per hour (2.0 gallons per hour) should be considered poor. Ordinary brick masonry construction can be expected to provide much better performance than this. Walls with rates of this magnitude usually result from workmanship which ignores industry recommendations, wall materials which are not compatible, or both.

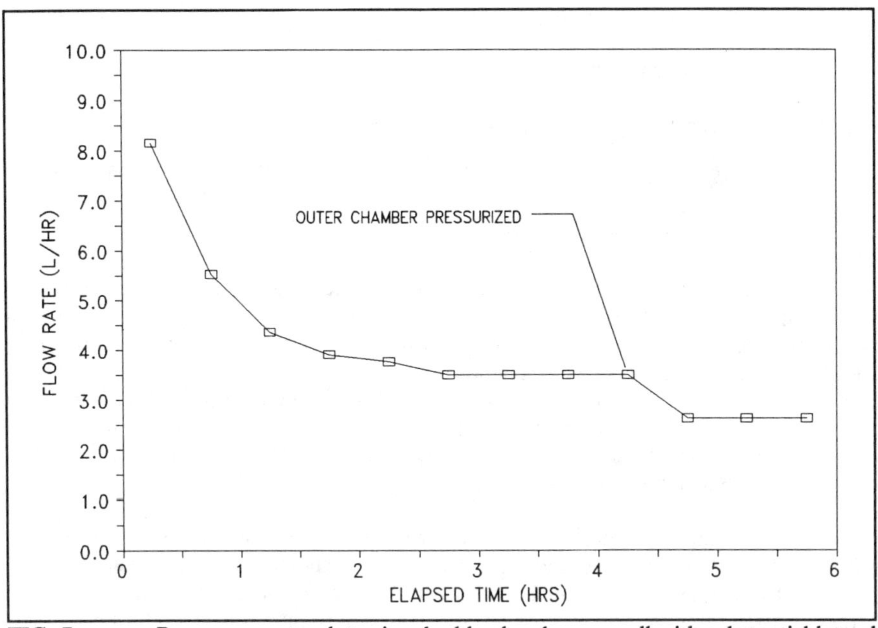

FIG. 7 --   Permeance rate plot using double chamber on wall with substantial lateral migration.

EVALUATING WORKMANSHIP AND MATERIALS

The permeance performance of masonry is influenced by many variables. The individual effects of variables such as workmanship or material compatibility cannot be distinguished from each other solely on the basis of the E 514 test. Inspection openings and sampling in conjunction with the E 514 test are used for evaluating workmanship quality and wall construction features. Observations should be made regarding:

- Fullness of both head and bed joints.

- Flashing and weeps.

- Extent and type of cracking present.

- Bed joint furrowing.

- Joint tooling.

- Mortar protrusions and cavity bridging.

- Grouting, parging, and dampproofing.

- Average joint widths and variation.

- Plumbness of the wall.

- Anchors and ties.

- Venting and pressure equalization.

- Overall visual condition.

- Presence of sealers or coatings

Case studies have shown that the visual appearance of a wall is not always a reliable indicator of how it will perform. Walls which exhibit visually good workmanship characteristics may not perform as well as walls with a poor visual appearance. Factors which can lead to less than optimal permeance, not detected by cursory visual examinations, include rolling or tapping of masonry units as they are placed, and bonding incompatibility between the masonry units and the mortar mix. Compatibility between the masonry materials is essential in achieving optimal masonry permeance performance. Pre-construction testing using E 514 is recommended for assessing material compatibility [8].

A study of three wallette samples was performed to determine the effects of mortar head joint fullness on water permeance performance. The three wallette samples were constructed by the same mason using uncoached techniques and identical masonry materials. The only variation between the samples was the degree of fullness of the mortar head joints. The first specimen was constructed with 100 percent full head joints, the second specimen had 50 percent full head joints and the third sample had 25 percent full head joints. The fullness of the head joints had a direct effect on the permeance performance of the test specimens. Figure 8 shows the leakage rates for these three wallettes as a function of time.

In another study, two brick wallette specimens were constructed by two different masons using identical masonry materials. The Type N mortar mix used was controlled by weighing the materials, including the water, to assure uniformity between mortar batches. The masons were instructed to construct the walls with full head and bed joints using their standard brick laying techniques. After the wallettes were constructed, the walls visually appeared different. One specimen exhibited good workmanship qualities, including uniform joint widths and proper tooling. The other specimen visually exhibited inferior workmanship qualities which included uneven bed joints, mortar joint width variations, premature tooling and variations in plumbness. Interestingly, results of the modified E 514 tests performed on the specimens were identical, with permeance rates of 0.95 liters per hour (0.25 gallons per

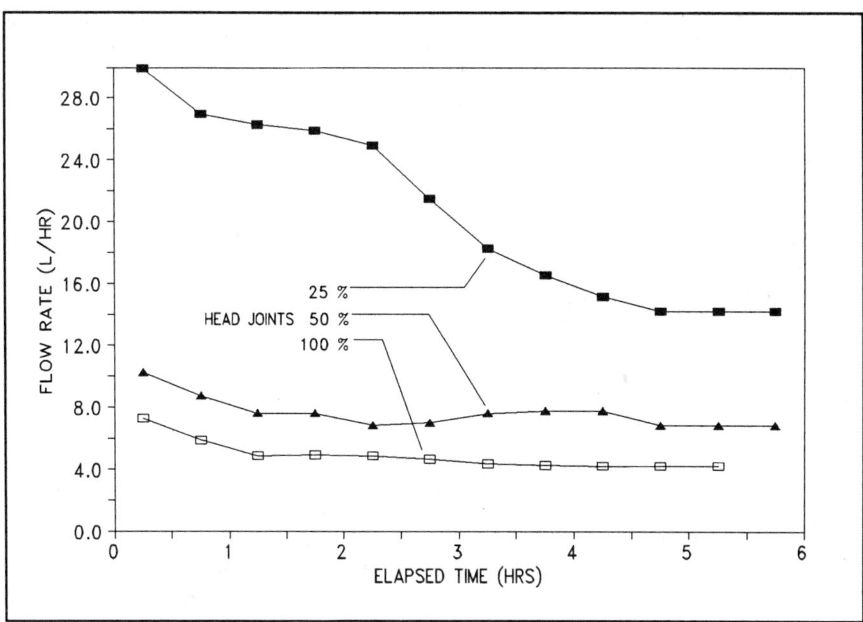

FIG. 8 --   Comparison of permeance rates for test walls with head joints filled 25, 50, and 100 percent during construction.

hour). The study shows that cursory visual examinations of masonry workmanship cannot be solely relied upon in evaluating permeance performance.

EVALUATING REPAIRS

The field modified version of ASTM E 514 can be a useful and effective tool in evaluating proposed repair methods for problem walls [9]. Remedial approaches like tuckpointing, surface grouting, and clear sealers all have known advantages and disadvantages, such as service life, initial costs, and life cycle costs. What is not always known is how effective each method will be on a given wall at improving the actual masonry water permeance performance.

Tests are used to quantify the actual performance of a wall both before and after repair methods are implemented. Using the test procedure in the evaluation of repair options can provide a building Owner with meaningful information which can be used in making informed decisions on the best repair approach to meet the projects goals and constraints.

Tests must be conducted prior to the implementation of any repairs in order to establish the net effectiveness. Test areas should be selected that are representative of the typical condition of the building and visually exhibit similar characteristics and workmanship qualities.

Once the base performance of the test areas is determined, the proposed repair methods can be implemented. The trial applications should be allowed sufficient time to cure prior to any additional testing. Follow-up tests should then be conducted using the same method and location as the initial tests. Comparing results from the initial and subsequent tests enables the engineer to evaluate the net effectiveness of each repair method. Lack of any improvement clearly indicates an incorrect or ineffective repair method. In evaluating various sealers, surface grouts, or other products and techniques, the test procedure is very effective and can save the project time and money. These tests determine initial levels of performance improvement [9]. Evaluating the long-term performance of repair methods requires periodic retesting. Tests conducted several years after the application of sealers frequently indicate a loss of effectiveness.

CONCLUSIONS

Expectations for the use of E 514 in the field should be confined to the determination of a performance property of the wall, which is permeance under controlled conditions. This property reflects the combined effects of several wall parameters, and is not a measure of any particular parameter in isolation unless a sufficient number of controlled tests or other investigations permit such a distinction. Likewise, the determination of a performance property of a wall should not be confused with an evaluation of building leakage and interior damage. The E 514 test, by itself, simply does not provide the information necessary for a thorough evaluation of leakage. The use of measurements from the E 514 test as the sole instrument for evaluating building leaks is a misapplication of the test.

If properly conducted and interpreted, E 514 is a valid quantitative and qualitative tool for the evaluation of masonry, even in the absence of an absolute rating of performance. Indeed, an absolute rating of performance may be impossible, and is specifically disclaimed in the text of the E 514 standard. This does not preclude: an engineering evaluation of the test results based on an empirical body of data from similar tests; the use of E 514 for comparative purposes; the use of E 514 as part of construction qualification and quality control program.

REFERENCES

[1]   Raths, C. H., "Brick Masonry Wall Nonperformance Causes," _Masonry: Research, Application and Problems, ASTM STP 871_, J. C. Grogan and J. T. Conway, Eds., American Society for Testing and Materials, Philadelphia, 1985.

[2]   Monk, C. B., Jr., "Adaptation and Additions to ASTM Test Method E 514 (Water Permeance of Masonry) for Field Conditions," _Masonry: Materials, Properties, and Performance, ASTM STP 778_, J. G. Borchelt, Ed., American Society for Testing and Materials, Philadelphia, 1982.

[3]   Krogstad, N. V., Monk, C. B., White, W. E., "Field Evaluation of Masonry Water Permeance," _Fifth North American Masonry Conference_, D. P. Abrams, Chairman, Urbana, Illinois, 1990.

[4]   Krogstad, N. V., "Masonry Wall Drainage Test - A Proposed Method for Field Evaluation of Masonry Cavity Walls for Resistance to Water Leakage," <u>Masonry: Components to Assemblages, ASTM STP 1063</u>, J. H. Matthys, Ed., American Society for Testing and Materials, Philadelphia, 1990.

[5]   Brown, M. T., "A Critical Review of Field Adapting ASTM E 514 Water Permeability Test Method," <u>Masonry: Components to Assemblages, ASTM STP 1063</u>, J. H. Matthys, Ed., American Society for Testing and Materials, Philadelphia, 1990.

[6]   Fishburn, C. C., "Water Permeability of Walls Built of Masonry Units," <u>BMS 82</u>, National Bureau of Standards, Washington, D.C., 1942.

[7]   Fishburn, C. C., Watstein, D., and Parsons, D. E., "Water Permeability of Masonry Walls," <u>BMS 7</u>, National Bureau of Standards, Washington, D.C., 1938.

[8]   Raths, D. C., "Preconstruction Brick Veneer Evaluation Testing," <u>Masonry: Materials, Design, Construction and Maintenance, ASTM STP 992</u>, H. A. Harris, Ed., American Society for Testing and Materials, Philadelphia, 1988.

[9]   Coney, W. B. and Stockbridge, J. G., "The Effectiveness of Waterproofing Coatings, Surface Grouting, and Tuckpointing on a Specific Project," <u>Masonry: Materials, Design, Construction and Maintenance, ASTM STP 992</u>, H. A. Harris, Ed., American Society for Testing and Materials, Philadelphia, 1988.

W. Dale Jones[1] and Mahendra B. Butala[2]

PROCEDURES AND FIXTURES FOR REMOVING, CAPPING, HANDLING, AND TESTING MASONRY PRISMS AND FLEXURAL BOND SPECIMENS

REFERENCE: Jones, W. D., Butala, M. B., "Procedures and Fixtures for Removing, Capping, Handling, and Testing Masonry Prisms and Flexural Bond Specimens," _Masonry: Design and Construction, Problems and Repair, ASTM STP 1180,_ John M. Melander and Lynn R. Lauersdorf, Eds., American Society for Testing and Materials, Philadelphia, 1993

ABSTRACT: Martin Marietta Energy Systems, Inc. (MMES), is currently assessing the safety of the Oak Ridge Y-12 Plant.[3] Many of the facilities at this plant are constructed using unreinforced hollow clay tile walls (HCTWs). In the spring of 1990 a testing program that includes in situ, laboratory, and shake table tests, and nondestructive examinations was initiated.

This paper presents the techniques used for removing, capping, and handling masonry prism specimens for testing normal and parallel to the bed joints. It also presents similar techniques used for removing, handling, and testing flexural bond masonry specimens. How these techniques can be used to obtain consistent test results and why they are required to meet relevant ASTM standards is described. These fixtures can also be used for fabrication of laboratory test specimens.

KEYWORDS: in situ masonry specimens, removal, capping, handling, testing, prism, flexural bond, clay tile

---

Many of the buildings at the Oak Ridge Y-12 Plant in Oak Ridge, Tennessee are constructed of unreinforced hollow clay tile walls (HCTW) infilled between steel or concrete frames. Although usually not considered as a part of the lateral structural resistance of these buildings, these infills will significantly change the response of the structure to seismic motions, since they provide the major component of lateral load resistance.

These structures do not meet present codes and standards for new construction and a literature review revealed that little, if any, information directly related to the performance of HCTWs to seismic motions exists [1]. Since one aspect of the safety of the plant, which is currently being assessed, depends on the seismic performance of these

---

[1]Senior Staff Engineer and Program Test Director, Center for Natural Phenomena Engineering, Martin Marietta Energy systems, Inc., Oak Ridge, TN 37831-8083

[2]Senior Structural Engineer, Structural and Architectural Design Department, Martin Marietta Energy Systems, Inc., Oak Ridge, TN 37831-7231

[3]managed by Martin Marietta Energy Systems, Inc. for the U.S. DEPARTMENT OF ENERGY under contract DE-ACO5-84OR21400

infills, a Hollow Clay Tile Wall Test Program was initiated by the
Center for Natural Phenomena Engineering (CNPE) of Martin Marietta
Energy Systems, Inc. (MMES) [1]. The approach for making the safety
assessment is risk based and, therefore, leads to the evaluation of HCTW
limit states. The CNPE test plan includes in situ, laboratory, and
shake table tests, and nondestructive examinations, coupled with a
parallel analytical effort. Nineteen types of in situ and laboratory
tests will be performed, ten of which will be both in situ and
laboratory.

Two major goals of the test program are to obtain consistent,
reliable test results and to obtain a statistically sound data base.
Achieving consistency in test results of small laboratory-built brick
prisms is difficult [2]. It follows that consistent results for the
larger laboratory-built prisms that are required are even more difficult
to obtain. Additional difficulty is encountered in removing and testing
in situ prisms. The prism sizes required are 2-ft wide by 4-ft high by
8- or 13-in. thick (610- by 1 219 mm by 203- or 330 mm). These sizes
were set because the walls of concern are either 8-in. thick (203 mm)
single-wythe or 13-in. thick (330 mm) double-wythe construction. The
8-in. (203 mm) walls are built using 12-in. long by 12-in. high by 8-in.
thick (305 mm by 305 mm by 203 mm) tiles, using a running bond with
cells laid horizontally. The 13-in. (330 mm) double-wythe walls are
also built using a running bond with cells placed horizontally. Here,
one 8-in. (203 mm) thick and one 4-in. (102 mm) thick tiles are placed
side by side in a course with a 1-in. (25.4 mm) collar joint between.
The tiles are offset such that the head joints do not align. Alternate
courses have the 8-in. (203 mm) tile set above the 4-in. (102 mm) tile,
making the collar joint discontinuous. Figure 1 depicts the
construction of a 13-in. (330 mm) thick wall.

Probably the most difficult task is to obtain undamaged specimens
from existing walls. Cavanagh et al. reported some difficultly in the
removal of wallettes from an old brickwork building, a number of them
being broken during cutting, handling, and transporting [3]. In the
initial phase of the program MMES had a similar experience in removing
undamaged 4-ft by 4-ft (1 219 mm by 1 219 mm) specimens for diagonal
tension (shear) testing. Figure 2 shows a specimen after removal.
Several problems were encountered. First, since no saw was available
for cutting completely through the 13-in. (330 mm) thick wall, it had to
be cut from both sides. This produced rough edges that later required
trimming. Next, the specimen was lifted out and set down using slings
and an overhead hoist. The specimen was then packaged for shipment to
the test site by encasing it with plywood held in place by steel banding
straps. After the specimen was received at the test site, it was
unpacked, trimmed, and then moved into the test fixture as shown in
Fig. 3. By this time some tiles had been completely jarred loose and
numerous cracks were observed in the mortar, especially in the head
joints. The specimen was so damaged that thin plastic banding straps
had to be used to hold the specimen together. Testing of three specimens was finally accomplished but with very questionable results [4].

FIXTURE DEVELOPMENT

These experiences led to the development of removal, handling,
capping, and shipping fixtures and procedures that should lead to
reliable and consistent test results [5]. The test program plan called
for the testing of compression prisms and flexural bond and diagonal
tension specimens. Prisms and flexural bond specimens were to be tested
normal and parallel to beds joints. These requirements led to the
design of five types of fixtures. The goal was to design and write a
procedure for each fixture type that could be used easily by technicians
and by craftsmen. A fixture was also designed so it could be placed on
the testing machine and be used for specimen alignment (the largest

Fig. 1. A 2-ft by 4-ft (610 mm
by 1 219 mm) laboratory-built
prism of a 13-in. (330 mm)
wall is prepared for testing
normal to the bed joint.

Fig. 2. A 4-ft square
(1 219 mm) specimen for
diagonal tension (shear)
testing has been removed and
is ready to be prepared for
shipping to the test site.

Fig. 3. A 4-ft square
(1 219 mm) diagonal tension
(shear) specimen was moved
several miles to the test
site, was uncrated and
trimmed, and is now ready for
testing.

specimen weighs approximately 900 lbs
(408 kg). Once aligned, the fixture
base could remain on the machine
during testing. Since the fixture
base was designed to remain on the
test machine, this accomplished another goal (i.e., the fixture could
be used to fabricate, handle, and test
laboratory-built specimens). This
is a highly desirable feature since,
as previously discussed, a major goal
of the CNPE test program is to obtain
a statistically sound data base.
Obtaining and testing numerous in situ
specimens would be very costly. This
cost not only involves labor but also
the impact on operations of the
facilities. To minimize this cost the
plan calls for the development of
correlation factors between laboratory-
built and in situ removed specimen test

results. To accomplish this goal, the testing of both types of specimens must be as nearly identical as possible.

<u>Prism Compression Normal to Bed Joints</u>

The fixture has been developed to the point where specimens have been successfully removed, capped, and are waiting to be tested. First, a wall is selected and the outline of the prism is marked on the wall. Then the tiles are removed from below the bottom and the sides of the first course, shown in Fig. 4. It was quickly learned that all tiles and cuts must be made by a saw to avoid damaging the specimen. A hydraulically powered, water-cooled chain saw with an abrasive, diamond impregnated blade was selected to make the cuts. Fattal and Cattaneo also suggest using a saw with a diamond or silicon-carbide cutting edge that is capable of cutting completely through the wall [6]. Next, a removal assembly is brought into place as shown in Fig. 5. The assembly has a fixture base, item 1, for receiving the prism, and a subbase, item 2, which is not physically attached to item 1 but which will be used to level item 1 as described later. Before lifting the assembly into place, weather stripping is attached to the fixture base to form a dam, shown in Fig. 6. To break bond, a thin plastic wrap is placed over the surface of the fixture base and between the weather stripping. The assembly is lifted into position and the subbase wedged into place so the bottom of the specimen is inside and below the top surface of the weather stripping dam shown in Fig. 7. Jacking screws supplied with item 1 are now used to raise the fixture to just below the bottom of the specimen. In performing this operation a bubble level is used to accurately level the fixture base. Next, a gypsum cement mix is poured into the dammed area and allowed to harden. The final saw cuts for removing the prism are now made and stabilizing bars are added. The assembly and the prism are lifted out of the wall and lowered to the floor, as shown in Fig. 8. The prism is now ready for capping its top surface.

Fig. 4. The first steps in removing a 2-ft by 4-ft (610 mm by 1 219 mm) normal compression prism from an 8-in. (203 mm) thick wall are completed.

Top Cap Preparation--Many attempts using ASTM Standard Methods of Sampling and Testing Brick and Structural Clay Tile (C 67) Section 6.2, Capping Test Specimens, and other recommended techniques, were tried without obtaining a flat surface that was within the desired tolerance of 0.003 in. (0.076 mm) in 16 in. (406.4 mm). The problem was probably compounded by the need to cap an uneven surface of 13 in. by 24 in.,

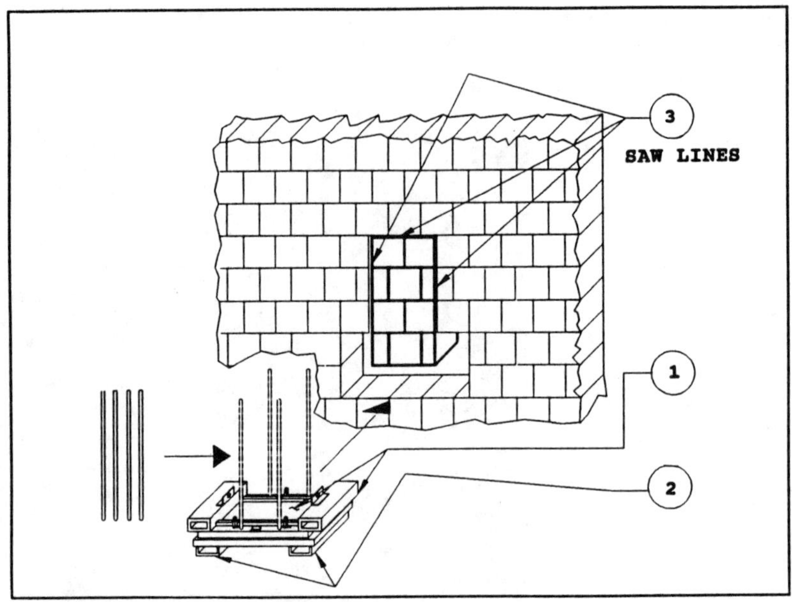

Fig. 5. A schematic showing the fixture base and subbase ready to be lifted into place under a 2-ft by 4-ft (610 mm by 1 219 mm) prism.

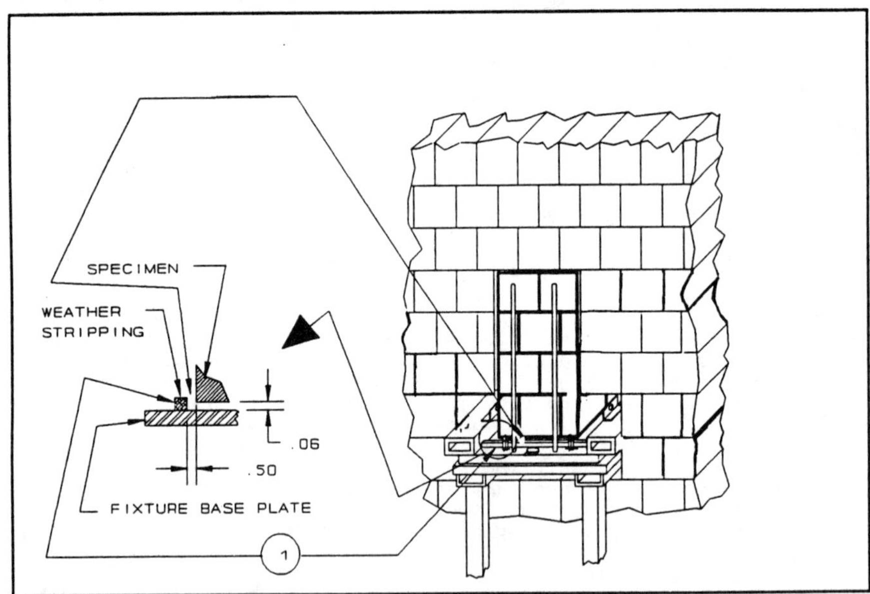

Fig. 6. A schematic showing the removal fixture positioned under a 2-ft by 4-ft (610 mm by 1 219 mm) prism. Details are shown of the formation of a weather stripping dam to hold the gypsum cement that will form the bottom cap.

Fig. 7. The removal fixture subbase is wedged into place under a
2-ft by 4-ft (610 mm by 1 219 mm) prism; the base is leveled with a
bubble level and then elevated with jack screws to just below the
bottom of the prism.

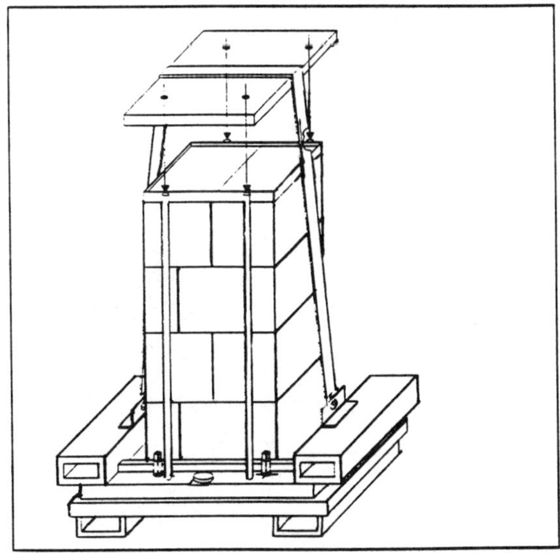

Fig. 8 A schematic of a 2-ft by 4-ft (610 mm by 1-219 mm) prism,
removed and capped, showing its preparation for shipment to the test
site.

(330 mm by 610 mm) which was required for
the 13-in. (330 mm) thick wall. Although a
stiff, flat machined plate was tried, heat
of hydration, shrinkage, etc., caused the
finished cap to deviate considerably from
the desired flatness. This effect was not
always apparent to the unaided eye but was
discovered while making flatness checks.
Figure 9 is a photograph of a finished
trial cap. This was the fifth cap tried
and the circles indicate the areas that
were out of tolerance. To obtain a flat
cap, machining was suggested but was ruled
out on the basis of cost. Since the bottom
cap was already formed level and perpendicular to the axis of the prism by the
procedure described above, it was decided
to relevel the specimen base with the
specimen (Fig. 8), and then form a level
top cap using the assembly shown in
Fig. 10. A dam is formed using precision
ground flat plates, item 1. One side of
the dam is leveled using jacking screws
(against and angle frame, item 2) and a
precision level. The opposite side of the
dam is leveled with respect to the first
side by placing the level across the
specimen. All this is done while keeping
the high point of the specimen less than
0.125 in. (3.175 mm) below the leveled dam.

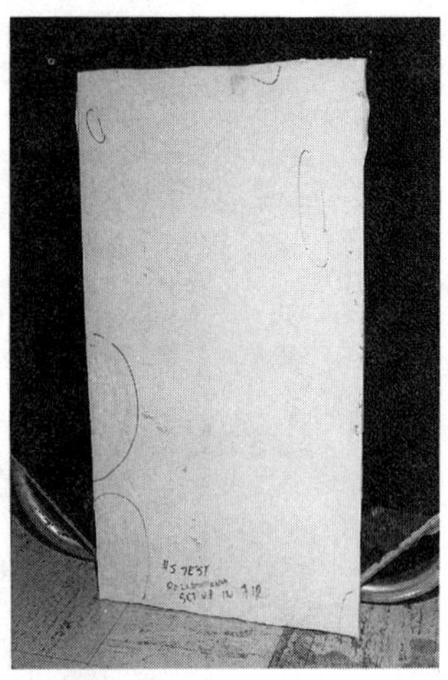

Fig. 9. A finished 13-in.
by 24-in. (330 mm by
610 mm) trial cap is
checked for flatness. The
circled areas indicate
those portions exceeding
flatness requirements.

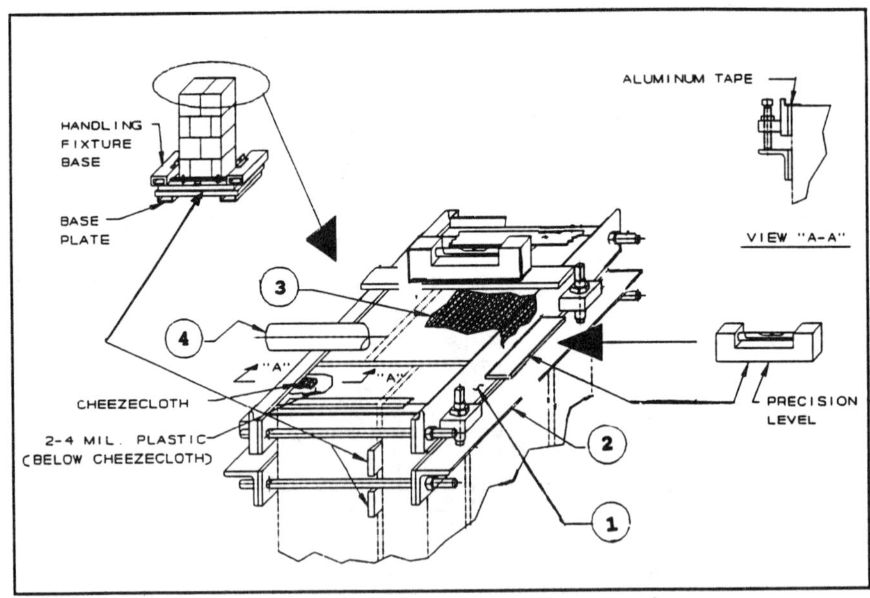

Fig. 10. A schematic of the fixture for making a prism's top cap.

The ends of the dam are formed with aluminum tape which provides a
runout for the cap. The dam is now poured full with the gypsum mixture.
A thin metal strip is used to remove air bubbles by using vertical
chopping motions into the mixture. The surface is lightly sprayed with
water and a damp cheesecloth, item 3, is placed across the gypsum and
dam, making sure that no wrinkles are left in the cloth. The cloth is
again lightly sprayed with water. A precision ground steel bar, item 4,
starting at one end, is rolled across the dam to squeeze out excess
gypsum and water. Several passes may be needed to remove all the water
that rises to the surface. Care is taken to keep the roller clean and
damp between each rolling operation. Finished caps made in this manner
are found to be well within the desired tolerance. This procedure
assures that the top and bottom caps are parallel to each other, and
that they are flat and perpendicular to the axis of the prism within the
accuracy of the precision level and/or the accuracy of the constructed
walls. Although ASTM C 67 only states that the caps will be approximately perpendicular to the axis
of the prism, other researchers
have suggested that a tolerance
of 1° is adequate [2]. For these
prisms, 1° could mean that the
caps could be out of parallel by
0.84 in. (21.3 mm).

<u>Shipping and Laboratory-
Built Specimen</u>--Figure 11 shows
an 8-in. (203 mm) thick normal
prism being loaded onto a truck,
which was then transported to a
nearby test facility and loaded
into a test machine. Inspection
at the test site revealed no
damage to the specimen. Figure 1
shows a laboratory-built normal
prism on a test machine. The
prism was built on the handling
fixture base and then set into
place using a forklift truck (the
steel tube sections used for
inserting the forks of the truck
can be seen on both sides of the
specimen).

<u>Prism Compression Parallel to Bed
Joints</u>

        Many other unusual problems
had to be overcome in the removal
and testing of prisms parallel to
bed joints. First, the prism
would have to be removed and then

Fig. 11. An 8-in. (203 mm) thick
normal prism, removed from a wall, is
loaded onto a truck for shipment to
the test site.

rotated 90° for testing. Secondly, since the cells run horizontally,
the capping procedure would have to be modified. Thirdly, the ends
would need to be capped while the prism was laying on its side, or it
could be rotated 90° capped on one end (top) then rotated 180° and
capped on the opposite end (now top). As previously noted, the
fabrication details of the 13-in. (330 mm) wall would make a prism very
fragile to handle because of the double-wythe construction. Here the
4- and 8-in. (102 mm and 203 mm) tiles are separated by a collar joint
which makes the top row very easy to knock loose. Figure 12 shows a
laboratory-built 13-in. (330 mm) parallel prism that was fabricated on
its handling base and then capped in this position.

<u>End Cap Preparation</u>--The capping fixture consists of two plexiglass plates connected by four threaded rods to form a parallelepiped. With this arrangement, the perpendicularity of the caps to the specimen, the thickness of the caps, and the parallelism of the caps to each other can be controlled. After alignment aluminum tape is used to seal the sides and the bottom gap between the plexiglass and the specimen. The gap is then filled with gypsum cement mix. Capping in this manner allows the airbubbles to migrate out the top gap. Caps thus formed meet the flatness tolerance as discussed above.

After the caps have hardened the specimen must be rotated 90° for testing. To accomplish this the capping fixture is removed and a steel baseplate (that later will be placed with the prism on the test machine) is attached to the handling base and placed flush against the cap. On the opposite end of the prism an aluminum plate and a soft material, such as a fiber glass mat, is pressed against the cap and the stabilizing rods are attached. Figure 13 shows the prism after being rotated 90°. The aluminum plate is on top and the baseplate that will be placed on the test machine is on the bottom. After the prism has been placed on the test machine, all parts of the handling fixture will be removed except the baseplate. The testing, fixture bases, and caps meet the intent of ASTM Standard Test Methods for Compressive Strength of Masonry Prisms (E 447).

Flexural Bond Fixtures

Two other fixtures similar to those described for prism compression have been designed for removing and testing flexural bond specimens. They have been successfully used to fabricate laboratory-built specimens but have not, as yet, been used for removing specimens from existing walls. The flexural bond specimen size selected for

Fig. 12. Shown here is a 13-in. (330 mm) thick prism, with the ends capped and to be tested parallel to the bed joints, that was laboratory-built on a handling and removal fixture.

Fig. 13. This shows the prism in Fig. 12, rotated 90° and prepared for shipping to the test site.

testing normal to bed joints is 2-ft wide by 6-ft long (610 mm by
1 829 mm) with the end bearing rods (to assure simple beam bending) 5 ft
(1 524 mm) apart. The width of the flexural bond specimen to be tested
parallel to bed joints has been changed to 3 ft (914 mm) to assure that
two bed joints are included in the specimen. The fixtures are designed
for accommodating third-point loading, Test Method A of ASTM Standard
Test Methods for Flexural Bond Strength of Masonry (E 518). Procedures
for grouting the end bearing rods and third-point loading rods to the
specimen, while holding them parallel and in a common plane
respectively, are included with the fixture design [5]. With this
arrangement there is no need to place leather shims (or other
compressible materials) between the rods and contacting surface of the
test machine. The fixtures are designed to axially compress the
specimen to prevent damage during handling, transporting, and placing
onto the test machine. Figure 14 shows a schematic of a specimen that
is prepared for shipping and for testing parallel to the bed joints.
Item 1 depicts the fabrication or removal base. Item 2 is a rod ready
for grouting to the specimen. As noted above, there are two bearing and
two loading rods. Parallelism of the rods are controlled by installing
them perpendicular to the base. Item 3 is a banding strap to hold the
specimen firmly on the base. Item 4 is to be attached to the base after
in situ removal or after laboratory fabrication of specimens. It is
also used to rotate the specimen 90° about its longitudinal axis.
Notice the two hollow steel tubular sections attached to items 1 and 4.
These features provide an easy method for using a forklift truck to
handle the specimen. After the specimen is rotated, item 4 is used to
set the specimen onto the test machine (Fig. 15.). After alignment all
parts of the fixture are removed except the rods.

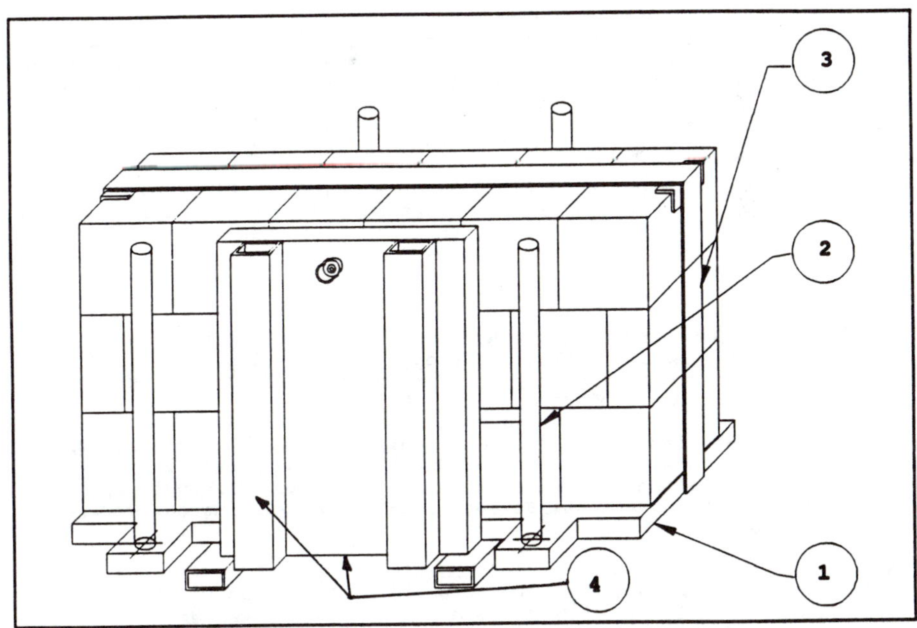

Fig. 14. A schematic of a 3-ft by 6-ft (914 mm by 1 829 mm) flexural
bond specimen prepared for shipping and for testing parallel to the bed
joints.

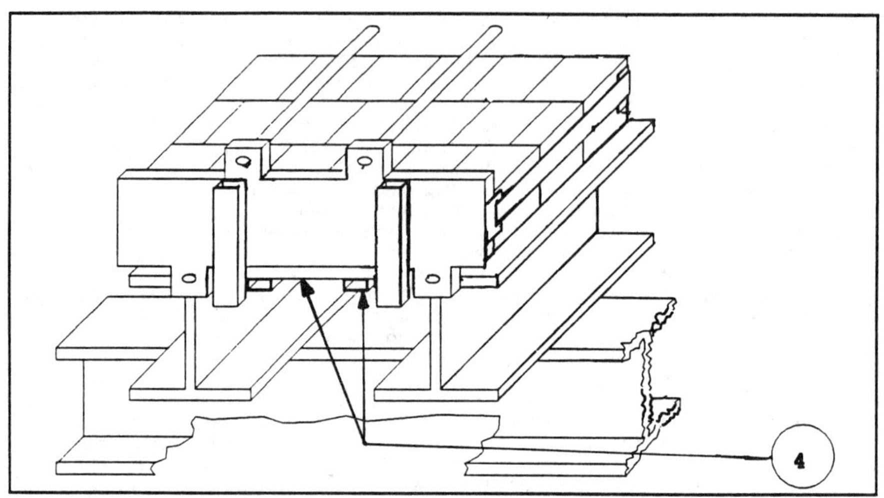

Fig. 15. The flexural bond specimen in Fig. 14 is shown here rotated 90°
and placed onto the test fixture.

CONCLUSIONS

Fixtures and procedures have been developed for removing and for
testing in situ (removed) prism compression and flexural bond specimens
that will enhance the ability to collect consistent and reliable test
data. These fixtures and procedures can also be used to fabricate and
test laboratory-built specimens, thus making it possible to better
correlate the test data from the two types of specimens.

The fixtures prevent damage during removal and shipping, and
provide features for controlling fabrication and capping tolerances much
better than do the current state-of-the-art procedures.

REFERENCES

[1]   Bennett, R.M. and Flanagan R.D., "Structural Testing of Clay Tile
       Walls: Testing and Evaluation Program," Y/EN-4325, Center for
       Natural Phenomena Engineering, Martin Marietta Energy Systems,
       Inc., Oak Ridge, Tennessee, May 1992.

[2]   Atkinson-Noland and Associates, Inc., "Preparation of Stackbond
       Masonry Prisms," Department of Civil, Environmental, and
       Architectural Engineering, University of Colorado, Boulder,
       Supported by: National Science Foundation, Washington, D.C.,
       May 1985.

[3]   Cavanagh, C., Webb, W.F. and Hodgkinson, H.R., "In-situ Lateral
       Load Tests on an Old Brickwork Building," British Ceramic Research
       Association Ltd.,Technical Note No. 345, December 1982.

[4]   Fricke, K.E., and Jones, W.D., "A Test Program to Determine the
       Structural Properties of Unreinforced Hollow Clay Tile Masonry
       Walls of the DOE Oak Ridge Plants," Masonry Society Journal
       9: 56-63.

[5]   Jones, W.D. and Hornyak, G.J., "Hollow Clay Tile Wall Test
        Program: Test Procedures for Removal and Handling of Test
        Specimens," Y/EN-4593, Center for Natural Phenomena Engineering,
        Martin Marietta Energy Systems, Inc., Oak Ridge, Tennessee,
        January 1992.

[6]   Fattal, S.G. and Cattaneo, L.E., "Evaluation of Structural
        Properties of Masonry in Existing Buildings," National Bureau of
        Standards, Washington, D.C., NBSIR 74-520, July 1974.

Strategies and Techniques

A. Rhett Whitlock[1], William B. Fairley[2], and Alan J. Izenman[3]

QUANTIFICATION OF MASONRY DETERIORATION THROUGH STATISTICAL MODELLING -- A CASE STUDY

REFERENCE: Whitlock, A. R., Fairley, W.B., and Izenman, A.J., "Quantification of Masonry Deterioration Through Statistical Modelling -- A Case Study," Masonry: Design & Construction, Problems & Repair, ASTM STP 1180, John M. Melander and Lynn R. Lauersdorf, Eds., American Society for Testing and Materials, Philadelphia, 1993.

ABSTRACT: Structures of all types experience physical damage due to numerous conditions or circumstances. Errors in design, defective materials, improper construction, or poor workmanship can lead to deterioration or failure of building structures or components. Natural aging processes, chemical attack, or the effects of natural disasters create damage. Determining the precise causes or mechanisms responsible for the condition is important and the procedures for doing so have been well researched and published. Also important, however, is quantifying the amount of damage. Quantification is important so that accurate estimates of repair costs can be established, budgeting for the repairs, and for scheduling.

This paper presents a case study concerning a condition survey of five nearly identical buildings constructed of fired clay masonry. Specifically, the surveying and estimation of spalled masonry units is discussed. In this case study, total damage is first assessed by means of two different and independent observational procedures, one of which suffers from incomplete data collection. Missing data are estimated using Poisson log-linear regression modelling of the relationship between damage assessments and a number of qualitative factors. The resulting two sets of predicted total damage estimates are then obtained through calibration techniques.

KEYWORDS: statistics, failure investigations, condition assessments, damage, quantity estimates, spall, regression, condition survey, modelling, calibration, deterioration, non-destructive evaluation (NDE)

Condition surveys are performed on existing structures to assess damage or deterioration. Several consensus documents [1 - 6] have been developed to aid engineers in the planning and execution of structural assessments (or failure investigations). Also, several papers have been published on this topic for masonry investigations [7 - 10]. All of these guidelines and papers focus on the methods of investigation, testing equipment and methods, and the identification of the causes of the problems -- all of which are useful, necessary, and very important;

---

[1]Senior Vice President, KCI Technologies, Inc., 8832 Rixlew Lane, Manassas, VA 22110.

[2]President, Analysis & Inference, Inc., P.O. Box 364, Swarthmore, PA 19081.

[3]Professor, Department of Statistics, Temple University, Philadelphia, PA, 19122.

however, for purposes of allocating funds for repair, estimating costs or establishing claims, total quantity estimates are crucial, yet this aspect of the process has been neglected in the literature.

Estimating damage quantities is often difficult because damage and deterioration extends beyond that which can be seen by the naked eye through visual surveys. It is necessary to couple visual surveys with some amount of physical testing to establish a relationship between observable damage and "true" damage. True damage is taken here to mean an amount, or quantity, of damage that can be stated with a reasonable degree of certainty.

The required level of effort depends on size of the structure, nature of deterioration, and types of material. For example, estimating the quantity of paint removal required as a result of vandalism in the form of graffiti is easily established by direct measurement, whereas, estimating the number and locations of severely corroded wall ties within a masonry wall requires considerable time and effort.

As every project is unique with respect to its problems, so is the optimum level of effort for a proper condition survey. Using adequate investigative techniques, tools, and analysis procedures, an optimum condition assessment will provide reliable data, increase the probability of quality repairs, and will increase the accuracy of cost estimates.

For large structures with relatively wide-spread damage, detailed testing and investigation to determine the precise amount of damage can be arduous and often cost prohibitive; therefore, visual assessments are commonly employed. Visual assessment is the most widely used and least costly non-destructive evaluation (NDE) technique available. Accurate qualitative information is provided through visual assessment, but quantitative estimates may be subject to subconscious bias, estimating errors, unobservable conditions, and oversight.

Just as situations or problems can suffer from the lack of sufficient study, they can be "studied to death." For example, it would be impractical to hammer-sound every square foot of a one million square foot parking structure to estimate corrosion related damage. Hence, visual assessments coupled with non-destructive testings, and in some cases semi-destructive testing, can provide data sufficient to establish correlations and statistically based estimates of total true damage.

This paper describes the application of a calibration procedure for estimating deterioration of a masonry structure.

PROJECT DESCRIPTION

The structures were five separate buildings located in The Bronx, New York. They were five to seven stories tall, with load bearing fired-clay tile (suitcase tile) exterior walls. Figure 1 is a map which shows the location of the five buildings.

The most significant type of damage on these buildings was spalling of the faces from the clay tile. It was generally agreed that the mechanism for the spalling was freeze/thaw yet there was disagreement regarding responsibility, which is not the subject of this paper. What became an important issue was <u>how much</u> spalling was present. Therefore, a detailed series of studies were undertaken to provide an estimate of total spalling.

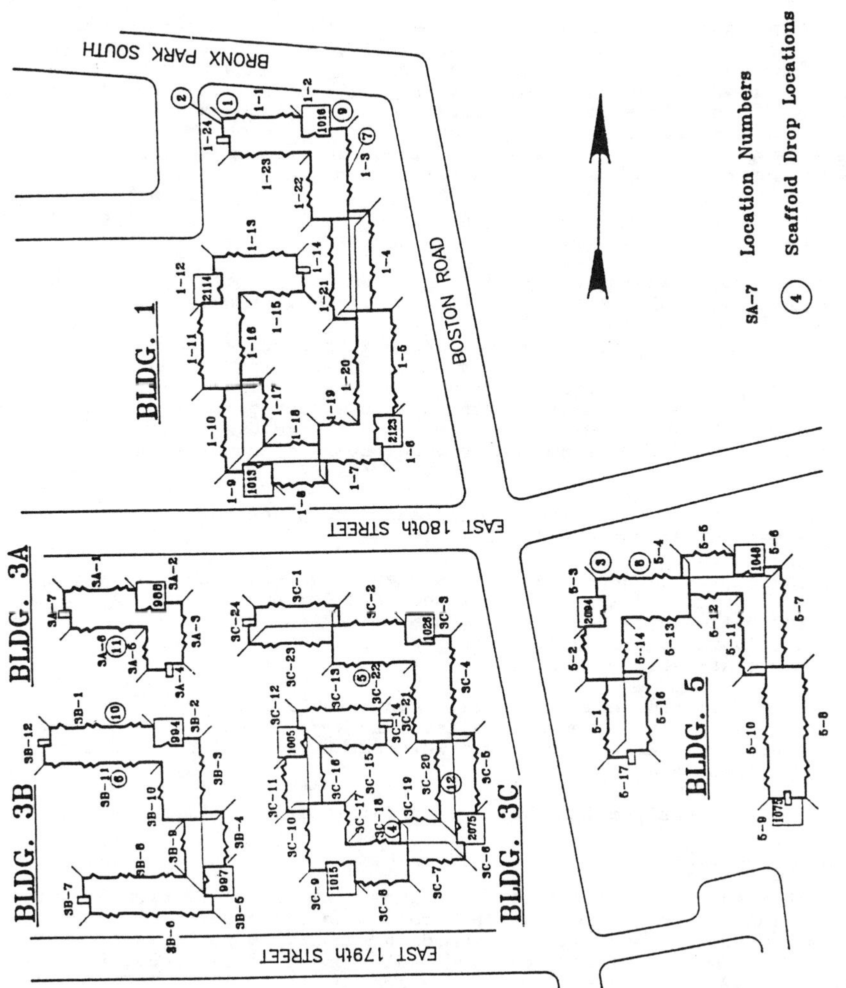

FIG. 1—Building Location Plan

FACADE SURVEYS

Three different types of condition surveys where performed in an effort to determine the amount of spalled (or spalling) tiles. These condition surveys consisted of counts from photographs, total visual counts made on site, and up-close scaffold drop survey counts.

Photographic Survey

A photographic survey was carried out to study the extent of spalling throughout the entire complex. This relatively inexpensive survey assessed spalling from large photographs of most of the 83 wall segments shown on Fig. 1. The photographs, however, also contained many tiles which could not be viewed sufficiently well to be judged as spalled or not spalled. These unobservable tiles (which resulted in extensive missing data) were due to either outside obstructions (such as trees, bushes, pipes, vehicles, etc.) or photo angles in photographing the wall segments. Figure 2 is a photograph of a typical wall elevation with notations regarding the count boundaries. Figure 3 is a photograph of a spalled tile just below a precast parapet coping.

The photographic survey was conducted with carefully selected definitions and rules for counting. The procedures were:

1.   Spall and tile counting boundaries in each photograph were established and defined such that: a) counting boundaries would terminate at edges or corners of buildings, and b) selected break points were chosen where more than one photograph was required to document a particular wall elevation. Care was exercised so that count boundaries from two or more separate photographs did not overlap.

2.   All count boundaries were delineated by black vertical lines marked on an acetate overlay (see Fig. 2).

3.   There were 24 separate count segments representing Building 1; 7 segments for Building 3A; 12 segments for Building 3B; 24 segments for Building 3C; and 16 segments for Building 5. Segment No. 15 for Building 5 does not exist due to a numbering error. In all, there were 83 count segments.

4.   To facilitate an accurate tile count, "Observable" and "Unobservable" tile were defined and were denoted on the acetate overlays. Unobservable tile were hatched in color on the acetate overlays (see Fig. 2).

The results of the photo survey are contained in Table 1.

Visual On-Site Survey

This survey consisted of an unaided examination of each surface of each building from ground level and recording the number of apparent spalls. This procedure was performed in one day by three inspectors. The findings were recorded for each wall segment of each building. A total of 1532 spalls was recorded and are contained in Table 1. Comparison of this amount to the photo survey total of 670 illustrates the wide disparity among the two procedures.

FIG. 2--Typical Wall Elevation with Counting Boundaries

FIG. 3--Spalled Tile at Parapet Coping

TABLE 1--<u>Photo and Visual Spall Counts and Rates by Building Location</u>

Bldg.- Loc.	Visual Spalls	No. of Units	Photo Spalls	Photo Observable Units	Visual Rate (per 1000)	Photo Rate (per 1000)
1-1	166	8,524	123	6,829	19.5	18.0
1-2	48	7,387	35	5,837	6.5	6.0
1-3	39	11,078	8	6,234	3.5	1.3
1-4	55	12,708	2	7,660	4.3	0.3
1-5	64	12,145	55	11,184	5.3	4.9
1-6	7	8,570	4	3,593	0.8	1.1
1-7	10	9,586	0	6,943	1.0	0.0
1-8	5	5,223	1	5,096	1.0	0.2
1-9	3	6,507	0	2,465	0.5	0.0
1-10	7	9,406	*0	0	0.7	...
1-11	5	9,225	0	6,541	0.5	0.0
1-12	25	7,450	1	3,822	3.4	0.3
1-13	77	9,033	12	8,327	8.5	1.4
1-14	41	7,008	40	3,544	5.9	11.3
1-15	0	7,263	0	4,765	0.0	0.0
1-16	37	10,370	12	6,085	3.6	2.0
1-17	7	6,327	4	4,192	1.1	1.0
1-18	21	8,610	3	2,129	2.4	1.4
1-19	13	5,419	15	3,835	2.4	3.9
1-20	19	12,718	7	10,465	1.5	0.7
1-21	8	13,211	3	9,342	0.6	0.3
1-22	2	8,486	0	6,461	0.2	0.0
1-23	5	8,671	0	8,095	0.6	0.0
1-24	28	6,879	0	3,522	4.1	0.0
Total	692	211,804	325	136,966		
			Pooled	Bldg. Rate	3.3	2.4
3A-1	26	7,052	15	5,823	3.7	2.6
3A-2	17	7,387	7	5,486	2.3	1.3
3A-3	4	7,972	6	5,111	0.5	1.2
3A-4	1	7,569	2	3,363	0.1	0.6
3A-5	2	5,472	0	4,003	0.4	0.0
3A-6	16	8,363	4	6,976	1.9	0.6
3A-7	0	6,732	1	2,107	0.0	0.5
Total	66	50,547	35	32,869		
			Pooled	Bldg. Rate	1.3	1.1
3B-1	18	11,469	6	10,151	1.6	0.6
3B-2	3	7,195	2	5,599	0.4	0.4
3B-3	16	11,687	7	9,540	1.4	0.7
3B-4	20	6,231	18	4,426	3.2	4.1
3B-5	9	7,423	3	3,191	1.2	0.9
3B-6	5	13,185	4	11,289	0.4	0.4
3B-7	9	7,068	1	3,581	1.3	0.3
3B-8	46	11,161	1	7,540	4.1	0.1
3B-9	5	5,780	0	1,502	0.9	0.0
3B-10	15	6,339	6	3,845	2.4	1.6
3B-11	15	12,881	0	6,842	1.2	0.0
3B-12	0	5,775	1	3,113	0.0	0.3
Total	161	106,194	49	70,619		
			Pooled	Bldg. Rate	1.5	0.7

TABLE 1--<u>Photo and Visual Spall Counts and</u>
<u>Rates by Building Location</u>
(Continued)

Bldg.-Loc.	Visual Spalls	No. of Units	Photo Spalls	Photo Observable Units	Visual Rate (per 1000)	Photo Rate (per 1000)
3C-1	28	6,560	0	5,868	4.3	0.0
3C-2	15	9,167	0	5,045	1.6	0.0
3C-3	17	7,387	3	3,782	2.3	0.8
3C-4	21	12,666	3	10,305	1.7	0.3
3C-5	19	7,524	4	6,789	2.5	0.6
3C-6	11	6,297	0	0	1.7	...
3C-7	6	7,756	3	6,064	0.8	0.5
3C-8	1	6,040	2	5,641	0.2	0.4
3C-9	0	7,660	0	2,207	0.0	0.0
3C-10	0	8,341	0	7,810	0.0	0.0
3C-11	0	6,156	0	3,284	0.0	0.0
3C-12	4	6,237	1	2,779	0.6	0.4
3C-13	3	7,788	1	7,109	0.4	0.1
3C-14	1	6,115	0	2,768	0.2	0.0
3C-15	3	6,210	0	4,636	0.5	0.0
3C-16	0	7,480	0	3,694	0.0	0.0
3C-17	0	4,707	0	3,530	0.0	0.0
3C-18	15	7,738	9	4,050	1.9	2.2
3C-19	2	4,065	0	3,448	0.5	0.0
3C-20	2	9,616	0	5,894	0.2	0.0
3C-21	3	9,401	0	6,078	0.3	0.0
3C-22	0	9,123	2	7,780	0.0	0.3
3C-23	0	8,088	0	5,055	0.0	0.0
3C-24	5	4,809	0	2,481	1.0	0.0
Total	156	176,931	28	116,097		
			Pooled	Bldg. Rate	0.9	0.2
5-1	4	9,406	3	8,653	0.4	0.3
5-2	7	6,104	1	5,385	1.1	0.2
5-3	6	7,330	0	3,838	0.8	0.0
5-4	211	11,046	129	10,208	19.1	12.6
5-5	89	6,168	61	5,427	14.4	11.2
5-6	51	7,318	30	4,161	7.0	7.2
5-7	39	13,766	*	*	2.8	*
5-8	17	16,969	*	*	1.0	*
5-9	2	7,251	0	3,784	0.3	0.0
5-10	8	17,474	2	8,489	0.5	0.2
5-11	0	8,945	0	5,319	0.0	0.0
5-12	15	8,544	2	6,053	1.8	0.3
5-13	0	7,452	0	3,600	0.0	0.0
5-14	6	7,924	2	2,148	0.8	0.9
5-16	2	7,488	1	4,499	0.3	0.2
5-17	0	6,095	2	1,990	0.0	1.0
Total	457	149,280	233	73,554		
			Pooled	Bldg. Rate	3.1	3.2
Proj. Total	1,532	694,756	670	430,105		
			Pooled	Proj. Rate	2.2051	1.5578

*   The tiles in these photographs were not observable.

<u>Scaffold Drop Spall Survey</u>

The scaffold drop survey was performed to establish highly reliable data. The name "drop" comes from the procedure of lowering individuals in a scaffold from the roof parapet of the building to examine in great detail (every brick in every course) portions of a wall segment (called a "drop" location) for spall damage. Figure 4 is a photograph of one of the twelve drops and shows the inspectors performing their work. The wall surveys at each drop location were performed identically. The load capacity of the scaffolding was limited to three individuals, two inspectors and one mechanic. The surveys began at the parapet elevation at each location. Each inspector was responsible for approximately 50% of the wall area at a given drop location. The survey was conducted by systematically progressing laterally left to right or right to left across the wall survey area observing each tile in every course, course by course, and floor by floor. Specially prepared data collection forms were used by each inspector. These forms illustrated the horizontal limits, floor line, and wall elevations for each of the twelve (12) drop locations. The inspectors noted physical conditions including spalls by symbols on the data collection forms.

FIG. 4--Photograph of Inspectors at a Scaffold Drop Location

The procedures for checking were methodically repeated at each drop location to ensure that a high level of reliability for agreement between the inspectors' data collection procedures and results was maintained. Non-destructive and slightly-destructive evaluations were performed to identify spalling. Slightly destructive is a term used to describe tests which cause relatively minor and repairable damage to a structure or component of a structure.

Typically, the inspectors' survey area and data collection overlapped at the midpoint of each drop. The overlap in data provided a limited form of double-checking. The inspectors would often switch sides on the scaffolding, re-survey a wall area, and double-check their

results against the other's data. The inspection forms were compared
and double-checked by the Principal Investigator on a daily basis.

    Occasionally the same wall area would be surveyed more than once.
For example, in a re-survey, the inspectors would begin at the midpoint
of the drop area and survey the wall in opposite directions.
Additionally, the inspectors would often re-survey a particular wall
area after returning from a break or in some cases the following day.
This systematic procedure of inspector's self checks, checking one
another, and close supervision by the principal investigator ensured
consistency and accuracy of data collection.

    Although this survey was very precise, it was expensive and time
consuming to conduct, and so the number of drop locations was limited to
12. The drop locations were selected to cover all types of wall
conditions in the complex, but were not chosen to be representative of
spall rates in the entire complex; indeed, the average spall rate (per
1000 bricks) in the 12 drop locations is considerably higher than that
in the entire complex (using data from either the photo or visual
surveys).

    As a measure of the relative expense of carrying out these
surveys, for every $1 of visual survey cost, the photo survey cost $10
and the scaffold survey cost $100. All three types of surveys are
common techniques in assessing damage, although it is not common that
all three be carried out to this extent within the same project.

    The scaffold survey permitted detailed inspection, sounding, and
exploratory investigation. Thus, the spall data obtained on the drops
are considered to be highly accurate.

<u>Gross Spall Rates</u>

    Based upon the results of the three procedures for observation the
gross spall rates were:

Spalls per Thousand Tile

Photo Survey	1.56
Visual Survey	2.21
Scaffold Survey	16.04

The above spall rates are based on the total observed spalls divided by
the total tile within the count area. The photo and visual surveys
display low spall rates because they do not include spalls that could
not be visually detected. The scaffold survey rate is high because the
areas chosen tended to be more severely damaged than an average area,
and practically all spalls within the area were identified -- even those
that were not detectable through close visual inspection.

The actual spall rates varied spatially, that is, orientation and height
above ground level were factors that affected spalling. North and East
elevations displayed more spalls than West and South, and locations near
the tops of the buildings displayed the highest concentration of spalls;
obviously, these conditions correlate to colder and wetter areas.

STATISTICAL MODELLING

<u>Photographic Model</u>

    The photographic counts were incomplete in the sense that
obstructions such as trees, cars, and other objects rendered some tiles
unobservable. Similarly, some photographs were taken at acute angles

making the determination of spalls impossible. To correct for the
missing data, it would be very simple to take the number of spalls
observed, divide by the number of observable tiles, and multiply by the
total tiles. This approach would yield 670/429,422 x 694,756 = 1084
spalls. This procedure may be inaccurate because it assumes that the
missing data have the same spatial attributes (factors) as do the
visible tiles. It could very well be that the observable and
unobservable areas were randomly distributed, or on the other hand, they
could have been located in areas which displayed either high or low
spall rates. Therefore, the data required correction (weighting), based
upon the locations of the missing observations.

To better assess this, a generalized linear Poisson regression
model [11,12] was employed using the unobstructed spall count data. The
fitted model was then used to estimate the number of spalls that would
likely be visible if all areas had been photographed. The spatial
factors are illustrated in Fig. 5 and were:

1.   Building Number: This factor had five levels corresponding to the
      five buildings. It was known from observation that buildings 1
      and 5 were more heavily spalled than the other buildings.

2.   Orientation: This factor had four levels corresponding to whether
      the wall segment faced north, east, south, or west. It was known,
      again, that the north and east orientations were more heavily
      spalled.

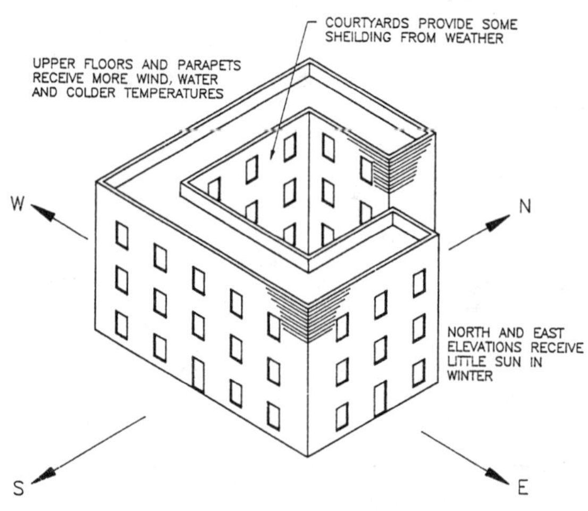

FIG. 5--Spatial Factors Which Affect Spalling

3.   Exposure: This factor had two levels corresponding to whether the
      wall segment was interior-facing (protected within courtyards and
      somewhat shielded) or exterior facing. Due to weather conditions,
      it was expected that exterior-facing wall segments might be more
      heavily spalled.

4.   Level: This factor had five, six, seven, or eight levels at each
        wall segment depending upon the number of floor-levels down from
        the roof parapet to the ground. Of the 83 wall segments, 1 had 5
        levels, 12 had 6 levels, 47 had 7 levels, and 23 had 8 levels. It
        was known from observation that the higher levels of the wall
        segments, such as top floors and parapets, were more heavily
        spalled than the lower levels.

The results of this model yielded a predicted Photo Spall count of
832 which is considerably less than previously predicted by simple ratio
and proportion. The corrected <u>overall</u> spall rate for the photo count is
832/694,756 x 1000 = 1.20.

Total Spall Prediction Through Calibration

The final step in this procedure is to compare the Photographic
and Visual counts to the Scaffold Drop counts. Table 2 contains the
results of the Drop, Photo, and Visual counts, at each drop location.
Linear regressions (see Figs. 6 and 7) were performed to predict
scaffold spall rates based upon both Visual and Photo counts. The
results were:

Predicted Spall Rate = 2.548 + 2.772(X), based on the Photo counts,(1)
and
Predicted Spall Rate = 2.038 + 1.757(X), based on the Visual counts(2)

In equation 1 and 2, X is the "observed" spall rate at any location.

Based on the overall rates of 1.20 and 2.205 for the entire project
based on Photo and Visual counts, respectively, the predicted "true"
rate per thousand would be:

2.548 + 2.772 (1.20) = 5.87 per thousand (based on the Photo survey)
2.038 + 1.757 (2.205) = 5.91 per thousand (based on the Visual survey)

Since the overall project contained about 700,000 tile, the two methods
yield a predicted total of 4100 spalled tile.

CONCLUSION

Prediction of total damage in terms of both quantity and costs can
be greatly enhanced using proper data gathering techniques along with
statistical modelling. Visual condition assessments alone may be
sufficient to identify the symptoms and, possibly, causes of problems,
but only rarely will they suffice for dependable quantity estimation.
There are many methods of statistical analysis that can be applied to
calibration and damage assessment; however, the models presented herein
are easily applied and should be applicable to damage or deterioration
estimation for all types of materials and structures.

The results clearly show that visual assessment alone can lead to
significant errors in estimating total damage, and thus calibration of
visual condition assessments with appropriate types and amounts of non-
destructive or slightly-destructive investigation is tremendously
important towards quantity estimation of total damage.

TABLE 2--<u>Drop, Photo, and Visual Counts and Rates of Spalls</u>

Drop No.	Drop Spall Count	Photo Spall Count	Visual Spall Count	Drop Tile Count	Drop Spall Rate (per 1000)	Photo Spall Rate (per 1000)	Visual Spall Rate (per 1000)
1	48	18	29	1,198	40.07	15.03	24.21
2	13	0	0	1,936	6.71	0.00	0.00
3	65	18	21	1,497	43.42	12.02	14.03
4	1	*	2	1,353	0.74	*	1.48
5	1	1	0	1,146	0.87	0.87	0.00
6	6	0	4	1,172	5.12	0.00	3.41
7	19	3	4	1,655	11.48	1.81	2.42
8	23	8	20	1,038	22.16	7.71	19.27
9	75	26	28	1,860	40.32	13.98	15.05
10	4	1	3	1,014	3.94	0.99	2.96
11	2	1	4	1,034	1.93	0.97	3.87
12	0	0	0	1,121	0.00	0.00	0.00
Total	257	76	115	16,024			
			Pooled	Drop Rates	16.04	4.74	7.18

* No photo data for this drop due to poor photograph.

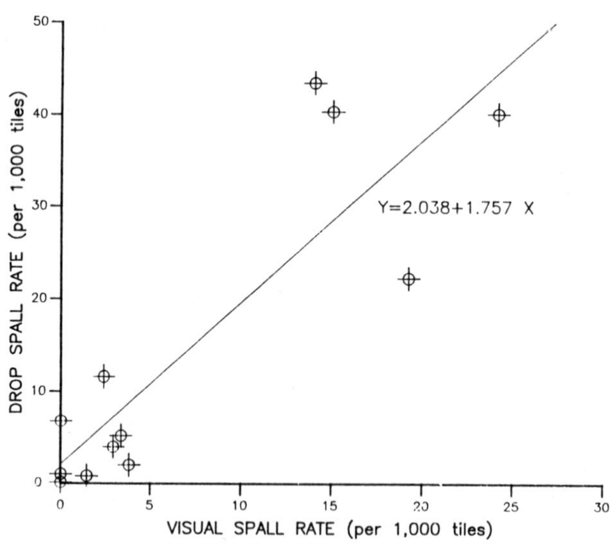

FIG. 6--<u>Calibration of Visual to Drop Spall Rates</u>

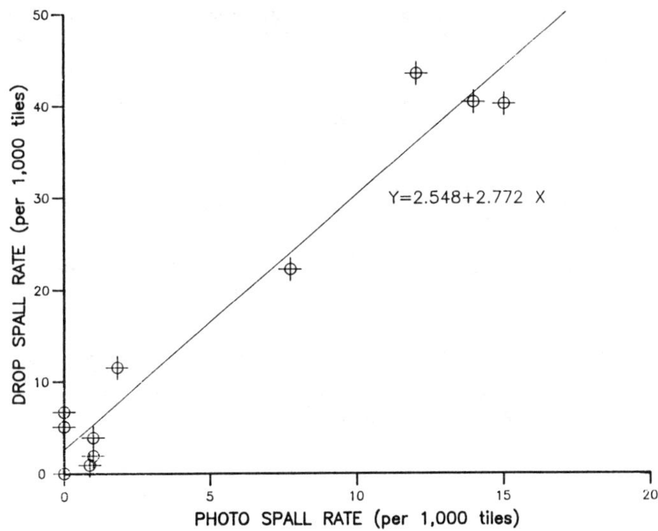

FIG. 7--Calibration of Photo to Drop Spall Rates

REFERENCES

[1]   "Guide for Making a Condition Survey of Concrete in Service," ACI
        201.1R-68, Revised 1984.

[2]   "Guide for Making a Condition Survey of Concrete in Service," ACI
        201.3R, 1986.

[3]   "Guidelines for Failure Investigation, ASCE Task Committee on
        Guidelines for Failure Investigation," American Society of Civil
        Engineers, 1989.

[4]   "Guideline for Structural Condition Assessment of Existing
        Buildings," ASCE 11-90, American Society of Civil Engineers, 1990.

[5]   "A Guide to Forensic Engineering and Service as an Expert
        Witness," Association of Engineering Firms Practicing in the
        Geosciences, 1987.

[6]   "Guidelines for the Professional Engineer as a Forensic Engineer,"
        National Society of Professional Engineers, Alexandria, Virginia.

[7]   Ali, M. M., "Failure Investigation of Masonry Structures,"
        Proceedings of the Second Canadian Masonry Symposium, Ottawa,
        Canada, June 1980, pp 481-491.

[8]   Monk, Jr., C.B., "Masonry Facade and Paving Failures," <u>Proceedings of the Second Canadian Masonry Symposium</u>, Ottawa, Canada, June 1980, pp 469-480.

[9]   Zetlin, Lev, "Investigation, Litigation and Prevention of Failures," <u>Fourth Canadian Masonry Symposium</u>, 1986.

[10]   Grimm, C.T., "Masonry Failure Investigations," <u>Masonry: Materials, Properties, and Performance, ASTM STP 778</u>. J.G. Borchelt, Ed., American Society for Testing and Materials, Philadelphia, 1982, pp. 245-260.

[11]   Numerical Algorithms Group (NAG) (1986), "The Generalized Linear Interactive Modelling System," Release 3.77. Oxford, U.K.: Royal Statistical Society.

[12]   Fairley, William B. and Izenman, Alan J., "Poisson Regression and Calibration: A Case Study of Masonry Deterioration in Building Structures," Analysis and Inference, Inc. and Temple University, Philadelphia, PA, April 1992.

Richard H. Atkinson[1], Michael P. Schuller[1]

EVALUATION OF INJECTABLE CEMENTITIOUS GROUTS FOR REPAIR AND RETROFIT OF MASONRY

REFERENCE: Atkinson, R.H., Schuller, M.P., **"Evaluation of Injectable Cementitious Grouts for Repair and Retrofit of Masonry,"** Masonry: Design and Construction, Problems and Repair, ASTM STP 1180, John M. Melander and Lynn R. Lauersdorf, Eds., American Society for Testing and Materials, Philadelphia, 1993.

ABSTRACT: Injection of grout into cracks may be used as a form of repair following a damaging event or as a means to strengthen multi-wythe masonry, enhancing composite action between the wythes. An experimental program has been conducted to evaluate grouting procedures, the suitability of different types of cementitious grouts for injection, and the effect of grout injection on structural behavior.

Over 30 separate mixes utilizing various cementitious components, aggregates, admixtures and different water/cement ratios have been evaluated. A standardized series of evaluation tests have been used to compare injectability, mix stability, and grout properties for each of the mixes. These grouts possess properties similar to the masonry being repaired, and would be available for widespread use as a material for masonry repair. An additional contribution of this research has been the application of nondestructive techniques as a means of measuring quality of the injection process.

KEYWORDS: masonry repair, grout injection, cementitious grout, admixtures, material behavior, nondestructive testing.

Unreinforced masonry buildings, many of historical and cultural importance, constitute a significant portion of the building inventory in the United States. Many of these buildings are structurally marginal for current use, suffering from the accumulated effects of inadequate construction techniques and materials, wind loadings, seismic forces, foundation settlements, and environmental deterioration. In addition to these factors, changed usage and more stringent seismic regulations have resulted in many masonry structures being designated as structurally deficient in their present condition. In California alone over 50,000 such buildings have been identified as being deficient and must, by state law, be brought into compliance with new regulations [1].

---

[1]Atkinson-Noland & Associates, Inc., 2619 Spruce Street, Boulder, Colorado, USA.

This paper provides details on a program investigating the efficacy of injection grouting techniques for repair and strengthening of old unreinforced masonry. The repair and strengthening of existing masonry buildings by injection grouting is a viable means to provide functional, durable, and safe structures without physically altering external aesthetics. The experimental and analytical program is part of a 3-year collaborative effort between the authors and P.B. Shing at the University of Colorado, L. Binda at the Politecnico di Milano, Milan, Italy, and P.P. Rossi at ISMES in Bergamo, Italy.

Injection of grout into cracks and voids in masonry has been used as a technique for repair or strengthening purposes, however there is little available information regarding the effect of grouting techniques on masonry behavior. The main purpose of this project is to provide an increased understanding of fundamental behavioral mechanics of older unreinforced masonry and to provide quantitative data on the effect of grouting on the strength and stiffness of old masonry. This particular paper reports on the development of cementitious grouts for masonry injection repair.

Several trial mixes were chosen based upon results of a grout evaluation program and used for repair and strengthening of a masonry test wall. Quantitative information regarding structural response before and after grout injection has been obtained through compressive loading of the specimen.

Nondestructive (NDE) techniques were used to monitor the quality of the grout injection process. Ultrasonic and mechanical pulse velocity techniques have proven useful for determination of grout penetration into cracks and voids. Results from these tests have shown a good correlation with injection quality and the effect of grout injection on masonry behavior.

BACKGROUND

Although injection of cementitious grout mixes for repair or retrofit of existing masonry has been applied by contractors and engineers on an individual job basis, very little information is available with which to select the optimum grout mix or to judge the efficiency of the repair. The City of Los Angeles has published a recommended procedure for repair of cracks in masonry walls detailing a specific mix design and grouting procedure [2]. Information regarding the basis for the mix design selection and on the adequacy of the resulting repair are not provided, however.

Considerably more work on this subject has been conducted in Europe. Paillere [3] has conducted an extensive study of grout mixes for injection into masonry including evaluation of cementitious materials and admixtures based upon various injectability criteria. Binda [4] has utilized grout mixes developed by Paillere to inject a number of masonry piers which were tested to quantify improvements in strength and stiffness. Binda also used ultrasonic NDE methods to evaluate grouting results. Tomazevic [5] has tested a limited number of full scale walls repaired by injection grouting.

While injection grouting of masonry is acknowledged as a potentially useful technique for the repair or retrofit of existing masonry the absence of published data on mix formulation and on the performance of the repaired masonry has made engineers and building officials hesitant to recommend or accept this method.

GROUT MIX EVALUATION

Grout Mix Components

Cementitious grouts were evaluated for use as possible materials for injection into older clay brick masonry. Old clay masonry is highly absorptive and can contain flaws in the form of small cracks or relatively large internal voids. The nature of old masonry thus requires a very fluid grout with fine particle size, both to penetrate into tiny flaws and to counteract absorption. At the same time the grout must be stable and resistant to segregation and shrinkage during initial set. It is also important for the grout to act as a binder, to tie together previously separated masonry components. Different types of cementitious materials, pozzolans, aggregate, and admixtures evaluated are described below.

Cementitious Materials--Portland cement was used as the main binding agent in all of the mixes. Portland cement, ASTM C-150 standard specifications for Portland cement, Types I, III, and a proprietary masonry cement, ASTM C-91 standard specification for Masonry Cement, were evaluated. Type I cement is readily available, relatively inexpensive, and commonly used in concrete construction. Type III cement is specially blended and more finely ground than Type I cement to produce a mix with a rapid strength gain. Masonry cement is sometimes used in masonry mortars and is ground very fine to provide a workable mortar. However it contains only 50% actual cementitious materials; the remaining 50% consists of a finely ground limestone filler. These different cements were chosen to represent a range of material fineness to determine the effect of particle size on grout properties. Approximate Blaine fineness values for the three cements used are as follows: Type I cement = 4000 cm^2/g; Type III cement = 5000 cm^2/g; masonry cement = 8000 cm^2/g.

Water--A grout mix with a high water/solids ratio is very fluid, has a low viscosity, and contains sufficient water to counteract the effects of rapid water loss during placement within highly absorptive masonry. However, grout mixes with excessive quantities of water are prone to rapid segregation during injection. Water/solids ratios ranging from 0.28 to 1.2 were evaluated to determine the optimum water content for injectable grouts.

Pozzolans--Pozzolanic fly ash and microsilica were used in some of the mixes to investigate their effect on mix properties. Pozzolans are normally used as a substitute for cementitious binders and have the advantage of having extremely fine particle sizes. An ASTM C-618 Class F fly ash was used in quantities ranging from 8% to 25% by weight of total solids. This particular fly ash contains 56% reactive silica, 24% aluminates, and has a particulate size approximately equal to that of normal cement. The microsilica additive used is a proprietary mixture formulated by W.R. Grace & Company and consists of extremely fine silica fume, dispersing agents and water in slurry form. The microsilica additive was added in quantities of 10% to 50% by weight of total solids. Microsilica is an extremely small particulate, with an average diameter of 100 times finer than normal cement particles, and consists of between 92% and 98% reactive silica.

The reactive nature of fly ash and microsilica requires the presence of free calcium hydroxide to produce calcium silicate hydrate binders. A portion of the required calcium hydroxide is produced by hydration of the cement within the grout; additional calcium hydroxide in the form of ASTM C-207 standard specification for Hydrated Lime for Masonry purposes Type S hydrated lime was added to some of the mixtures to ensure complete reaction of all pozzolanic materials. The use of lime by itself as a stabilizing agent was also investigated. Lime quantities ranging from 5% to 33% by weight of total solids were used.

   Aggregate--Aggregate in the form of a fine silica sand was used in mixes intended for injection of larger voids and cavities. The addition of sand helped to control grout shrinkage and resulted in a more economical mix, however the size of injectable cracks was limited by the size of the aggregate. Number 70 (0.212 mm maximum particle size) silica sand was chosen for use and is similar to what is recommended by the City of Los Angeles [2]. The fine sand aggregate was used in quantities ranging from 55% to 77% of total mix volume.

   Admixtures--A superplasticizer was used in nearly all of the mixtures as a fluidizing agent and helped to disperse particles of cement and other fines. The superplasticizer used during this project was manufactured by Prokrete, has a modified napthalene sulphonate formaldehyde base, and was used in percentages up to 2.33% by weight of cement.
   One problem inherent to grouting of masonry is severe plastic volume change as excess water within the grout is rapidly absorbed by the surrounding masonry. Expansive admixtures have been quite useful for reducing volume change in plastic masonry grouts with a high water content. Two types of expansive admixtures were considered in this study, both of which are manufactured by the Sika Corporation. Grout-Aid is formulated for use with masonry grouts, whereas Intraplast-N is intended for use with normal concretes and grouts. The main reactive constituent of each admixture is a fine aluminum powder which reacts with the alkaline cement paste to form tiny bubbles of hydrogen gas, offsetting total volume reduction due to water loss. Expanding agents were used in quantities of 1% by weight of cement.

Evaluation of Grout Mix Components

   Grout injectability is recognized as an important property, however it is not sufficient for injection grouts to simply fill voids within the masonry. The hardened grout must also act as a binder to tie together portions of masonry that were previously separated by cracks or voids. Cementitious grouts chosen for injection must be compatible with the masonry being injected, possess adequate compressive and tensile strengths and, most importantly, must bond well to the surrounding masonry.
   A standard program for specimen preparation, grout mixing, and specimen injection has been developed and used throughout this project to provide a consistent base for evaluation of grout mixtures. Standardized tests were used whenever possible to evaluate various aspects of injectable grout mixes so that a rational assessment of the relative merits of mix constituents could be undertaken. Parameters evaluated and the tests used during evaluation are summarized below.

   Injectability--This parameter encompasses many different characteristics, including viscosity, particle size, fluidity, and cohesion, but basically implies the suitability of a given mix to flow without segregation within small cracks and voids. Several tests, as shown in Figure 1, were used to provide a relative measure of injectability:

* The amount of time taken for a measured quantity of grout to flow through the Marsh Funnel viscometer provides a measure of viscosity. The test was conducted in accordance with API Recommended Practice 13 (b) [6] as shown in Figure 1 (a), is very simple and quick, and provides a reasonable measure of grout fluidity.

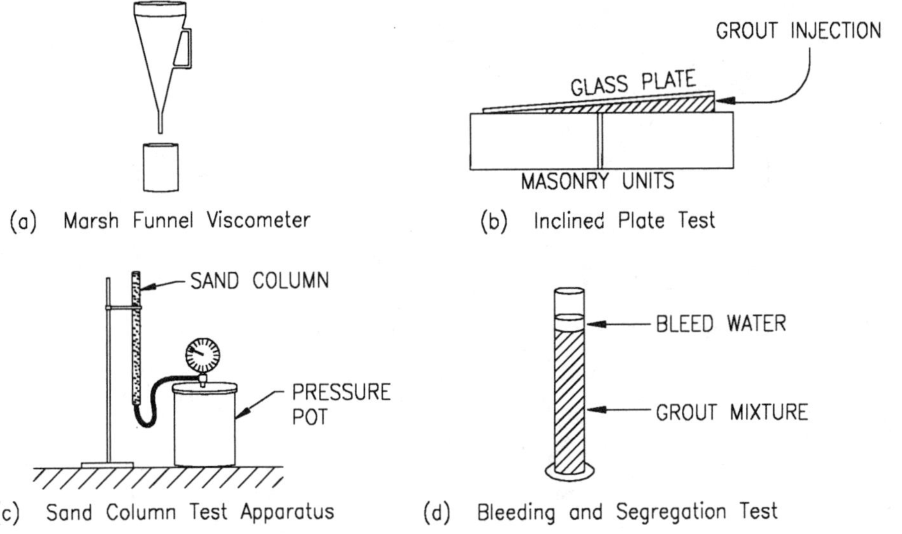

Figure 1--Apparatus used for evaluation of grout mix fluidity,
injectability, and stability.

- The inclined plate test, first developed by Hooton and Konecny [7], utilizes a sloping glass plate to simulate a converging crack. The glass plate assembly is clamped to brick masonry units to expose the grout to absorptive masonry units as shown in Figure 1 (b). The "crack" opening is 3 mm (0.125 inch) at the injection port, with the plate sloping to a zero crack width at the opposite end. The minimum penetrable crack opening is calculated based upon the observed grout penetration.
- Measurement of grout penetration into a finely graded sand mixture provides a measure of the ability to penetrate small voids. Grout was injected into a sand column apparatus, described in French standard NF P 18-891 [8] and shown in Figure 1 (c), using a normal injection pressure of 0.5 to 0.7 atm (8 to 10 psi). The sand gradation was chosen to simulate an average void size of 0.2 to 0.4 mm (0.008 to 0.016 inch).

Stability--Mix stability implies the ability of the mixture to remain homogeneous without continual mixing. "Unstable" mixtures exhibited rapid settlement and segregation of the constituents during and after injection, leading to inferior penetration and incomplete filling of voids during grouting. Grout mix stability for the first 60 minutes following mixing was measured for each of the trial mixes using ASTM Test Method C 940-87, "Test Method for Expansion and Bleeding of Freshly Mixed Grouts for Preplaced-Aggregate Concrete in the Laboratory," for bleeding and expansion of grouts, as shown in Figure 1 (d).

Material Properties--Specimens for grout tensile splitting and compressive tests were obtained by cutting segments from an injected column of sand. The bond strength between masonry units and injected grout was determined using the bond wrench, as described in ASTM Test

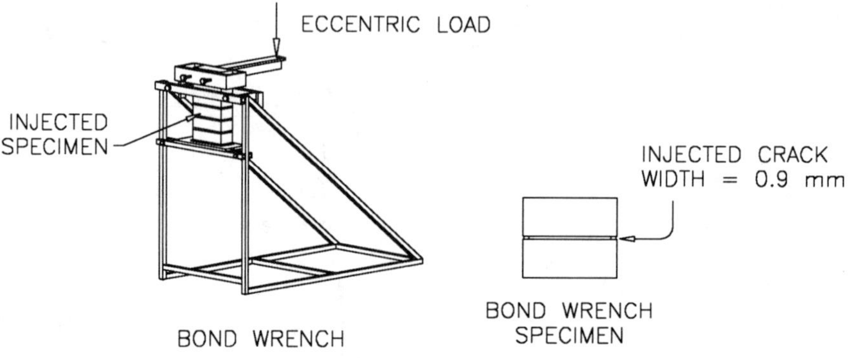

Figure 2--Bond wrench test apparatus and typical specimen for
            determination of bond strength of injected grout.

Method C 1072, "Method for Measurement of Masonry Flexural Bond
Strength." The bond wrench apparatus shown in Figure 2 was used to
subject an injected "crack" to flexural stresses, providing an
indication of flexural tensile bond strength. Specimens for bond wrench
testing consisted of a small stack of reclaimed molded clay brick units
(circa 1910), each separated from the adjacent unit with four small
metal spacers. In this way, a constant and uniform "crack" was
simulated for injection. A crack opening of 0.9 mm (0.035 inch) was
used throughout the majority of this study; this opening was increased
to 2.4 mm (0.095 inch) for some of the grout mixes containing
aggregates. Eight bond wrench specimens were injected for each grout
mix and tested at an age of 28 days.

     Guidelines for specimen cleaning, saturation, and injection
summarized here are based upon recommendations provided by the City of
Los Angeles [2] for injection of cementitious grouts into masonry.
Grout was injected into the simulated crack of the bond wrench specimens
using either surface mounted or drilled injection ports. Surface
mounted ports were most convenient, yet could only be used on relatively
wide cracks. A hollow-core drill bit with vacuum chuck attachment is
recommended for drilling ports into finer cracks: the vacuum removes
dust and cuttings during drilling rather than forcing these materials
into the crack opening. Cracks were sealed at the surface of the
masonry using either a vinyl spackle paste or polyester resin epoxy.
The masonry surface was sealed on some of the specimens with a
commercial masonry water sealant prior to grouting to aid in cleanup of
spilled grout and prevent staining of the masonry. Immediately prior to
grout injection, the crack was flushed with water at low pressure to
remove any remaining debris and saturate the surface of the crack.
Grout mixing was conducted using a high-shear mixer operating at
approximately 3500 rpm for 3 minutes, followed by additional high-shear
mixing at 10-minute intervals until the grouting operation was
completed. Grout was injected into the specimens at a pressure of 0.5
to 0.7 atm (8 to 10 psi); ports which flowed freely were capped and
pressure maintained until grout would no longer flow through the crack.
Specimens were cured in air for 28 days, then prior to bond wrench
testing the crack sealants were removed: vinyl spackle was removed with
water and a fiber brush; polyester resin epoxy required softening by

heat application with a torch prior to removal with a scraper. The
cleaned specimens were allowed to dry for at least one day before
testing.

Results of Grout Mix Evaluation

Over 30 separate combinations of cementitious materials,
pozzolans, aggregate, and admixtures for varying water/cement ratios
were evaluated. Results from several grout mixes, selected to provide a
representation of the variation in results for each of the components
investigated, are presented here.

The plots shown in Figures 3 and 4 show the effect of varying
grout mix parameters on Marsh funnel time, sand column penetration and
time, percent bleeding, crack penetration, tensile splitting strength,
and bond strength for 10 of the grout mixes. A mix containing only Type
I Portland cement with a water/cement (w/c) ratio of 0.75 is included in
each of the plots as the baseline for comparison: all other mixes
contain additional items as listed.

Viscosity, as measured by the Marsh funnel viscometer, is shown in
Figure 3 (a). An increase in the w/c ratio or the addition of
superplasticizers both served to reduce viscosity. In general, the
addition of ultra-fines such as lime, fly ash, and microsilica had the
effect of increasing viscosity because an increased quantity of water
was required to fully coat the very fine particles. It is important to
note that all mixes containing ultra-fines required the addition of
superplasticizer to permit flow through the funnel within a reasonable
time. Addition of Grout-Aid expanding admixture also increased the
viscosity of the grout mix. Grout mixes containing sand aggregate were
not able to flow through the Marsh funnel. The time for flow of water
through the Marsh funnel was measured at approximately 28 seconds; this
can be considered the practical lower limit for Marsh funnel time and is
shown in Figure 3 (a) as a dashed line.

Sand column penetration was measured for each of the mixes as the
amount of time required to penetrate the full 36 cm height of the column
(Figure 3 (b)). For those mixes where full penetration was impossible,
the maximum height penetrated is listed. The results shown in this plot
show the same general trends as the Marsh funnel results. Mixes
containing ultra-fines were expected to be more injectable due to their
smaller particle size, however most mixes with ultra-fines were unable
to penetrate into the sand column due to a large increase in viscosity.
Mixes utilizing plain cement and water, or with the addition of
superplasticizers, proceed quite rapidly through the column. It appears
as if the base w/c ratio of 0.75 utilized for these mixes must be
increased when ultra-fines are to be used to provide proper fluidity for
injection. Grout mixes using sand aggregate were not able to penetrate
the sand column.

The addition of ultra-fines tended to increase viscosity, however
use of these materials provided a definite positive effect on bleeding
and segregation as shown in Figure 3 (c). Ultra-fines increased the
overall water demand of the mix, and, for the most part, reduce bleeding
to negligible levels. The addition of Grout-Aid had the same effect of
reducing bleeding, and it is theorized that formation of numerous tiny
bubbles during the expansion process may serve to impede downward
migration of the particles. It should also be noted that, in addition
to reducing bleeding, Grout-Aid produced an expansion of over 9%, as
measured in an unconfined container. The majority of the expansion
occurred between 15 and 60 minutes following mixing. Mix 28, containing
Portland cement, lime, fly ash, and a fine #70 (0.212 mm maximum
particle size) sand aggregate appears to be well graded and displayed
almost no segregation during the test.

The glass plate test provided a comparison of the minimum
penetrable crack opening for the different grout mixes as shown in
Figure 3 (d). Results from this test show that both fluidity and
particle size had an effect on crack penetration. In general, those

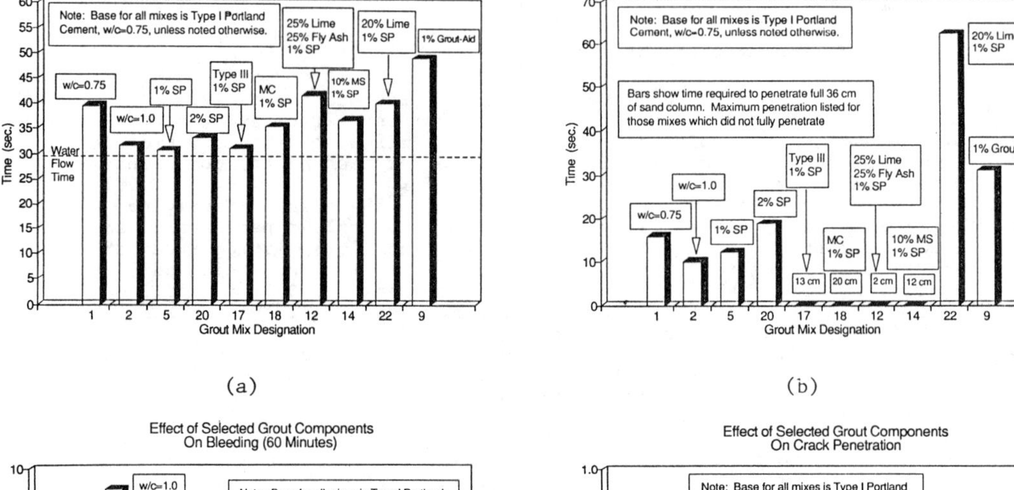

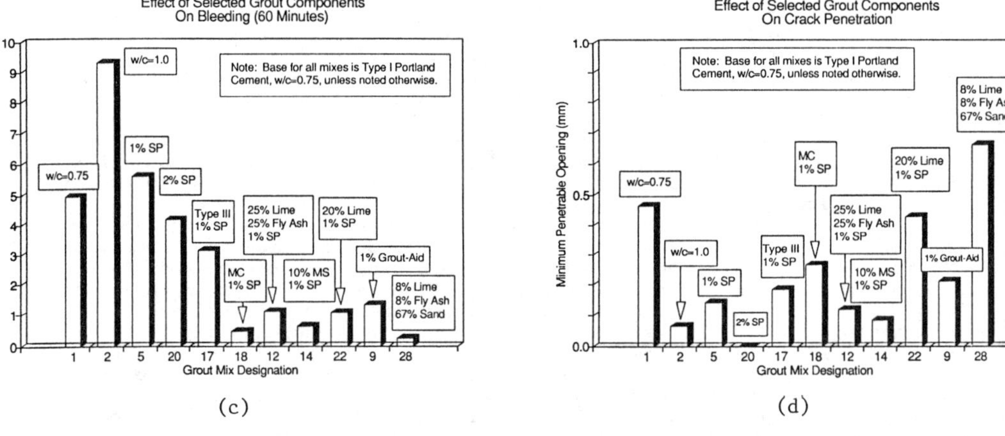

Figure 3--Representative results showing the effect of various grout
components on: (a) Marsh funnel time; (b) sand column
penetration; (c) bleeding and segregation; (d) crack
width penetration. Note: SP = superplasticizer, MC =
masonry cement, MS = microsilica.

mixes which were more fluid were also able to penetrate further into th
simulated crack. Increasing fluidity by increasing the w/c ratio or
adding superplasticizer greatly reduced the opening which was penetrate
by the grout. Addition of ultra-fines had the same effect of providing
a grout with enhanced penetration capabilities. The majority of the
mixes were able to penetrate cracks on the order of 0.1 to 0.2 mm (0.00
to 0.008 inch). The addition of fine sands to mix 28 increased the
average particle size, which in turn greatly increased the minimum
penetrable crack. This mix was able to penetrate an opening with a
width of only 0.7 mm (0.03 inch).

Cylindrical specimens for tensile split tests were obtained by
cutting segments from the injected sand column. Results from this
testing (Figure 4 (a)) indicate that in addition to the type of
cementitious materials used, overall mix injectability also had an
effect on tensile strength. Fluid mixtures were able to fully penetrat
spaces between sand grains, producing a dense column with enhanced
tensile strength. Microsilica was especially effective at forming
strong bonds during the hydration process and had the greatest tensile

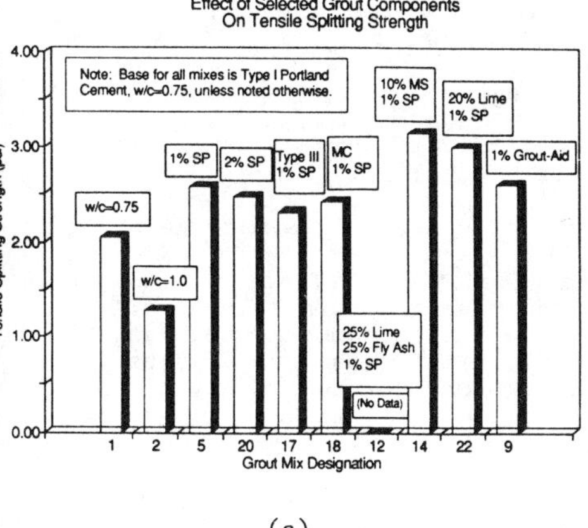

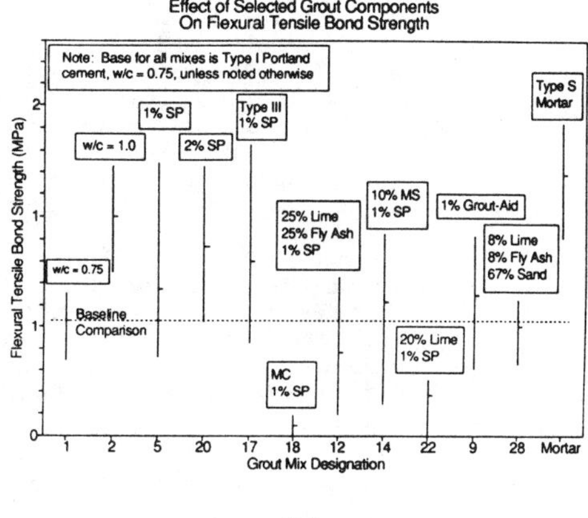

(a)   (b)

Figure 4--Representative results showing the effect of various grout components on: (a) tensile splitting strength; (b) bond strength. Note: SP = superplasticizer, MC = masonry cement, MS = microsilica.

strength of these mixes. Expansion of the Grout-Aid mixture was effective in offsetting normal shrinkage and provided an increase in tensile strength of approximately 25% over the base mix.

The data in Figure 4 (b) shows a similar effect, where both injectability and mix components have an effect on the bond strength at the grout/masonry interface. This plot shows the range in bond strength values for each mix, in addition to the mean value. Bond strength values for a modern type S mortar and the same old masonry units are included in this plot for comparison. Injected specimens show bond strengths from 0.7 to 1.0 MPa (100 to 150 psi), whereas the average bond strength of the specimens using type S mortar was 1.2 Mpa (175 psi).

It is interesting to note that the strength of bond between units and grout is not only a function of the mix components, but is also affected by overall mix injectability. Mix fluidity proved to have a direct effect on bond strength: the enhanced penetration capabilities of mixtures with high water content or with the addition of superplasticizers resulted in an increase in bond strength of 28 to 90 percent over the base mixture. Use of Type III cement also produced a mix with excellent bond characteristics, whereas bond strength of the masonry cement mix was nearly zero. Of the pozzolanic admixtures, only the microsilica additive provided an increase in bond strength; lime and fly ash proved to be especially poor. The addition of Grout-Aid expansive admixture was also beneficial, effectively counteracting plastic shrinkage of the injected grout. This mix had an increase in bond strength of 25% over the base mix.

Summary--The simple tests discussed above were effective at determining the effect of varying mix parameters on injectability, stability, and material properties of cementitious grouts for masonry injection. Highlights of the evaluation are summarized below.

- Increasing the water/solids ratio enhances the capability of the grout to penetrate small cracks and voids, however mixtures with a high water/solids ratio exhibited excessive segregation and bleeding.

- Mix stability can be improved with the addition of ultra-fines or water retaining agents: lime, fly ash, silica fume, and Grout-Aid all worked well to control bleeding and segregation.
- A water/cement ratio of 0.75 to 1.0 is optimal for a mix containing only Type I cement. Further evaluation is needed to determine the optimal w/c ratio for mixes containing other components.
- Type III cement and microsilica increased stability, tensile strength, and bond strength of the grout. Masonry cement, lime, and fly ash generally provided poor grouts.
- Addition of superplasticizer increased injectability of the mixture; a quantity of 2% by weight of cementitious materials is optimal.
- Addition of expansive admixtures had a beneficial effect on mix stability by reducing settlement and bleeding to nearly zero, with the added benefit that expansive action combats the high amount of plastic shrinkage encountered in mixes with a large w/c ratio, increasing the strength of bond to the masonry.

MASONRY COMPONENT TESTS

The second phase of this research effort is to quantify the effects of repair by injection grouting on overall masonry behavior. This phase will evaluate grout mix components and general injection procedures, provide information for analytical models, and allow refinement of nondestructive techniques.

A series of direct shear specimens and masonry piers will be tested. The general procedure will be to damage the specimens by an initial overload followed by injection grouting repair. The specimens will be loaded a second time following injection to determine the efficacy of the repair. Following these initial tests, a number of full-scale, multi-wythe shear wall specimens will undergo the same general procedure to determine the effect of injection grouting on masonry seismic behavior and resistance to lateral loads.

Preliminary results from this second phase study involved repair of a large-scale compressive specimen are shown in Figure 5. This particular specimen consisted of a 2-wythe masonry wall constructed of old clay masonry units and high-lime mortar with approximate dimensions of 1.52 m (5 feet) tall by 1.22 m (4 feet) wide. Compressive behavior in the original state was approximately linear up to the maximum applied stress of 200 psi. The wall was subjected to compressive overloads, which caused damage in the form of vertical cracks in addition to vertical and horizontal cracking from in situ shear test reactions. A reduction in the initial compressive modulus and an initial stiffening behavior were observed during compression loading of the damaged wall. The test wall was repaired by injection grouting with a mix consisting of Type I cement, microsilica, superplasticizer, and Grout-Aid. Compressive behavior in the repaired state was noticeably improved, with an increase in initial compressive modulus, and marked reduction of the stiffening behavior. Injection grouting effectively restored the initial compressive modulus (measured as the secant modulus from zero to 200 psi) from a value of 68% of the original modulus in the damaged state, to 89% of the original modulus in the repaired state (an improvement of 31%), indicating that injection grouting can be used to restore damaged masonry to near its original condition. An even greater improvement in the initial tangent modulus was achieved. Repair by grout injection increased the initial tangent modulus of the damaged wall by 123%, effectively eliminating the stiffening behavior noted for the damaged wall.

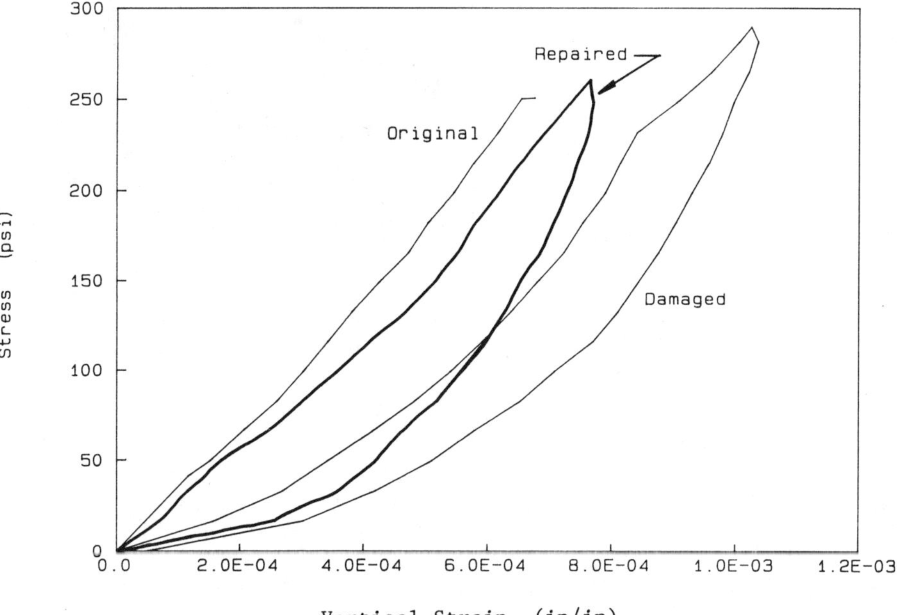

Figure 5--Compressive behavior of masonry test wall for the original and
damaged states, and following repair by grout injection
(145 psi = 1.0 MPa).

NONDESTRUCTIVE EVALUATION TECHNIQUES

Ultrasonic and mechanical pulse velocity techniques have been used
to identify flaws, cracks, and voids present within masonry [9] and as a
means of evaluating repair following grout injection [10]. A standard
procedure for measuring direct (through-wall) and indirect (surface)
horizontal and vertical pulse velocity was established during this
program to: (a) evaluate areas prior to injection for location of
cracks and voids; (b) determine the extent of grout penetration
following injection; and (c) provide a means to measure the efficacy of
the injection.

The general procedure involved measurement of pulse velocity in
the original (as-built) condition, in the damaged state, and following
repair by injection. Damage in the form of cracks and other flaws
hinder transmission of the stress wave, resulting in a reduction of the
measured pulse velocity. Filling of these cracks with grout during
injection effectively restored the velocity to its original value.
Results from these tests have shown a good correlation with both
injection quality and the measured structural effect of grout injection
on masonry behavior. The two-wythe masonry test wall described
previously provided a good initial assessment of the nondestructive
techniques. Horizontal and vertical indirect velocity tests showed
that, for the most part, the injection process was effective at filling
larger cracks. These tests were also consistent with visual
observations: locations where very fine cracks were not filled with
grout showed no improvement in ultrasonic pulse velocity.

The direct, or through-wall, technique proved to be an especially
effective indicator of injection quality for the multi-wythe wall. A
dense gridwork of through-wall measurements is shown in Figure 6 as a
three-dimensional surface plot where pulse arrival time is displayed on

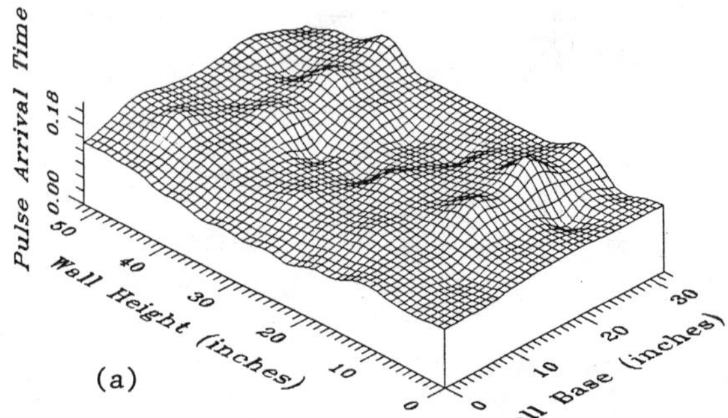

Wall 7, Through—Wall UPV, Original Condition

Figure 6--Three-dimensional surface plot of through-wall ultrasonic pulse arrival time: (a) original, as-built condition; (b) damaged condition; (c) following repair by injection of cementitious grout (1 inch = 25.4 mm).

Condition	Mean Pulse Arrival Time (microsec.)
Original	148
Damaged	357
Repaired	111

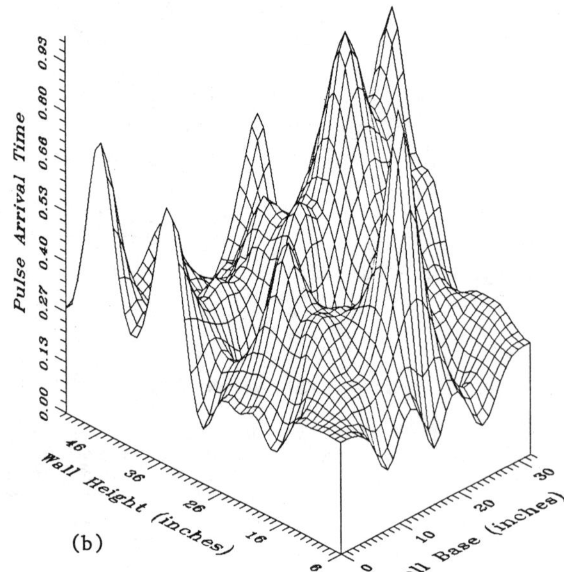

Wall 7, Through—Wall UPV, Damaged Condition

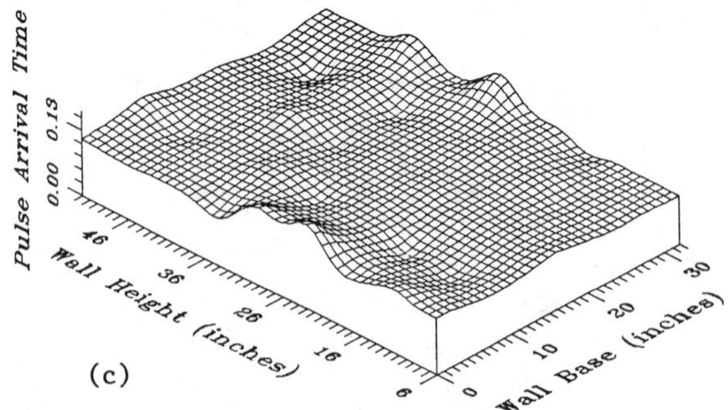

Wall 7, Through—Wall UPV, Repaired Condition

the vertical axis. A greater arrival time is depicted as a "peak" on
the plot, indicating lower pulse velocity and hence an area of reduced
masonry quality. Damage in the form of cracks and delaminations is
clearly evident in Figure 6 (b) when compared to the original condition
in Figure 6 (a). The condition of the wall was effectively restored by
grout injection as shown in Figure 6 (c). It is interesting to note
that not only was the grouting operation successful in repairing the
damage, but that an overall improvement over the as-built state was also
realized. The mean pulse arrival time in the repaired state is actually
25% less than the velocity in the original state, indicating that the
overall void area was reduced by the injection grouting process.
Evaluation of nondestructive techniques and in situ material tests will
continue to further validate their application as part of the masonry
repair process.

CONCLUSIONS AND RECOMMENDATIONS

Results from this study have shown that injection of cementitious
grouts is an effective technique for repair and retrofit of damaged or
deficient masonry. Careful formulation of the grout provided a mixture
with good flow properties, able to penetrate into crack widths as small
as 0.002 inch, and exhibiting little segregation. Material property
tests have indicated that the compressive, tensile, and bond strength of
grout injected into fine cracks is affected by overall mix
injectability, in addition to individual mix constituents. In general,
grouts which displayed better flow into cracks and voids also displayed
greater bond strengths. The use of nondestructive techniques are
recommended during the repair process to verify grout penetration and to
provide information on the effectiveness of the repair. Additional
tests will be conducted during the next phase to further investigate the
effect of injection grouting on masonry structural behavior.

REFERENCES

[1]   Earthquake Engineering Research Institute Newsletter, Vol. 22, No.
      4, April 1988, p. 4.
[2]   City of Los Angeles Rule of General Application RGA #1-91, "Crack
      Repair of Unreinforced Masonry Walls With Grout Injection," May
      22, 1991.
[3]   Paillere, A. M., Buil, M., Miltiadou, A., Guinez, R., Serrano,
      J.J., "Use of Silica Fume in Cement Grouts for Injection of Fine
      Cracks," Third Int. Conf. on Fly Ash, Silica Fume, Slag and Natu-
      ral Pozzolans in Concrete, ACI SP 114, Vol. 2, 1989.
[4]   Binda, L. and Baronio, G., "Performance of Masonry Prisms Repaired
      by Grouting under Various Environmental Conditions," Masonry In-
      ternational, Oct. 1989.
[5]   Tomazevic, M. and Anicic, D., "Research, Technology and Practice
      in Evaluating, Strengthening, and Retrofitting Masonry Buildings:
      Some Yugoslavian Experiences," Proceedings: International Seminar
      on Evaluating, Strengthening, and Retrofitting Masonry Buildings,
      The Masonry Society, Oct. 1989.
[6]   Recommended Practice 13 (b), Standard Procedure for Field Testing
      of Drilling Fluids, American Petroleum Institute.
[7]   Hooton, R.D., Konecny, L., "Permeability of Grouted Fractures in
      Concrete," Concrete International, July 1990.
[8]   "Special Materials for Hydraulic Concrete Structures, Materials
      for Grouting Concrete Structures, Sand Column Grouting Test in Wet
      and Dry Media," L'Association Francaise de Normalisation (AFNOR),
      NF P 18-891.

[9]   Kingsley, G.R.; Noland, J.L.; Atkinson, R.H.; "Nondestructive
       Evaluation of Masonry Structures Using Sonic and Ultrasonic Pulse
       Velocity Techniques," Proceedings: 4th North American Masonry Con-
       ference, Los Angeles, 1987, Paper 67.

[10]   Berra, M., Binda, L., Anti, L., Fatticcioni, A., "Utilization of
       Sonic Tests to Evaluate Damaged and Repaired Masonry," Proceedings
       of the Conference on Nondestructive Evaluation of Civil Structures
       and Materials, University of Colorado, Boulder, May 1992.

G. Gabriel Cole and Paul L. Kelley[1]

CLADDING AN EXISTING BUILDING WITH MASONRY VENEER

REFERENCE: Cole, G. G. and Kelley, P. L., "Cladding an Existing Building with Masonry Veneer," <u>Masonry: Design and Construction, Problems and Repair, ASTM STP 1180</u>, John M. Melander and Lynn R. Lauersdorf, Eds., American Society for Testing and Materials, Philadelphia, 1993.

ABSTRACT: Brick is highly regarded as a durable, low-maintenance cladding material. As a result it is often the material of choice for recladding an existing building. Masonry recladding projects are, however, oftentimes problematic. It is challenging to detail and install a masonry veneer, particularly if the building was originally clad with another material. Owners also typically have high expectations for the recladding, especially if the previous cladding failed prematurely. The projects are further complicated if the building is occupied and interior operations must be maintained throughout the construction process. These factors have led to substantial cost and schedule over-runs on several projects. As a result, many owners, designers, and contractors are reluctant to engage in recladding projects. Recladding need not be intimidating. A sound technical approach and a carefully selected and managed project team can successfully undertake and complete a recladding project. With illustrations from a recently completed project, this paper explores the major technical challenges, design constraints, material considerations, and coordination requirements that are encountered in a recladding project.

KEYWORDS: brick masonry, recladding, flashing, renovation, relieving angles, dimensional tolerances, trade coordination, project team

---

Owners of institutional buildings, such as hospitals and colleges, demand durable cladding systems for their buildings, since they are typically expected to be in service for a long period of time. In New England, brick masonry is the material by which all other cladding systems are judged for durability. There are numerous masonry-clad institutional buildings in New England that have weathered the elements for over 100 years with modest maintenance.

Due to its reputation, brick masonry is often selected by institutional owners as replacement cladding for existing buildings. This is particularly true when cladding systems have failed prematurely or unexpectedly.

---

[1]Senior Engineer and Associate, respectively, Simpson Gumpertz & Heger Inc., Consulting Engineers, 297 Broadway, Arlington, MA 02174

Design and construction of masonry as replacement cladding is considerably different from that for new construction. Recladding projects have the following general characteristics that distinguish them from new construction projects:

1.   The detailing and installation of the new masonry cladding is challenging and frequently problematic. This is true even on buildings with similar original cladding.

2.   The building owners typically have high expectations for the new cladding. Since they have had to suffer with the failure of one cladding system, they explicitly set out requirements concerning durability, leak resistance, and future maintenance requirements.

3.   The projects are particularly challenging on occupied buildings where care must be taken to maintain interior operations and mitigate exposure of occupants to weather, noise, dust, or construction chemicals.

The following case study of a recently completed project illustrates the many challenges inherent in a masonry recladding project.

CASE STUDY

The subject building is a medical research institute in the Boston area. The institute is housed in a 1924 vintage 14-story building originally constructed to house garment manufacturing firms. The building has a concrete frame and is clad on the exterior with cast stone panels (Fig. 1). The owner renovated the building into medical laboratories and office space between 1984 and 1986. These renovations concentrated primarily on interior elements of the building. Although new windows were installed, and the exterior was patched and painted, the cast stone was not significantly altered during the renovations.

Leakage through the outside walls plagued the building after occupancy in 1986. Repeated attempts to address the leakage problems failed. In late 1989, a large piece of the exterior cast stone fell off of the building. No one was injured by this incident, but it prompted the owner to hire a design team to conduct an investigation and develop a remedial solution for the exterior walls.

The design team was led by an architect who was responsible for communications with the building owner, design of the new facade, participation in local design reviews, and management of the project team. The design team included a building technology consultant and structural engineer.[2] The construction team, including the construction manager and key subcontractors assisted the design team throughout the design process.

The detailed field investigation indicated that the majority of the facade had deteriorated to the point that it could not be reliably repaired and that its performance was unpredictable. The design team proposed three schemes for recladding:

<u>Glass fiber reinforced concrete (GFRC)</u>--This was a replication alternate. Molds would be taken from the existing facade elements and GFRC panels would be manufactured to duplicate the existing as closely as possible.

---

[2]Author Cole served as project engineer for building technology and author Kelley served as project engineer for structural.

FIG. 1--View of case study building with original cast stone
cladding.

<u>Brick masonry</u>--This option would be sympathetic to the lines of
the original facade, but would have its own distinctive architectural
style. Materials would be selected in a manner to "tie" the building to
others in the surrounding area.

<u>Metal panels</u>--This option would be radically different from the
existing, but would probably be the most economical alternative.

Brick masonry was recommended by the architect and accepted by the
owner, since it was the most reliable and proven system that would
satisfy local building design review concerns to maintain the original
character of the building facade.

<u>Project Constraints</u>

The owner imposed several constraints on the design team, the most
significant of which was the need to maintain interior operations. The
laboratories inside the building perform vital medical research that
cannot be moved to other buildings. Therefore, the existing windows
cannot be removed, and all construction operations must take place on
the exterior.

In order to accomplish this, a fast-track approach was used
whereby the design team would rapidly develop demolition documents, and
would develop the new cladding design while the demolition was underway.

Predesign/Schematic Stage

     The first task of the design was to organize the analysis from the investigation and extract information necessary for the new design, including:

     Determination of why the existing system failed--The original cast stone depended on a "barrier" concept of waterproofing and, as such, had no concealed cavities or built-in drainage system. Over the years water worked its way into the system and deteriorated the facing material and the attachment devices.

     Identification of any components which could be reused--The building technology consultant identified one small area around the entrance of the building where the original cladding was sound and, therefore, suitable for incorporation into the new facade. The building technology consultant determined that all other exterior components, with the exception of the windows, were severely deteriorated and unsuitable for continued use. Even the original relieving angles on the building could not be reused, because they were heavily deteriorated and poorly attached to the building.

     Determination of "typical" physical configuration--Fortunately, one floor of the building was empty and this afforded an opportunity to do some heavy demolition on the interior which allowed for examination, measurement, and detailing of concealed elements within the wall system without disrupting building occupants.

Demolition strategy

     The demolition procedure was challenging because of the need to leave the windows in place and minimize the impact on interior operations. Unfortunately, the windows were attached to the cast stone facade which was to be removed. The windows were also large, with an average frame size of 4.37 m (14 ft 4 in) long by 2.39 m (7 ft 10 in) high and weight of 545 kg (1200 lb). To further complicate the situation, the window assembly sill consisted of several decorative pieces screwed together, with no continuous structural member to allow the window to hang from the jambs and head.

     The structural engineer developed the following procedure for progressive demolition and resupport of the window:

     1.   Remove the cast stone panels along the window jambs and on the same day install steel clips to tie the window jambs to the concrete building columns (Fig. 2).
     2.   Install a temporary tubular support made of aluminum in front to the window sill (Fig. 3). The tube has tabs which extended beneath the window to pick up the gravity load (Fig. 4). Once this tube is in place, remove the cast stone spandrel beneath the window.
     3.   Install a permanent steel tube, that the waterproofer previously has covered with membrane flashing (Fig. 5). Once this tube is in place, remove the temporary aluminum tube for reuse in another location.

     Due to the complexity of this procedure, the design and construction team built a mock-up to evaluate and refine the design. The mock-up was extremely useful because it identified several difficulties with the installation of the design. The design team modified the design considerably to accommodate these concerns during the full-scale construction.

     Since the demolition was done before the recladding and interior operations were to be maintained, it was necessary to waterproof the

FIG. 2--New attachment clips between concrete column and window jamb.

building structure after removal of the cladding. The design team reviewed several options and ultimately selected a "peel and stick" membrane. This material provided a high degree of waterproofing protection as a temporary membrane during demolition. Ultimately, this material was incorporated into the final design as the cavity waterproofing membrane. The material has a relatively low permeability, and the building technology consultant performed a detailed analysis of the wall system to assure themselves that the potential for internal condensation is minimal.

As the demolition proceeded, the distressed condition of the existing facade became more apparent. In many areas the cast stone was not tied to the structure in any way, but was simply held in place by friction, mortar, and sealant. This lack of lateral restraint led to outward displacement of the stones, particularly at the parapet level. The deterioration of the concrete frame was also more severe than indicated by the initial field survey (Fig. 6). The owner established an allowance for restoration of the frame, including cleaning and painting of reinforcing steel, concrete patching, and epoxy injection of cracks in spandrel beams. This work was performed on an as-needed basis during the demolition process.

<u>Cladding Design Philosophy</u>

FIG. 3--Temporary aluminum support tube mounted in front of
window sill.

FIG. 4--Tabs on temporary aluminum support tube extend
beneath window sill frame.

     The design team developed a traditional brick masonry curtain wall
design based on previous experience and relevant industry literature
[1][2]. The design included the following basic components:

     Support scheme--Since the existing relieving angles could not be
reused, the structural engineer designed a new relieving angle system to
support the masonry at each floor level. The structural independence of
each floor became critical at later stages, when the schedule required
the upper three floors to be clad with brick before the rest of the
building.
     The demolition process revealed that the original concrete was of
relatively low strength (13.8 to 20.7 MPa (2000 to 3000 psi)) and of

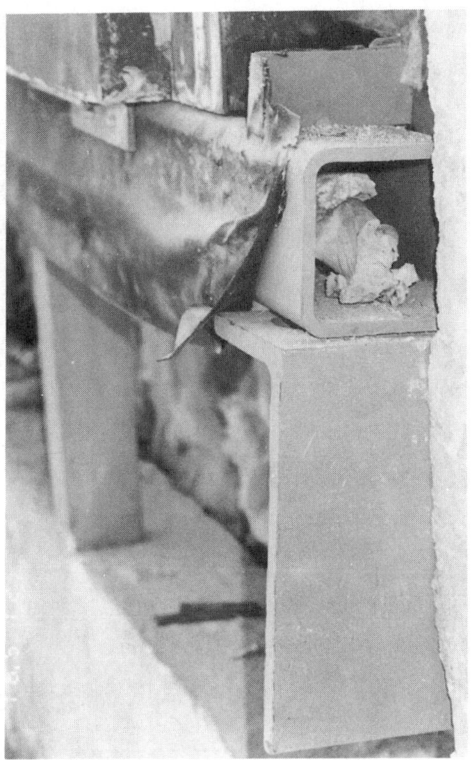

FIG. 5--Permanent steel tube in place beneath window sill
with uncured neoprene membrane flashing.

inconsistent quality. The face of the concrete frame varied
considerably and in many areas appeared to have been hammered off during
the original construction. In response to this variability, the
structural engineer developed a separate hanger fixture that was
installed in advance of the relieving angle to provide adjustability for
the relieving angle (Fig. 7).
     The hanger strap system made it possible to install the expansion
bolt anchorage a minimum of 127 mm (5 in.) above the bottom of the
spandrel. This avoided edge distance problems which would have occurred
had the angle been attached directly to the spandrel beam.
     To provide vertical adjustability, a proprietary punch and washer
system was used with the relieving angle (Fig. 8). This system allowed
approximately 13 mm (1/2 in.) of vertical adjustment during final
installation of the relieving angles.
     The design team selected an L 203 mm X 152 mm X 13 mm (L 8 in. X 6
in. X 1/2 in.) as the typical relieving angle, and an L 203 mm X 203 mm
X 13 mm (L 8 in. X 8 in. X 1/2 in.). These are unusually large
relieving angles. The 152 mm (6 in.) to 203 mm (8 in.) horizontal legs
are fairly common with insulated cavity wall construction; the 203 mm (8
in.) vertical leg was dictated by the hanging strap detail and field
conditions which required the horizontal leg of the angle to project as
much as 64 mm (2-1/2 in.) below the bottom of the spandrel.
     Because of the large unit weight of the angle and the desire to
erect the angles without a crane, the angle length was limited to 1.5 m

FIG. 6--Deterioration of concrete frame and corrosion of reinforcing steel.

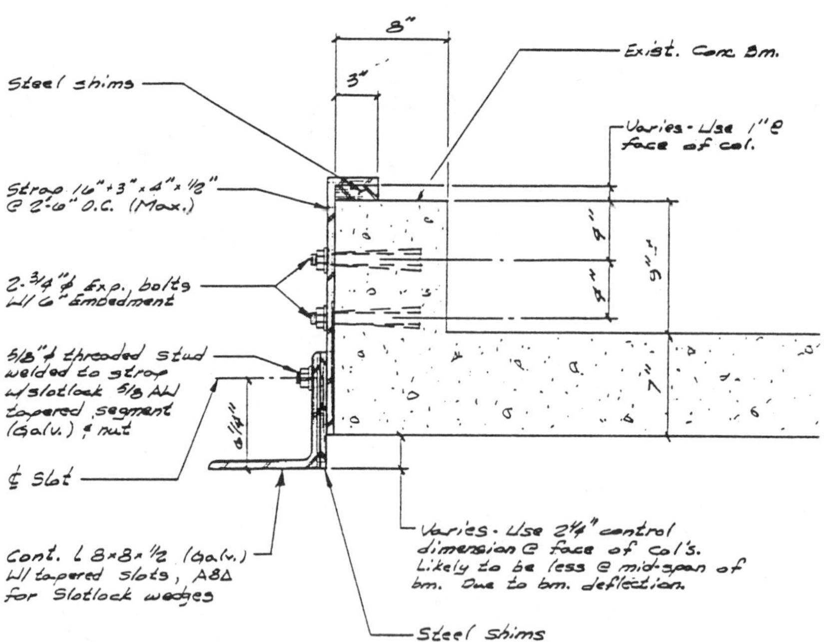

FIG. 7--Relieving angle attachment detail.

(5 ft.). Relieving angles and hanger straps were typically hot-dip galvanized after fabrication. In some instances, where field conditions required substantial modification of angles after the initial fabrication, the angles were shot blasted and painted with a zinc rich primer to eliminate the two to three week delay for hot-dip galvanizing.

FIG. 8--Relieving angle installation in process showing wedge
shaped holes and washers that allow for vertical adjustment of
angle.

The structural engineer specified expansion bolt sizes and
spacings to keep bolt forces low so that the setting torque could be
reduced to avoid damage to the old low strength concrete frame and to
allow reduced edge distance where the optimum edge distance could not be
achieved. The bolts were typical 19 mm (3/4 in.) diameter wedge anchors
embedded a minimum of 152 mm (6 in.). The installation torque was
limited to 203 N·m (150 lbf·ft). Using standard torque wrenches, the
erector torqued each bolt and the field monitor verified all bolts when
each floor was finished. Bolts that could not reach the minimum torque
were replaced with a section of galvanized threaded rod set in ceramic
epoxy.

Ties--The design team developed a tie layout based on a maximum
spacing criteria of 457 mm (18 in.) in either direction. The new
masonry facade included several articulations intended to provide
architectural effect. These articulations, combined with the
irregularities of the building frame resulted in large cavities and/or
applied dead load on the tie. This required careful analysis and
selection of ties. Several unusual tie details were required,
including:
     1. Install an extra row of ties directly above special shaped
soldier brick (Fig. 9). These brick overhang the relieving angle and
tend to roll outward until one or two additional courses have been
installed. The extra row of ties provides restraint during construction
to prevent outward movement or rolling of the soldier course.
     2. Install lateral reinforcement in every fourth to sixth course
at corners and projected column covers (Fig. 10).
     3. Install field shaped stainless steel bar stock to provide
"long-reach" ties at corners (Fig. 10).
     4. Prefabricate brick assemblies for projecting corners (Fig.
11). This moves the center of gravity of these units onto the relieving
angle, thereby facilitating construction and minimizing the chance of
movement of the brick in the future.

Movement control--In addition to soft joints below relieving
angles at each floor, vertical expansion joints were placed at regular
intervals to allow for future movement of the wall system. The existing

FIG. 9--Tie layout on spandrel panel beneath windows.

FIG. 10--Lateral reinforcement in masonry at corner of column
cover. Arrow identifies stainless steel bar stock used for
long tie at corner.

floor slabs had deflected considerably over time. Although the
potential for future creep deflection was small, the structural engineer
was concerned that the reloading of the building during recladding might
result in displacement. To avoid problems with the masonry, the design
team located vertical joints on both sides of the spandrels. This broke
the wall system into rectangular sections consisting of column covers
and spandrel panels with no re-entrant corners.

Cavity waterproofing and flashings--The peel and stick membrane

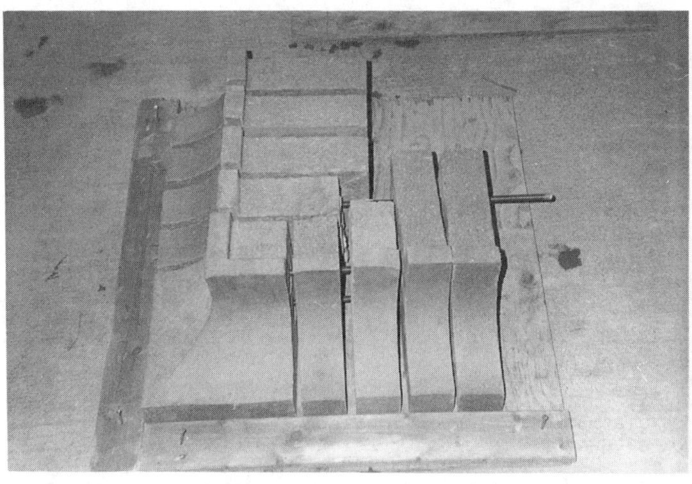

FIG. 11--Prefabricated brick assembly under construction. Stainless steel threaded rod that was embedded into brick with ceramic epoxy.

applied during the demolition phase was left in place to function as cavity waterproofing. Initially, there was some hesitation to count on this material, since it had been exposed to sunlight for more than six months. With few exceptions, this material weathered the exposure well.

An uncured neoprene sheet was installed beneath the windows to capture any window frame leakage. This flashing lapped over the top of the peel and stick membrane such that any water leaking through the

FIG. 12--Spandrel prior to installation of brick showing window sill flashing (arrow a), peel and stick cavity waterproofing (arrow b), and lead coated copper through wall flashing (arrow c).

window frame would be captured and directed down into the cavity (Fig. 12).

Lead coated copper through wall flashings were installed over the top of the relieving angles at each floor. These flashings extended out past the exterior face of the masonry with an exposed drip edge (Fig. 12).

The design team required that the wall be watertight without reliance on the external sealants. To help assure the integrity of the waterproofing system, the field monitor selected several panels at random for water testing prior to installation of sealants.

Adjustability and Tolerances--Adjustability was a major factor in the design of all details. The construction manager and subcontractors sought to avoid custom fabrication of materials. Based on the data from the predesign investigation, the design team developed details and set tolerances which appeared to encompass the range of variations on the existing building.

Constructability and schedule--During design, the design team was assisted by the construction manager and likely subcontractors who were hired as consultants. The construction manager monitored the project budget and advised on construction techniques and sequencing. The input of the masonry subcontractor was particularly useful due to the great number of special shapes contemplated for the project; this assistance was crucial to understanding lead times for various materials and allowed early ordering.

Technical Problems During Construction

Several technical problems arose during the construction process which slowed the progress of the job. Some of these items required substantial modification of the original design to compensate for unexpected variations in the construction of the original building. The most significant construction problems were the following:

Building lean--Shortly after demolition began at the top of the building, the construction team discovered that the concrete frame was highly irregular. The owner retained a surveyor to establish dimensional control lines on the building and map the contours of the concrete frame using lasers mounted on the roof and referenced to control points on the ground. The surveyor discovered that one end of the building leaned approximately 203 mm (8 in.) to one side.

The design and construction teams evaluated several options to deal with the lean ranging from constructing the wall to conform to the lean to constructing a plumb wall. Ultimately, the teams agreed that a hybrid solution would best satisfy the concerns for structural integrity, constructability, and appearance. The design was modified to include 127 mm (5 in.) of lean (roughly 6 mm (0.25 in.) per floor) and to accommodate the balance of the lean by shimming and/or trimming the relieving angles. This prompted the structural engineer to develop more precise shimming tolerances and to design alternate details for areas where the required shimming exceeded the tolerances. In some areas, horizontal legs were increased to 203 mm (8 in.). In other areas, the horizontal legs of the angles were cut down in the shop. Where transitions between angle sides were required, the steel erector used a hand-held band saw to field cut the leg.

Shimming of the typical spandrel straps was increased to 32 mm (1-1/4 in.); associated fastener bending was accommodated by the friction developed by torquing the threaded stud on the strap. At columns, where the angle was fastened directly to the concrete without a hanger strap, shimming at the expansion bolts was held to 19 mm (3/4 in.) to limit bending of the bolt. Where larger shim dimensions were required, a heavy bar spacer was inserted into the gap. Individual instances of shim tolerance violation were dealt with by special

FIG. 13--WT sections used to support angle at expressed
columns.

supplemental connections applied under the angle or by field drilling
through the angle and installing additional bolts. Shims were required
to have full bearing against the back of the relieving angles. This
required thorough monitoring and education of the ironworkers,
particularly regarding how to deal with shimming at irregularities in
the building frame. The shimming did, however, proceed smoothly after
the first few weeks. In a few localized cases near the top of the
building a special assembly consisting of two wide tee (WT) sections
allowed the angle to be pushed out 127 to 178 mm (5 to 7 in.) away from
the wall (Fig. 13).

     Tie length--Since the new wall did not completely follow the lean
of the building, the cavity increased in size near the top of the
building. This created a problem for the masonry ties as well as the
relieving angle. The structural engineer designed special tie anchorage
of light gauge metal framing so that the unsupported tie length could be
kept below 127 mm (5 in.).

     Alignment of flashing drip edge--Once the angle adjustment was
complete, the sheet metal through wall flashings were installed. These
flashings were held in place with a piece of butyl-tape, but had a
tendency to displace. Since the drip edge stuck out beyond the future
face of masonry, it interfered with the string lines used by the
ironworkers and masons. On the first few floors constructed, the
flashing waved in and out, creating an irregular line on the building
that was unacceptable to the architect. The architect agreed to allow
this work to stay in place, since it was up high on the building, with
the understanding that the contractor would resolve the problem on
future floors. The construction manager expended considerable effort to
coordinate the trades and develop procedures to alleviate the problems.
In particular, the sheet metal workers left the drip edge open at
corners and did not solder these areas until after the masonry
installation was complete. This allowed the mason to pass the string
lines through the open corners. The construction manager also secured
the flashing with C-clamps until the mason completed the first few
courses of brick.

<u>Waterproofing details</u>--The concept of the waterproofing design depends on the back-up or cavity waterproofing providing the primary waterproof seal. The design team had difficulty communicating the importance of the cavity waterproofing to the construction team, and spent considerable effort educating them that the cavity waterproofing was the most important element and that it was not simply a back-up that could be compromised when necessary.

Non-Technical Construction Issues

In addition to the technical issues outlined in the previous section, the project team was beset by a series of additional problems that hindered the job progress, including:

<u>Use of Volatile Organic Compounds (VOC)</u>--To guard against migration of the biological materials used for research, the owner operates the building HVAC system at a significant negative pressure. On several occasions materials from outside (i.e. dust, chemicals, etc.) were drawn inside through openings in the building envelope. In some cases these contaminants damaged experiments in progress. Although this was a nuisance and resulted in wasted effort, the occupants were more concerned about the impact of this exposure on themselves. Since they are medical researchers, many are aware of the potential ill affects of exposure to chemicals and were particularly sensitive to the use of VOC containing materials, including primers, solvents, adhesives, paints, and other common construction materials. Occasionally, the construction manager had to halt portions of the work to address these concerns. In some cases modified work procedures, such as method of application and handling and disposal of rags and brushes, were sufficient to minimize the exposure to the interior occupants; in other cases the design team selected new materials with a reduced VOC content.

<u>Trade jurisdiction</u>--The building is located in an area of New England with very strong trade unions. Even though the construction manager and respective subcontractors reviewed the construction documents, the project experienced several jurisdictional problems. In particular, the ironworkers claimed that they had to install any sheet metal work that met the existing windows. This created problems because most of this sheet metal was architectural closure. The iron workers were not, however, accustomed to such finish work.

<u>Strike/Labor Problems</u>--The project schedule was made more difficult when the ironworkers went on strike for six weeks. The strike occurred when they were approximately half-way through erection of the relieving angles. The strike halted the masonry work and forced the job into winter months, increasing project costs for protection and heating.

Results

The construction manager substantially completed the project in February 1992 (Fig. 14). To date the project has performed well and provides the building occupants a weathertight envelope. The pace of the project was brisk and many compromises, such as the architect's acceptance of the wavy flashing on upper floors, helped bring the project to completion. The design team was, however, careful not to make any compromises that lessened the durability or integrity of the design and, in some cases, asked the construction manager to remove work that did not conform to the construction documents.

The project construction cost was about $125 per square foot of gross wall area, 33% of which is existing windows. About 40% of the construction cost is attributable to demolition of the cast stone facade and resupport of the windows. Change orders accounted for 8% of the construction cost (2% building lean, 3% strike, 3% miscellaneous). Design services and the owner's project overhead added an additional $20

FIG. 14--New masonry facade.

per gross square foot (16% of the total construction cost).

CONCLUSIONS

Facade replacement is more complex than new construction from both the design and construction perspectives. Much of the information necessary to detail the work is hidden until demolition. Problems and unforeseen circumstances are inevitable on renovation projects, but proper planning and control can ease their impact.

Based on the lessons of the case study and similar projects, the methodology should include the following elements:

Project team members--Communication and coordination are key elements of the renovation process, and each of the involved firms should have the necessary resources to support the group effort. The team must carefully divide tasks to eliminate needless repetition, yet avoid omission.

Timing of entry of the various team members is also important. The project team, including the general contractor or construction manager, and key subcontractors should assemble early in the design phase. This runs contrary to traditional competitive bidding strategy, but the economic loss due to elimination of competitive bidding is offset by design efficiency and reduced extras and delays when the project team can interact throughout the design and construction phases.

All project team members must clearly communicate their expectations and limitations early in the process to assure acceptance of unusual techniques, materials, and details. Phasing of the work and coordination of the different trades may constrain the design significantly. Design changes cannot be avoided in renovation, and they are best accomplished through the cooperative effort of all team members.

Field assessment procedure--Condition appraisal must be thorough and is most reliable if done up close from a swing stage or mobile work platform. Binocular surveys from the ground are usually not accurate enough for design work. The field assessment must identify the condition of elements, the cause of distress, and the range of as-built dimensions. The construction documents should record this data accurately so all project team members have the same understanding of the existing conditions. When possible, a single team member should coordinate all building dimensions.

Detail flexibility and reasonable tolerances--Tolerances should encompass the full range of variation on the building, yet should not place unreasonable demands on the field workers. Strict observance of tolerances will serve as a guide to unusual job conditions; if the contractor advises that he/she cannot comply with a particular tolerance it likely represents an unanticipated condition in the field that warrants the designers special attention. Change order allotments for dealing with these conditions typically need to be greater than for new construction, particularly if some key elements will not be uncovered until the project begins.

Mock-ups--The use of mock-ups is imperative, particularly if any of the repairs are innovative. The mock-ups should encompass all typical conditions and should be conducted before completion of design phase. Mockups on the building can be invaluable in the development of final details. The installing contractors frequently suggest ingenious modifications to details. The mock-up should include objective test criteria to assess the efficacy of the procedures.

Field monitoring--Facade renovation can involve many trades including wreckers, carpenters, ironworkers, masons, waterproofers, and sheet metal workers. The presence of a full time monitor who promptly reports all variances in the specifications and job tolerances promotes high quality work and identifies conflicts before they become major problems. Monitoring is critical for elements that impact public safety such as steel erection and masonry installation.

Work in occupied buildings--The project team must assess the impact of all elements of design, material selection, and construction procedure on interior occupants. To maintain occupant cooperation and satisfaction, the construction manager or general contractor must keep schedules up-to-date and promptly address complaints about noise, dust, odors, leakage and other problems. Material selection must include careful consideration of the potential exposure of building occupants to irritating and harmful construction materials.

The most important element for success of the project is effective communication and mutual respect among project team members. Recladding, like all renovation, is by nature a process that requires feedback and change throughout design and construction. Without effective communication, even the best designs are destined for trouble.

ACKNOWLEDGEMENTS

We thank the principals and associates of Simpson Gumpertz & Heger Inc.

for their support in writing this paper. We also thank Imai/Keller Architects, Inc., Barr & Barr Inc., and New England Medical Center Hospitals with whom we worked on the case study project.

REFERENCES

[1]   Brick Institute of America, <u>Technical Notes on Brick Construction</u>, Brick Institute of America, Reston, VA, 1988.

[2]   Gumpertz, Werner H. and Bell, Glenn R., "Engineering Evaluation of Brick Veneer/Steel Stud Walls, Part I - Flashing and Waterproofing," <u>Proceedings, Third North American Masonry Conference</u>, Arlington, TX, June, 1985.

Bruce S. Kaskel,[1] and Rocco C. Romero[2]

FAILURES OF INTERIOR MASONRY WALLS SUBJECTED TO LATERAL AIR PRESSURE

REFERENCE: Kaskel, B.S., and Romero, R.C., **"Failures of Interior Masonry Walls Subjected to Lateral Air Pressure,"** Masonry: Design and Construction, Problems and Repair, ASTM STP 1180, John M. Melander and Lynn R. Lauersdorf, Eds., American Society for Testing and Materials, Philadelphia, 1993.

ABSTRACT: Interior masonry partition walls in three different buildings were found to fail when subjected to significant unanticipated lateral loads due to air pressures. These walls were built around the 1970's and were empirically designed. Governing codes then, as now, commonly required that an interior partition be designed to sustain a lateral load of 5 lbs. per sq ft (24.4 kg per sq meter). In one case, loads of 15 lbs. per sq ft (73.2 kg per sq meter) were recorded by instrumentation of the walls.

The investigation and its findings are presented for each of the three case studies. Lessons learned from these examples about lateral loads and the design of masonry partition walls are presented to assist in future designs. This paper also reviews empirical design and minimum design load provisions in current codes and standards to examine whether they accurately account for such load cases.

KEYWORDS: interior non-loadbearing walls, empirical design, lateral loads, air pressure, investigation, codes.

Recent investigations of distress to interior masonry partition walls revealed failures caused by lateral loads from unanticipated air pressures. Three buildings that were built in the late 1960's and early 1970's were examined. In one case, a concrete masonry wall was leaning inward over 6 in. (15.2 cm). In a second case, portions of a masonry wall collapsed. In a third case, masonry walls, although intact, could be moved laterally at the top by hand.

The masonry walls that failed in each of these three buildings had several similarities:

<u>Interior walls</u> -- The walls that failed were all built inside the building and were not exposed to the elements. The first wall was in the penthouse of an office building; the second wall was in the garage of a convention center; and in the third case, walls separated an elevator bank and a return air shaft in an office building.

---

[1]Consultant, Wiss, Janney, Elstner Associates, Inc., (WJE), 29 North Wacker Drive, Chicago, IL

[2]Engineer/Architect I, Wiss, Janney, Elstner Associates, Inc., (WJE), 29 North Wacker Drive, Chicago, IL

<u>Non-loadbearing</u> -- All walls were constructed of non-reinforced masonry with nominal horizontal joint reinforcement for shrinkage control. The walls did not receive in-plane loads other than their own weight.

<u>Air pressure loads</u> -- All three walls failed due to unanticipated air pressure differentials. In the first case, wind loads on the exterior wall was transmitted to the interior masonry wall; in the second case winds were channeled through openings in the exterior wall; and in the third case, return air flow and elevator movements caused lateral loads on the walls.

<u>Anchorage</u> -- The walls were built tight to adjoining construction without special attention given to the anchorage details, especially at the top of the walls.

<u>Empirical design</u> -- The span length to wall thickness ratio of the walls are based on empirical design criteria for interior non-loadbearing masonry.

CASE STUDY 1

On the morning following a spring wind storm, building personnel at this 22 story office building noticed mortar and debris on the penthouse floor. They saw that the 16 ft (4.9 m) tall by 75 ft (22.9 m) long concrete masonry wall was severely damaged with significant step and horizontal cracks. The top of the wall was leaning inward about 6 in. (15.2 cm) and was wedged tight to the underside of roof slab (Fig. 1).
Upon arrival of the investigative team, the wall was quickly braced to prevent further displacement, and then carefully dismantled. In the interim, the investigative team observed the following conditions:

1. The masonry wall was intended to serve as a fire-rated back-up to the exterior aluminum curtain wall. The curtain wall spanned from the penthouse floor to the roof and was not tied or supported by the concrete masonry wall. A distance of about 6 in. (15.2 cm) separated the two walls.

2. The aluminum curtain wall was designed with open gutter joints between sections of the curtain wall panels. Originally, foam material had been installed in these joints to reduce air leakage. At the time of the investigation, this foam material had deteriorated and daylight could be seen through many of the gutter joints.

3. The nominal 6 in. (15.2 cm) thick concrete masonry wall spanned 16 ft, 4 in. (5 m) from the floor to the roof slab, an h/t ratio of 33.

4. The masonry wall was built from the inside. The top masonry unit was about 1/2 in. (1.3 cm) below the underside of the concrete roof slab and the joint was packed with mortar on the interior face shell of the concrete masonry.

5. Anchors were not provided at the top of the masonry wall.

The investigation concluded that the failure of the wall was caused by a pressure differential between the two sides of the masonry wall. This pressure differential was caused by wind loads on the exterior curtain wall. The open curtain wall joints allowed the air pressure in the space between the masonry wall and the curtain wall to equalize with the exterior pressure. This in turn caused a pressure difference across the interior masonry wall (Fig. 1). The mortared joint at the top of the wall was incapable of restraining the wall from moving and displaced inward until the wall became wedged to the underside of the roof slab.

Had adequate support at the top of the wall been provided, the wall may not have failed. However, as will be shown later in this paper, even if adequate support had been provided at the top of the wall, the calculated flexural stresses in the wall would still exceed allowable stresses for non-reinforced masonry.

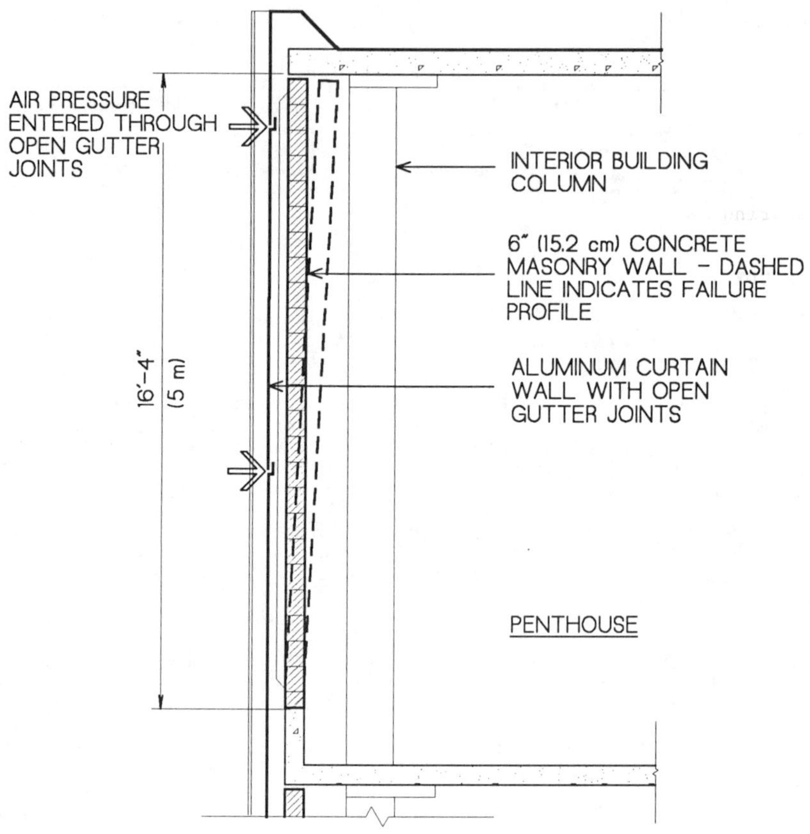

Fig. 1 -- Case Study 1: Wall Section

The wall was subsequently rebuilt with masonry to comply with the fire enclosure requirements of the building. Due to space and weight constraints, 6 in. (15.2 cm) masonry was the only realistic replacement. Steel wide flange column braces, spanning from floor to ceiling, were introduced in the plane of the masonry wall at 10 ft (3 m) centers, to gain increased load resistance. The column braces were designed to be independent of the building's structural frame. This permitted movement of the building's structural frame without introducing gravity load transfer through the bracing system.

The wall was rebuilt with nominal 6 in. (15.2 cm) thick non-reinforced concrete masonry, which was designed to span horizontally between the column braces. The rebuilt wall had a l/t ratio of 20. The wall was checked for calculated flexural stresses. This analysis indicated that the rebuilt wall had a significantly greater load resistance than the original wall, both by reducing the masonry span length and by spanning the wall parallel to the masonry bed joints.

CASE STUDY 2

Interior masonry walls in the garage area of a large convention hall were constructed as partitions between driveway ramps and as an enclosure to conceal pipe shafts (Fig. 2). The entrances to the driveways were open to the exterior. Portions of the interior masonry walls were built adjacent to these entrance areas (Fig. 3).

During a significant wind storm, a 90 ft (27.4 m) long wall section along one driveway ramp collapsed. As an immediate precautionary measure, the other walls in this area were braced. The investigation team examined the walls in the areas where the collapse occurred.

The investigation yielded the following conclusions about the design and construction of these walls:

1. The interior walls were subjected to exterior winds that were channeled through the driveway entrances into the building.

2. The governing building code required interior walls to be designed to withstand a minimum lateral load of 5 lbs. per sq ft (24.4 kg per sq m). Exterior walls at this building, were designed to withstand a minimum wind load of 20 lbs. per sq ft (97.6 kg per sq m).

3. The wall that collapsed was 20 ft, 5 in. (6.2 m) high. It was built as a composite wall consisting of nominal 4 in. (10.2 cm) brick and nominal 4 in. (10.2 cm) block up to a height of 16 ft, 6 in. (5 m). Above that height the wall was built with nominal 8 in. (20.3 cm) concrete block (Fig. 2). The h/t ratio for this wall, based on solid masonry construction, is 31.

4. The wall was originally designed for support at the top, with clip angles at 3 ft (0.9 m) on center. During construction, these clip angle supports were deleted and replaced with a compressible filler material between the top of the masonry wall and the structure, for a construction cost savings. This change eliminated the lateral support at the top of the wall. Without this support, the wall was essentially cantilevered from the floor, the full height of 20 ft, 5 in. (6.2 m).

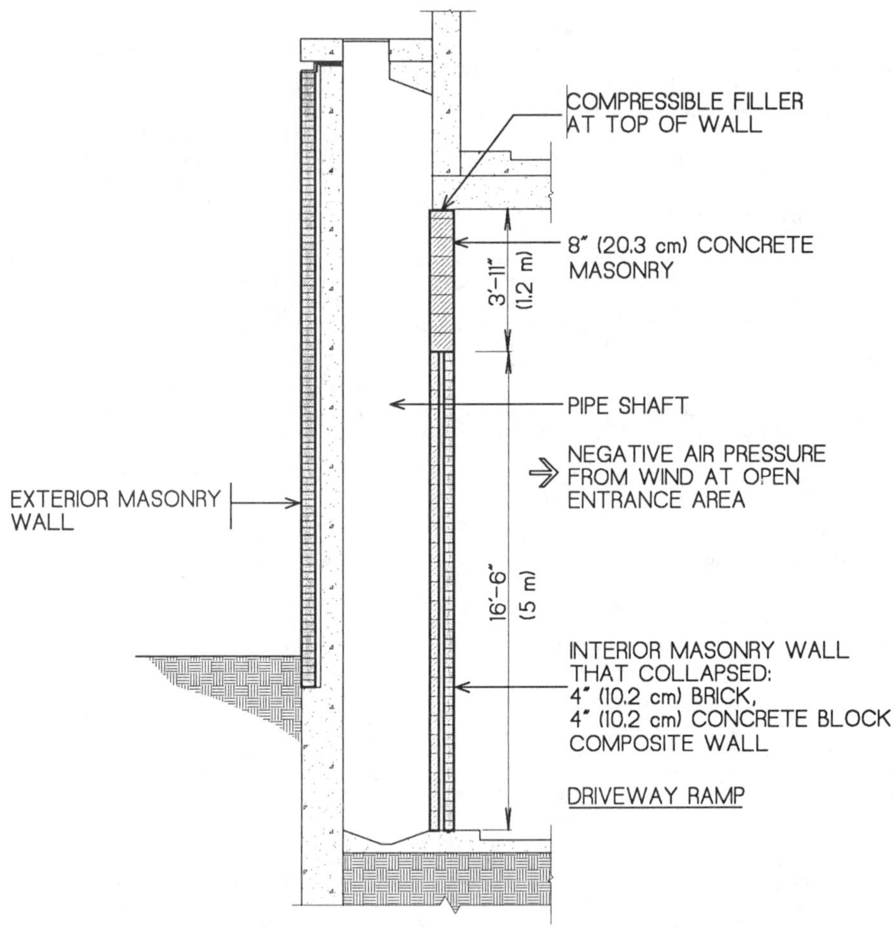

Fig. 2 -- Case Study 2: Wall Section

The investigation concluded that the collapse of the interior wall was caused by exterior wind loads, which were channeled through the open driveway entrance area. The 20 ft, 5 in. (6.2 m) tall wall that collapsed was intended to span vertically but was not adequately anchored at the top.

The collapsed wall was subsequently rebuilt. The other interior walls in the building were recommended to receive vertical supports to reinforce the walls. These supports consisted of steel double angle braces that spanned from the floor to the structure above. The existing masonry walls were bolted to these retrofit braces.

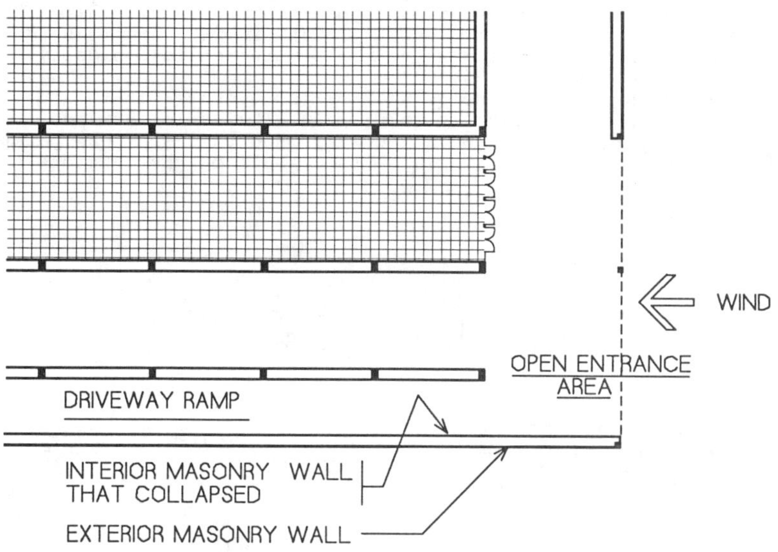

Fig. 3 -- Case Study 2: Partial Building Plan

CASE STUDY 3

During routine work in the elevator shafts of this high-rise office
building, maintenance personnel discovered that in several locations
masonry walls which separated the elevator shaft from an adjacent return
air shaft could be moved laterally by hand. Furthermore, the top of
these walls leaned toward the air shaft when the building's return air
fans were in service, as shown in Fig. 4. Due to concern that these
walls might collapse into the shafts, several of these walls were
removed and replaced with gypsum board and metal stud walls. An
investigative team was contacted to examine the conditions and to
recommend repairs to the masonry walls that remained.

The investigative team found the following conditions:

1. The building engineers had installed pressure transducers in the
walls at several locations and made measurements of the air pressure
differences between the air shaft and the elevator shafts. These
measurements found that air pressure differences up to 15 lbs. per sq ft
(73.2 kg per sq m) occurred.

2. The nominal 6 in. (15.2 cm) thick concrete masonry wall spanned
10 ft, 8 in. (3.3 m) vertically, an h/t ratio of 21.

3. The walls were supported by perpendicular masonry walls at three of the four corners of the air shaft. At the fourth corner where the walls abutted a building column, the walls were not mechanically anchored to the column (Fig. 5).

4. The masonry walls were built from the elevator shaft side. The top masonry units were typically about 1 in. (2.5 cm) below the underside of the floor beam above the wall, and the joint was packed with mortar only on the elevator side face shell of the concrete masonry.

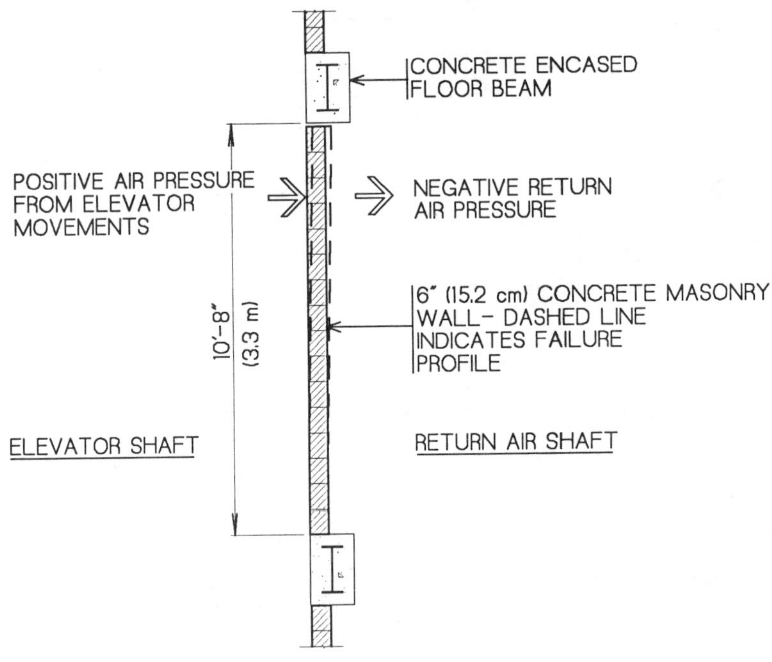

Fig. 4 -- Case Study 3: Wall Section

The investigation concluded that the movement of the walls was caused by the air pressure difference between the air shaft and the elevator shaft. The masonry walls, which were inadequately braced at the top and at the corner adjacent to the building column, were not designed to withstand the lateral loads created by these air pressures.

Since the majority of the remaining masonry walls were not cracked or distressed, the decision was reached to brace the walls in place with vertical steel angle braces. The braces were bolted to the walls and anchored to the top and bottom floor beams. The wall was analyzed as an non-reinforced masonry wall spanning parallel to the bed joints, between braces. Calculated flexural stresses were held to allowable stress values.

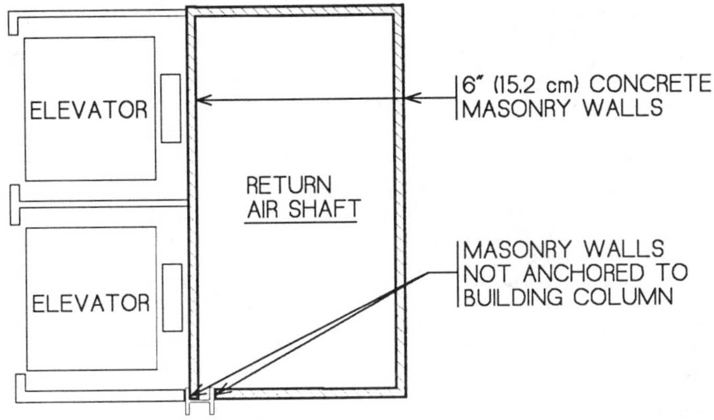

Fig. 5 -- Case Study 3: Partial Building Plan

DESIGN OF NON-LOADBEARING INTERIOR MASONRY WALLS

Empirical Design -- All three case studies where built in the late
1960's and early 1970's. At that time, empirical design of masonry
walls was accomplished using ANSI A41.1-1953 (R1970) "American National
Standard Building Code Requirements for Masonry." [1] Empirical design
of masonry wall was also described in the masonry industry literature
[2][3].

Empirical design is accepted for certain cases in the current
masonry design standard, "Building Code Requirements for Masonry
Structures", (ACI 530-88/ASCE 5-88) [4]. Empirical span length to wall
thickness ratios are provided for non-bearing walls in Table 9.5.1 of
this standard (Table 1).

TABLE 1 -- Wall Lateral Support Requirement

Construction	Maximum l/t or h/t
Non bearing walls	
Exterior	18
Interior	36

The span length to wall thickness ratio of the original wall constructions for Case Study 1, 2 and 3 were 33, 31 and 21, respectively. Therefore, based on empirical design requirements and the span length to wall thickness ratio, the original wall construction in all three case studies would be considered adequate for interior non-load bearing walls.

Engineered Design -- The alternative to empirical design is the engineered design of non-reinforced masonry based on flexural analysis allowing tension in the mortar joints. Flexural analysis requires identification of the loads on the wall. Design loads are commonly selected based on minimum requirements of the governing building codes. A review of two current model building codes reveal the following minimum requirements for interior partitions:

Uniform Building Code (1991 Edition) [5], Sec. 2309. (b):

> **"Interior Walls.** Interior walls, permanent partitions and temporary partitions which exceed 6 ft in height shall be designed to resist all loads to which they are subjected but not less than a force of 5 lbs. per sq ft (24.4 kg per sq m) applied perpendicular to the walls."

Standard Building Code (1991 Edition) [6], Sec. 1203.5:

> **"Interior Wall Loads.** Interior walls, permanent partitions, and temporary partitions shall be designed to resist all loads to which they are subjected but not less than 5 lbs. per sq ft (24.4 kg per sq m) applied perpendicular to the walls, except for decorative screen walls."

A flexural analysis of the three case study wall conditions was performed for comparison with the empirical design requirements. Flexural analysis was based on the wall's vertical span assuming lateral support at the top and bottom of the wall. As discussed in the three case studies, lateral support of these walls were found to be inadequate at the top. The flexural analysis (Table 2) shows that even if adequate lateral support were provided at the top of the walls, flexural stresses would have exceeded allowable stresses.

Flexural analysis was performed at two load levels: 1) 5 lbs. per sq ft (24.4 kg per sq m) minimum lateral load, per the model building codes, and 2) a 15 lbs. per sq ft (73.2 kg per sq m) lateral load indicative of the determined lateral load in Case Study 3. Although actual loadings were not determined at the time of the failures of walls in Case Study 1 and 2, this loading represents an approximation of the load conditions on these walls at that time.

For this analysis, the masonry was assumed to have been built with type N portland cement/lime mortar.   Flexural tension stresses were reduced by the weight of the masonry at mid-height of the wall. Allowable flexural tension values provided in "Building Code Requirements for Masonry Structures", (ACI 530-88/ASCE 5-88)[4] are listed for comparison.

TABLE 2 -- Flexural Analysis of Walls

Case Study	Allowable Flexural Tension[1] (psi)	Calculated flexural tension values (psi)	
		5 psf load (24.4 kg per sq m)	15 psf load (73.2 kg per sq m)
1	19	33	113
2	19	32	113
3	19	13	48

[1]Allowable flexural tension values for hollow concrete masonry with type N portland cement/lime mortar from ACI 530-88/ASCE 5-88.

Although all three case studies had inadequate lateral support at the top of the walls, even if the top of the walls were adequately supported, this analysis shows that the calculated flexural tension stresses exceed allowable flexural tension stresses. The allowable flexural tension stress is exceeded both at the minimum code lateral load of 5 lbs. per sq ft (24.4 kg per sq m) and at the higher load level which approximates actual loads on these walls. The sole exception is case study 3, which met the allowable flexural tension values at 5 lbs. per sq ft (24.4 kg per sq m) lateral load.

CONCLUSIONS

In conclusion, the three case studies indicated common problems with the design of interior masonry walls:

1. The walls were based on empirically designed span to thickness ratios, for interior non-load bearing walls. These empirical standards implicitly assume that the lateral loads on interior walls are slight.

2. The governing building codes required a minimum lateral load of only 5 lbs. per sq ft (24.4 kg per sq m) on interior partitions. Without knowledge to the contrary, the designer is not likely to select a greater load.

3. The actual lateral loads in all three case studies likely exceeded 5 lbs. per sq ft (24.4 kg per sq m). In the one case, lateral loads as high as 15 lbs. per sq ft (73.2 kg per sq m) were recorded.

4. The masonry walls in all three cases were built up to the underside of the structure, but were not adequately braced at the top. In one case, adequate anchorage was designed at the top, but was not provided in order to save money. Had these anchors been installed, the failures may not have occurred.

RECOMMENDATIONS

The design of interior masonry partition walls is dependent on
reliable design information in masonry standards and in the model
building codes. Current masonry standards should require that non-
bearing walls subjected to lateral loads which exceed 5 lbs. per sq ft
(24.4 kg per sq m), should be considered "exterior" walls for empirical
design, regardless of whether the walls are on the exterior of the
building. Model building codes should also indicate conditions where
lateral loads on interior partition walls may exceed 5 lbs. per sq ft
(24.4 kg per sq m).
Designers should recognize that in circumstances such as the
following, the interior masonry walls may be subjected to lateral loads
greater than 5 lbs. per sq ft (24.4 kg per square m):

1.   Interior walls built in close proximity to an exterior wall
where load sharing between exterior and interior walls is possible.

2.   Interior walls built near exterior wall openings.

3.   Interior walls subjected to mechanical air pressure loads or
loads from elevator air movements.

A more accurate means of determining lateral loads in such
circumstances is a subject worth further study. Designers should
consider engineered design of interior masonry walls, especially when
lateral loads are significant.
Finally, it is important that both designers and builders recognize
that positive anchorage of interior masonry walls is necessary.
Placing mortar between the top of the wall and the building structure
was shown to be insufficient in the three case studies explored in this
paper. For vertically spanning walls, adequate mechanical anchors at
the top of the walls are essential.

ACKNOWLEDGEMENT

The authors appreciate the review and helpful comments of John F.
Seidensticker and J. A. Wintz, III, both of Wiss, Janney, Elstner
Associates, Inc. (WJE).

REFERENCES

[1] "American Standard Building Code Requirements for Masonry," (ANSI
    A41.1), American National Standards Institute, New York, 1953
    (1970).

[2] "Empirical Design of Concrete Masonry Walls", _National Concrete
    Masonry Association Technical Note 73_, NCMA TEK-73, 1975, National
    Concrete Masonry Association.

[3] "Empirical Design of Brick Masonry", _Brick Institute of America
    Technical Notes on Brick Construction 42_, Brick Institute of
    America, 1976.

[4] "Building Code Requirements for Masonry Structures (ACI 530-88/ASCE
    5-88)", American Society of Civil Engineers, New York, 1988.

[5] _Uniform Building Code_ (1991 Edition), International Conference of
    Building Officials, Whittier, CA, 1991.

[6] _Standard Building Code_ 1991 Edition, Southern Building Code Congress
    International, Inc., Birmingham, AL, 1991.

Russell J. Kenney [1] and Richard S. Piper [1]

URETHANE FOAM INJECTION AS A METHOD OF REMEDIAL REPAIR FOR MASONRY CAVITY WALLS

REFERENCE: Kenney, Russell J., and Piper, Richard S., "Urethane Foam Injection as a Method of Remedial Repair for Masonry Cavity Walls," _Masonry: Design and Construction, Problems and Repair, ASTM STP 1180_, John M. Melander and Lynn R. Lauersdorf, Eds., American Society for Testing and Materials, Philadelphia, 1993.

ABSTRACT: In 1981, various tests were performed by the authors to determine if the injection of a urethane foam into the cavity of a typical cavity wall system would substantially alter the properties of the wall system and, more important, if the normal adhesive properties of the polyurethane would bond the veneer outer leaf sufficiently to prevent loosening of the outer leaf during periods of excessive movement. Tests included both laboratory tests on cavity wall panels and samples taken from the wall systems of projects where the cavity walls were repaired using the foam injection method.

Evaluations were made as to the practicality of the repair method and long-term durability of the urethane foam. One building approximately 21 years old, six buildings 5 to 9 years old, and two buildings 1 to 2 years old were monitored to see if the foam material injected into the cavity could have a long-term effect on dimensional stability.

The increased structural capacity of similar cavity wall systems when filled with foam to form a composite structure was substantial. The increase in bond and the bonding of each veneer unit to the substrate was also a benefit of this repair method.

This repair method offers a relatively inexpensive method of remedial repair of cavity wall systems, and can have wider applications particularly in areas prone to earthquake.

KEYWORDS: bond, corrosion, deflection, delamination, masonry veneer, polyurethane, urethane foam, water penetration, remedial repair, wall cavity, leaf

---

[1]President and Architect, respectively, R. J. Kenney Associates, Inc., P.O. Box 1748, Plainville, MA 02762

INTRODUCTION

The authors were involved with a number of buildings, eight to be precise, in the early 1970s until recently, that experienced problems with inadequate or improperly placed masonry ties or ties affected by acid washing or high chloride level. The use of light gague corrugated ties that lacked sufficient corrosion resistance, as well as improper placement of these ties, has made many wall assemblies subject to failure at loads well below the original design. The walls developed cracks or deformation. The present paper shares the experience gained in remedial repair of these walls by the use of urethane foam.

Initially, laboratory work was done to determine the effectiveness of the foam injection process, excellent results were obtained, the work was extended to repair of buildings with cavity wall systems having various types of defects.

The success of this repair was primarily attributed to the development of a technique of injection of urethane foam in a controlled programmed manner, using fiber-optic borescope to visually monitor the foam expansion to make sure that there is uniformity in expansion and injecting foam in a limited area at a time and employing delayed time between injections.

A substantial bond develops between the veneer, substrate, urethane, and results in a composite wall system that possesses excellent flexibility. Under excessive movement the composite wall permits its movement without debonding or losing a single brick. Not only was increased strength in composite unit observed, but also observed were increased thermal resistance and reduced water infiltration.

URETHANE FOAM

A two-part polyurethane foam was chosen for testing in the wall system. The foam had an average density of 2.3 lb/ft^3 (36 kg/m^3) and was dispensed using commercially available foam pour equipment.

The foam, a PAPI (polymethylene, polyphenylisocyanurate), was tested for physical properties and expansion characteristics in a confined cavity. The air quality monitoring equipment used was a fast alarm organic isocyanurate monitor.

The urethane foam used in test panel was only used on two of the eight buildings, a similar density urethane was used on the remaining six buildings comprised of three different foam manufacturers. Similar test results were achieved with the similar density foams.

Figs. 1 and 2 show the foam being injected into a cavity space with clear plexiglas on one side so the expansion properties could be observed.

When confined, the urethane foam, which is very temperature sensitive, tends to expand until the outer edges harden. This, however, does not always mean expansion has stopped. The center of the newly formed foam may be much higher in temperature and still expand causing substantial pressure to the abutting veneer or substrate.

It is the expansion problem and the temperature sensitivity of the foam that limits the amount of area that can be foamed at each interval. Generally, 18 in. (0.45 m) vertically and 24 in. (0.61 m) horizontally is the practical limit.

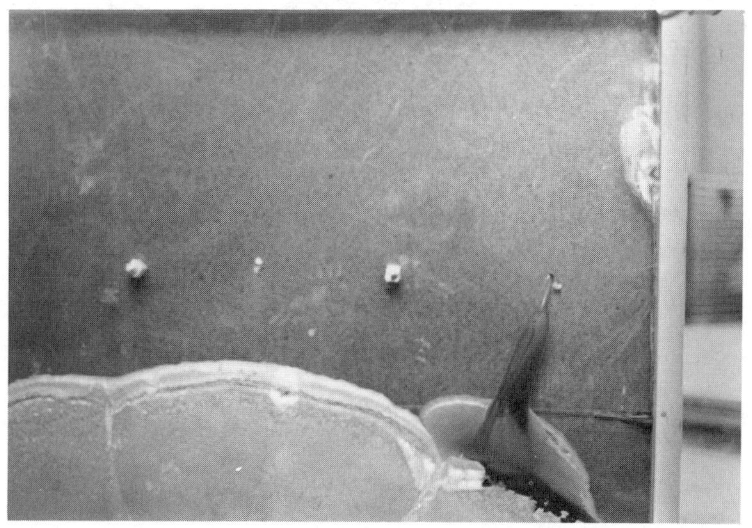

Fig. 1 -- Foam being injected into cavity

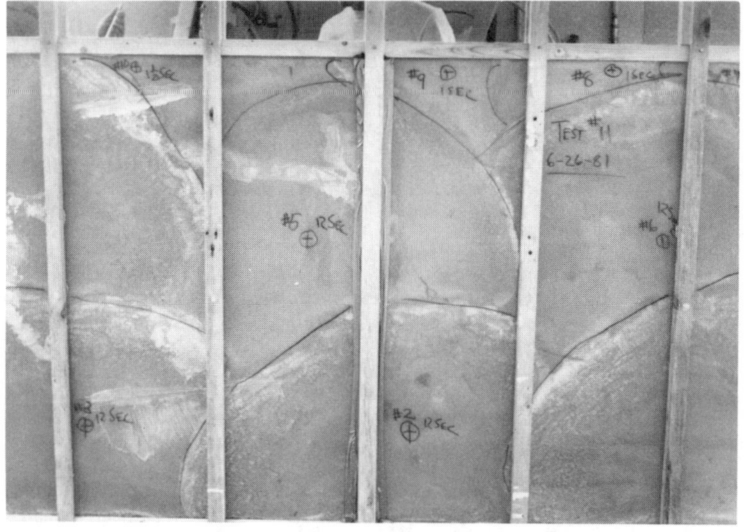

Fig. 2 -- Fully foamed test panel

Equipment, however, can be programmed to inject only a known amount of foam, and delayed time between injections included in the program.

Actual field conditions normally include the use of a fiber-optic borescope to visually monitor the foam expansion which can be altered by blockages in the cavity, by foam liquid temperature, air temperature, and by the surface temperature of the adjacent materials.

BOND STRENGTH

The pressure of the foam expanding against the brick or block veneer and block or sheathing substrate causes the foam to densify at the surface and forms a relatively dense, high strength urethane that bonds to the abutting materials.

In cases where the mortar joints are very thin, the expansion of the foam may seep to the exterior or into the block back-up voids. Pin holes in the mortar joints of the veneer may also allow the foam to appear on the exterior.

The bond between the foam and concrete block veneer and concrete block substrate depends on the properties of the foam, density being the predominant factor.

With this particular foam, the bond strength exceeds 9.0 psi (62.1 kPa) in tension and no failure occurred at the masonry interface. The failure was always in the foam. Bond tests were performed using a Dillion Tensile Tester with the specimens restrained and the plates bonded to the exterior leaf.

Bond test results of samples with split-face block veneer and steel stud gypsum sheathing substrate depends on the fasteners holding the gypsum sheathing to the studs and/or the condition of the bond between the gypsum core and the paper cover.

Fig. 3 shows a section of a wall with 4x8x16 in. (10x20x40 cm) masonry block back-up and the condition of the foam between the brick and block. A close examination of the photograph shows the dense skin that forms at the block and brick surfaces.

TYPICAL BOND TESTS (wall section Fig. 4)

Test Section #1   (Lower left hand side)

Total Load = 1,044 lb (475 kg)
Area = 64 sq. in. (413 cm^2)
Bond Strength = 16.31 psi (112.5 kPa)

Test Section #2   (Center bottom)

Total Load = 492 lbs. (223.5 kg)
Area = 64 sq. in. (413 cm^2)
Bond Strength = 7.68 psi[*] (52.9 kPa)

[*] - Only 69% of area bonded. Some mortar spillage prevented full bonding.

Fig. 3 -- Portion of test panel. Note dense surface
of foam in contact with masonry surfaces.

Fig. 4 -- Test panel

Fig. 5 -- Water spray at top of test panel

WATER TESTS

Test panels, 8x8 ft (2.43 x 2.43 m) with window openings, were erected using standard brick veneer, 2 in. (50 mm) air space, and 4x8x16 in. (100x200x400 mm) concrete masonry unit substrate. The panels were flashed at the bottom of the assembly and over the window lintel.

Foam was injected in a 18x24 in. pattern via 1/2 in. (13 mm) diameter holes drilled in the mortar joints. A fiber-optic borescope was used to monitor the expansion of the foam. The rate and quantity of the foam injected was controlled by a timer which was programmed to delay injection sequences to prevent over expansion.

At age 60 days, and 24 hours after foaming, water was applied to the wall panels via a bar at the top of the wall at a rate of 5.0 gallons per square foot (200 l/m^2) per hour for 2 1/2 hours (Fig. 5).

Immediately after the water tests, seven samples were cut out of the panel at various locations and the foam was tested for water migration.

Increase in water content over the equilibrium moisture levels were determined by drying the urethane foams in a forced air oven at 105 +/- 5°C for 24 hours.

Test Results

Water did not penetrate the foam in the cavity. It did, however, penetrate locations where mortar spillage prevented the foam from completely filling the cavity.

Observations

It is apparent that the foam can effectively seal a cavity and substantially increase its water resistance. The foam can also be used to create end dams at the flashing levels to prevent water from flowing off the end of the flashing.

This system, however, has limitations and is not always successful under conditions where excessive mortar is in the cavity and the flashing is inadequate.

Under these conditions, the cavity should be foamed and the bricks or blocks at the flashing level removed, the excess mortar dislodged, new flashing installed, and the bricks or blocks replaced. The foamed cavity will allow brick or block removal to a much greater extent than under unfoamed conditions. Removal of 13 linear feet (4 m), without shoring on projects with support angles at every floor, and no continuous loads transmitted through the support angles, was successful without any veneer distress.

The physical properties of the foam with the dense outer skin are substantially different than water resistance properties of a cut section of urethane insulation board, which has been reported to allow migration of water due to pressure differential between the exterior and interior.

Vapor transmission properties of the injected urethane foam in the cavity were not determined. The injection process causes denser skin at the outer surface thereby reducing vapor transmissions. Studies by others [1], reported accumulation and distribution of moisture in urethane insulation due to thermal gradient being in the same direction as the vapor pressure gradient.

However, none of the buildings constructed or repaired using this method of remedial repair have experienced any dimensional stability problems or reports of detrimental effects on the long-term thermal performance of the exterior wall system.

Further investigation duplicating the condition of the in-place foam with its dense outer shell needs to be studied to determine the effects on the foam due to thermal and vapor pressure gradients.

Other authors have reported the experimental use of similar systems for improving the water resistance of cavity walls on buildings [2].

COMPLETED PROJECTS

Since 1981, there have been several projects where the exterior wall systems were repaired or partially repaired using the foam injection method. One prior project, in 1971, used a foam cavity but the cavity was foamed during the initial construction rather than injection of foam into a confined cavity. No deterioration of the foam in monitored projects has been noted.

Projects	Date	Exterior	Cavity	
Arlington, MA*	1971	Brick Veneer **	60-70	mm
Lynn, MA	1983	Brick Veneer ***	40-70	mm
Salem, MA	1984	Brick Veneer **	50-60	mm
Malden, MA	1984	Brick Veneer **	40-50	mm
Salem, MA	1984	Brick Veneer ***	40-50	mm
Shrewsbury, MA	1985	Brick Veneer ***	50-70	mm
Revere, MA	1987	Block Veneer **	40-70	mm
Fall River, MA	1990	Brick Veneer **	40-50	mm
Parsippany, NJ	1991	Brick Veneer ***	40-50	mm

* Urethane foam was poured into the cavity during the
  construction. The authors have examined the building from
  time to time since 1979.
** Block substrate
*** Gypsum sheathing substrate

These buildings have been examined every six months, for any signs of expansion due to the foam cavities. There are numerous other buildings which utilized injected urethane foam in the cavity to increase thermal properties and reduce water infiltration. No failures due to thermal expansion have been reported in these cases. Both the waterproofing and increased thermal resistance were performed under controlled injection methods.

Observing these projects, it is evident that foam injection works best with cavities of at least 1 1/2 in. (38 mm) in width. Cavities as small as 1 in. (25 mm) can be successfully foamed depending on the amount of mortar spillage in the cavity. If the wall system is split-face block with steel stud gypsum substrate, the condition of the sheathing (water damage) may be the determining factor.

On projects where the substrate was concrete masonry block, field tests of the bond between the substrate and the veneer were similar to laboratory tests at around 9.0 to 11.0 psi (62.1 to 75.8 kPa).

The bond on projects with gypsum substrates was limited by the fasteners securing the gypsum sheathing to the steel studs and the condition of the sheathing.

On one particular project, most of the corrugated ties on the wall with the steel stud/gypsum sheathing back-up were improperly installed as shown by Fig. 6. The project also had a severe corrosion problem due to the improper use of an acid cleaner. See Fig. 7.

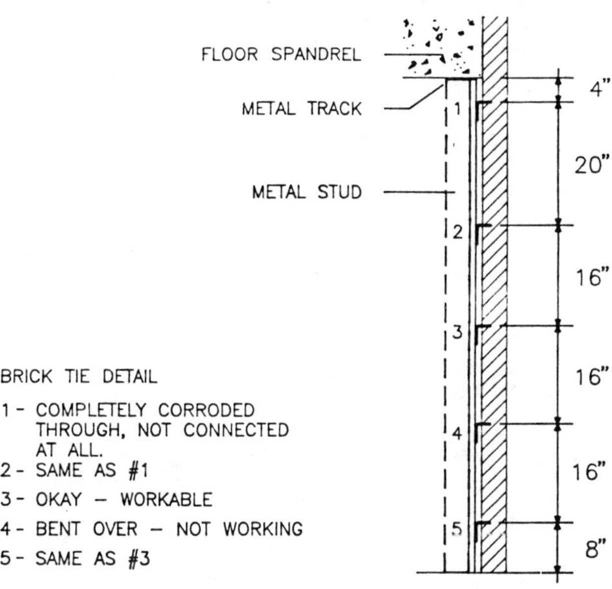

Fig. 6

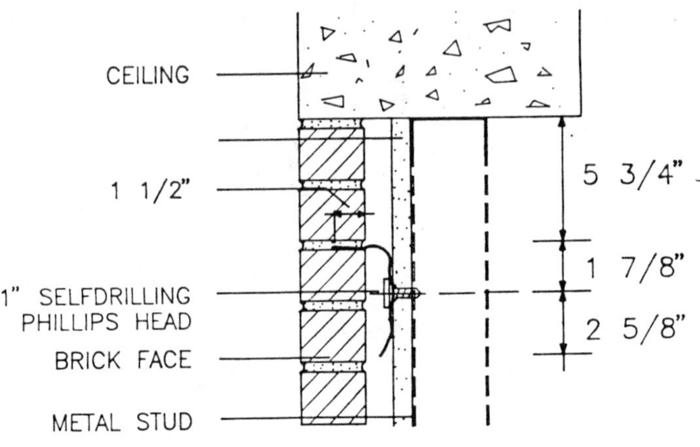

Fig. 7

SIMULATED WIND LOAD TEST

Various walls with both block substrates and with steel stud gypsum substrates were fabricated. The 4x4 ft (1.20 x 1.20 m) walls were constructed using 20 gauge (0.9 mm) 3 5/8 in. (92 mm) steel studs, 1/2 in. (13 mm) gypsum sheathing, 2 in. (50 mm) cavity, and standard brick veneer. Adjustable wall ties, 3/16 in. (5 mm) diameter rectangular ties with drip, were placed 16 in. (0.4 m) on center horizontally and 16 in. (0.4 m) on center vertically. Mortar meeting the proportional requirements of ASTM C270, Specification for Mortar Unit Masonry, Type S was used for the brick veneer and block substrate.

The test walls were restrained at the top and bottom via steel channels. A chamber was constructed in front of the test assembly which contained the air bag. The air bag assembly was open at the specimen face in order to apply pressure.

No special instructions were given to the experienced masons erecting the brick veneer walls. It was noted during the erection of the brick work that there was a tendency to pull the ties forward engaging the tie in a position to resist tension but not compression.

The purpose of the test was not to simulate the actual field conditions but to evaluate the performance of the wall with and without the cavity foamed with a two-part polyurethane foam with an average in-place density of 2.3 lbf^3 (36 k/m^3).

Dial indicators and strain gauges were installed as shown on Figs. 10 and 11. Pressure was applied via an air bag with a calibrated pressure gauge of 0.1 psi (0.7 kPa) accuracy.

TEST SET UP

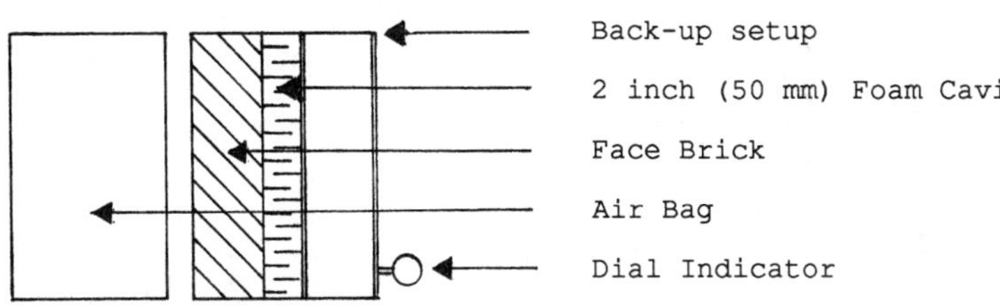

LOCATION OF DIALS ON BACK OF PANEL

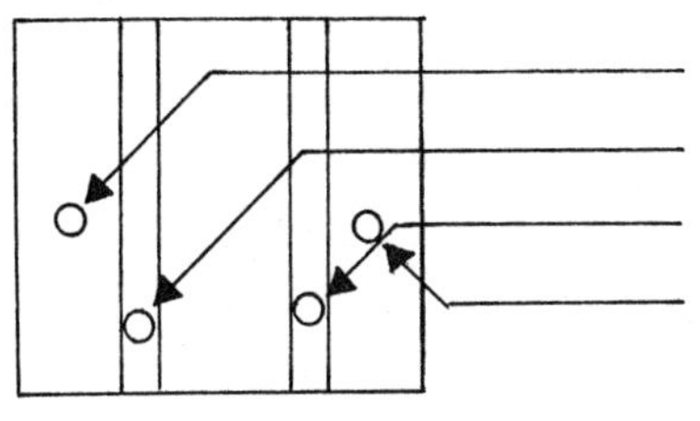

Fig. 8

Table 1 -- <u>Foamed Cavity</u>

Air Bag Pressure		Dial Indicator #1		Dial Indicator #2		Dial Indicator #3		Dial Indicator #4	
psi	Pa	in	mm	in	mm	in	mm	in	mm
0.50	3,447	0.0005	0.013	0.0005	0.013	0.0010	0.025	0.0010	0.025
0.75	5,171	0.0013	0.033	0.0014	0.036	0.0015	0.038	0.0015	0.030
1.00	6,895	0.0023	0.058	0.0026	0.066	0.0025	0.064	0.0025	0.064
1.10	7,584	0.0028	0.071	0.0031	0.079	0.0028	0.070	0.0030	0.076
1.20	8,274	0.0035	0.089	0.0038	0.097	0.0030	0.076	0.0040	0.102
1.30	8,963	0.0040	0.102	0.0043	0.109	0.0035	0.089	0.0045	0.114
1.50	10,342	0.0053	0.135	0.0055	0.140	0.0040	0.102	0.0050	0.127
1.60	11,032	0.0058	0.147	0.0060	0.152	0.0045	0.114	0.0055	0.140
1.70	11,721	0.0065	0.165	0.0068	0.173	0.0050	0.127	0.0060	0.152
1.80	12,411	0.0073	0.185	0.0075	0.191	0.0055	0.140	0.0065	0.165
1.90	13,100	0.0081	0.206	0.0087	0.221	0.0060	0.152	0.0070	0.178
2.00	13,790	0.0091	0.231	0.0091	0.231	0.0070	0.178	0.0075	0.191
2.10	14,479	0.0099	0.251	0.0099	0.251	0.0075	0.191	0.0080	0.203
2.20	15,168	0.0114	0.290	0.0116	0.295	0.0090	0.229	0.0090	0.229
2.30	15,858	0.0128	0.325	0.0126	0.320	0.0100	0.254	0.0100	0.254
2.40	16,547	0.0141	0.358	0.0138	0.351	0.0110	0.279	0.0110	0.279
2.50	17,237	0.0158	0.401	0.0152	0.386	0.0125	0.318	0.0120	0.305
2.60	17,926	0.0188	0.478	0.0176	0.447	0.0150	0.381	0.0130	0.330
2.70	18,616	0.0217	0.551	0.0201	0.511	0.0190	0.483	0.0150	0.381
2.80	19,305	0.0249	0.632	0.0222	0.564	0.0215	0.546	0.0170	0.432
2.90	19,995	0.0298	0.757	0.0257	0.653	0.0270	0.686	0.0190	0.483
3.00	20,684	0.0405	1.029	...	...	0.0370	0.940	0.0220	0.559

TEST SET UP

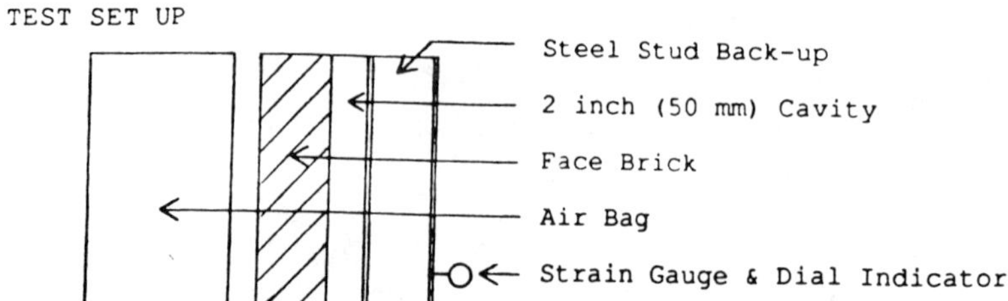

LOCATION OF DIALS ON BACK OF PANEL

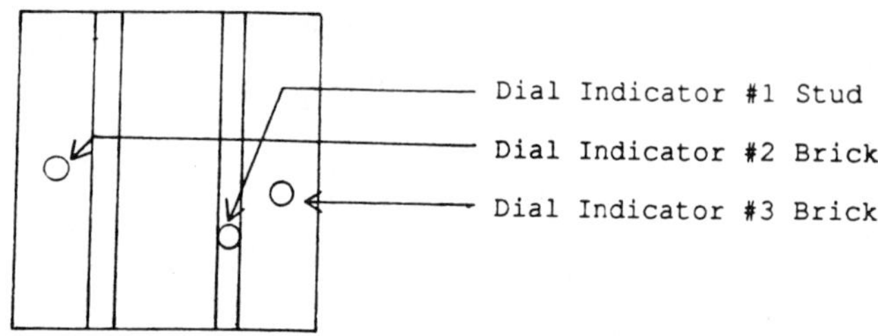

Fig. 9

TABLE 2--Air cavity

Air Bag Pressure		Dial Indicator #1		Dial Indicator #2		Dial Indicator #3	
psi	Pa	in	mm	in	mm	in	mm
0.10	689	0.0005	0.013	0.0010	0.025	0.0020	0.051
0.20	1379	0.0006	0.015	0.0020	0.051	0.0020	0.051
0.30	2068	0.0007	0.018	0.0030	0.076	0.0045	0.114
0.40	2758	0.0023	0.058	0.0060	0.152	0.0080	0.203
0.50	3447	0.0042	0.107	0.0090	0.229	0.0130	0.330
0.60	4137	0.0063	0.160	0.0120	0.305	0.0170	0.432
0.70	4826	0.0102	0.259	0.0190	0.183	0.0230	0.584
0.80	5516	0.0141	0.358	0.0240	0.610	0.0280	0.711
0.90	6205	0.0183	0.465	0.0310	0.787	0.0340	0.864
1.00	6895	0.0229	0.582	0.0380	0.965	0.0400	1.016
1.10	7584	0.0273	0.693	0.0440	1.118	0.0480	1.219
1.20	8274	0.0314	0.798	0.0510	1.295	0.0540	1.372
1.30	8963	0.0360	0.914	0.0570	1.448	0.0600	1.524
1.40	9653	0.0407	1.034	0.0650	1.651	0.0670	1.702
1.50	10342	0.0454	1.153	0.0710	0.803	0.0740	1.880
1.60	11032	0.0518	1.316	0.0810	2.057	0.0830	2.108
1.70	11721	0.0576	1.463	0.0880	2.235	0.0910	2.311
1.80	12411	0.0636	1.161	0.0960	2.438	0.0990	2.515
1.90	13100	0.0694	1.763	0.1040	2.642	0.1060	2.692
2.00	13790	0.0757	1.923	0.1120	2.845	0.1150	2.921
2.10	14479	0.0824	2.093	0.1210	3.073	0.1240	3.150
2.20	15168	0.0888	2.256	0.1300	3.302	0.1320	3.353
2.30	15858	0.0961	2.441	0.1390	3.531	0.1420	3.607
2.40	16547	0.1026	2.606	0.1470	3.734	0.1500	3.810
2.50	17237	0.1100	2.794	0.1570	3.988	0.1600	4.064
2.60	17926	0.1178	2.992	0.1680	4.267	0.1700	4.318
2.70	18616	0.1228	3.119	0.2010	5.105	0.3800	9.652

Test results are shown on Tables 1 and 2, and Figs. 8 and 9, for both the wall with the foamed cavity and the unfoamed test specimen. The deflection of the studs and brick veneer is also listed.

At 1.0 psi (6.8 kPa) of pressure, the foam cavity specimen had deflection of the brick veneer of 0.0026 and 0.0025 in. (0.066 and 0.064 mm). At the same pressure, the unfoamed specimen had brick deflection of 0.0380 and 0.0400 in. (0.965 and 1.016 mm).

At 2.0 psi (13.6 kPa) of pressure, the foamed specimen had brick deflection of 0.0091 and 0.0075 in.(0.965 and 0.19 mm), while the unfoamed specimen had deflection of 0.112 and 0.115 in. (2.8 and 2.9mm).

At a pressure of 2.7 psi (18.5 kPa), the unfoamed specimen failed. Deflection of the brick veneer at that time was 0.201 and 0.380 in. (5 and 9.5 mm). Failure occurred in the five mortar joints up from the bottom of the panel, ties bent, ties with mortar on top penetrated into the sheathing. Residual strains were left in the studs.

The foamed specimen was tested just beyond 3.0 psi (20.4 kPa) of pressure when the air bag ruptured. No permanent deformation occurred and the strain gauge showed no residue stress after loading ceased. Brick deflection at 3.0 psi (20.4 kPa) was 0.022 in. (0.569 mm) at dial indicator #4.

Figs. 10 and 11 show the substantial difference between similar wall sections subjected to the same stress when the cavity is foamed versus an unfoamed cavity.

BOND TESTS -- BRICK VENEER TO GYPSUM SHEATHING AND CONCRETE MASONRY UNITS

One of the main purposes of the research was to determine if a brick cavity wall could be filled with a polyurethane foam under controlled conditions and achieve an adequate bond between the brick veneer, the gypsum sheathing, or concrete masonry units due to the normal adhesive properties of polyurethanes.

After the simulated wind-load test and subsequent water tests of the 8x8-ft (2.43 x 2.43 m) wall sections with block back-up, samples were saw cut and bond tests were performed to determine the bond between the brick-foam-block, and gypsum sheathing.

In the case of the gypsum sheathing, oversized washers were installed over the fasteners at the sheathing to stud connection. Prior testing had shown that the bond between the foam and the gypsum sheathing far exceeded the resistance of S-12 fasteners to hold the sheathing in place, and all test failures were due to the sheathing pulling through the fasteners. In these cases, 2 in. (50 mm) diameter washers were used so that failure would occur either in the foam itself or the bond of the paper covering the gypsum core.

When failure occurred due to lack of bond to the sheathing or debonding of the paper covering the gypsum core, the failure range was relatively narrow, ranging from 6.59 psi to 7.43 psi (45 to 51 kPa).

Specimens of the cavity wall with block back-up, when tested for bond strength between the brick-foam-block assembly, ranged from 9.0 psi to 11.0 psi (62 kPa to 76 kPa) with the failure always due to failure of the foam rather than debonding at either the block or brick-foam interface. Tests on actual projects resulted in higher values for the concrete substrates and lower values for the gypsum substrates, especially where the sheathing was damaged.

Simulated Wind Load Test
Panel One   Foamed Cavity

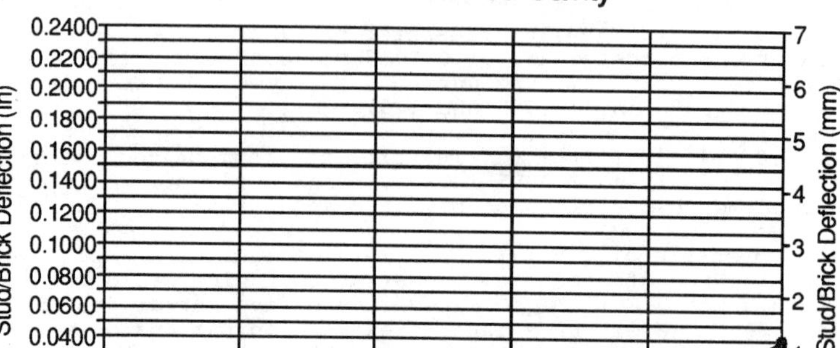

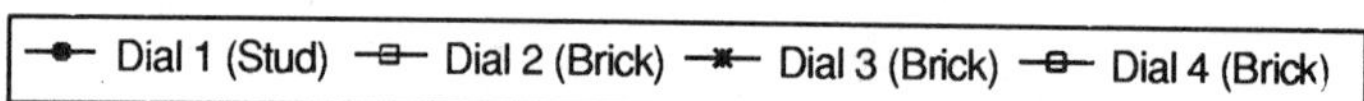

Fig. 10

Simulated Wind Load Test
Panel Two   Air Cavity

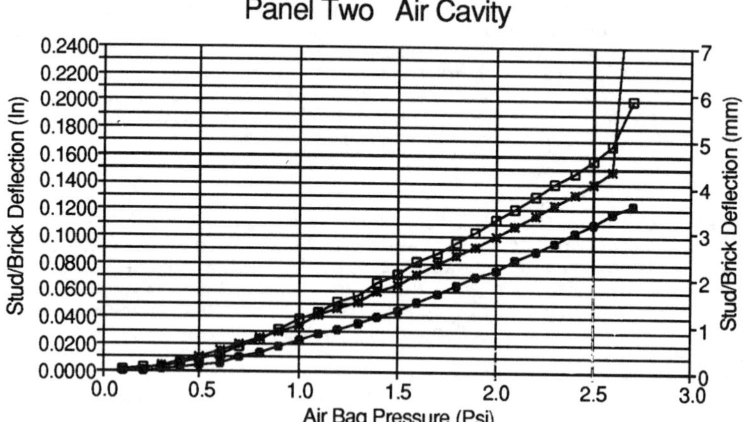

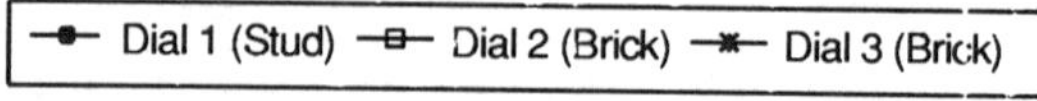

Fig. 11

CONCLUSIONS

This repair method offers a relatively inexpensive method of remedial repair of cavity wall systems. Depending on the wall construction, it can offer substantial increases in wind-load resistance and substantially greater bond of the veneer to the substrate.

The bond is such that excessive movement should not cause dislodging of the veneer.

The best feature of this repair method is the bond that develops between the veneer and the substrate, especially masonry substrates, and the movement the assembly can be subjected to without damage or loss of any of the veneer.

Increased water resistance, thermal properties, and decreased air infiltration are additional benefits of this repair system.

There are, however, conditions that may reduce the effectiveness or even render the use of this repair method impractical, if:

- building paper is used in the cavity

- the blowing agent of the foam is restricted by State Regulation

- the sheathing, in the gypsum-sheathing substrate, is damaged,

the use of this repair method is generally not practical for structural purposes.

REFERENCES

[1] Schwartz, N. V., Bomberg, M., and Kumaran, M. K., "Water Vapor Transmission and Moisture Accumulation in Polyurethane and Polyisocyanurate Foams," in <u>Water Vapor Transmission Through Building Materials and Systems: Mechanisms and Measurement, ASTM STP 1039</u>, American Society for Testing and Materials, Philadelphia, 1989, pp. 63-72.

[2] Beasley, K. J., "Water Leakage Through Brick-Clad Walls", Proceedings, Third North American Masonry Conference, Paper 81, University of Texas, Arlington, June 1985.

Frederick A. Herget[1], and Robert W. Crooks[2]

DETERIORATION AND STABILIZATION OF BEREA SANDSTONE ON THE HAMILTON COUNTY COURTHOUSE

REFERENCE: Herget, Frederick A., and Crooks, Robert W., "Deterioration and Stabilization of Berea Sandstone on the Hamilton County Courthouse," _Masonry: Design and Construction, Problems and Repair, ASTM STP 1180,_ John M. Melander and Lynn R. Lauersdorf, Eds., American Society for Testing and Materials, Philadelphia, 1993.

ABSTRACT: A rehabilitation program is underway for the 114 year old Hamilton County Courthouse (Indiana). The exterior masonry of the Courthouse is composed of Berea Sandstone and hydraulically pressed face brick. Poor details and workmanship along with improper maintenance have led to the extensive deterioration of the exterior masonry. Much of the sandstone is delaminating in thin layers and falling from the building. This is believed to have been caused by cyclical freezing of the saturated stone. Repair documents have been prepared utilizing proven methods and materials. These repairs are discussed in detail.

KEYWORDS: sandstone, deterioration, delamination, water infiltration, cyclical freezing, improper maintenance, repairs

---

The Hamilton County Courthouse was constructed in 1878 in Noblesville, Indiana. It is listed on the National Register of Historic Places. The exterior masonry consists of sandstone, hydraulically pressed face brick and lime mortar. The brick were manufactured in Philadelphia. The sandstone was quarried by The Cleveland Stone Company in Berea, Ohio near Cleveland. Promotional literature by this supplier in 1887 indicated the stone used for the Courthouse was Berea Sandstone. The Cleveland Stone Company supplied sandstone for hundreds of buildings from Boston, MA to Omaha, NE in the late 1800's including such prominent structures as the Parliament Buildings in Ottawa, Canada and the State Capitols of Michigan and New York.

---

[1]Professional Engineer, ARSEE Engineers, 14500 E. 136th St., Noblesville, IN 46060
[2] President, ARSEE Engineers, 14500 E. 136th St., Noblesville, IN 46060

The Hamilton County Courthouse is currently undergoing a major rehabilitation effort in conjunction with the construction of a new Judicial Building across the street. A significant portion of this work involves the rehabilitation of the exterior masonry which has extensive deterioration on all elevations of the building. Infrequent and improper maintenance practices in the past have contributed greatly to the deterioration.

WALL SECTIONS AND DETAILS

The exterior masonry of the Courthouse is composed primarily of sandstone with vertical bands of brick as shown in Figure 1. Horizontal stone bands encircle the perimeter of the structure at its base and at the 2nd floor level. This detail at the 2nd floor has been identified as the horizontal belt course. The main entrance is to the south which projects out and includes a balcony from which speeches reportedly were once presented. This entrance including the balcony slab and railings is constructed completely of sandstone and common brick back-up. Most of the exterior walls of the structure are load bearing. The 1st thru 3rd floors are constructed of shallow brick arches supported by cast iron joists and beams which bear into the common brick back-up. The roof and attic floor is supported by 4.5 m (15 ft) deep wrought iron trusses which also bear on the brick masonry.

Fig. 1 -- Southeast elevation of the Courthouse.

Few of the original drawings for the building were available.
Those that were available appeared to be schematic drawings by the
original Architect, Edwin May. The drawings provided little information
regarding wall sections or masonry details. A borescope was used to
determine typical wall sections and help understand how the stone and
brick masonry interrelated to form the exterior walls. The majority of
the walls were found to be 4 wythes thick with the exterior wythe
constructed of hydraulically pressed face brick and the remainder of
common brick. The face brick were laid in running bond with blind
headers installed every 6th course into the back-up. Much of the stone
was found to be a veneer with many of the pieces extending back into the
brick masonry less than 3 cm (1 in) in depth. The back of some of the
vertical stone bands actually coincided with the exterior face of the
brick. Few of the sandstone pieces extended back into the brick masonry
more than one brick wythe or approximately 10 cm (4 in). The exceptions
were at window sills and at the second floor horizontal belt course.
These stones ultimately took the place of flashings for water
infiltrating the wall and contributed to the deterioration of the
sandstone masonry. Examples of two of the typical wall sections
observed are shown in Figure 2.

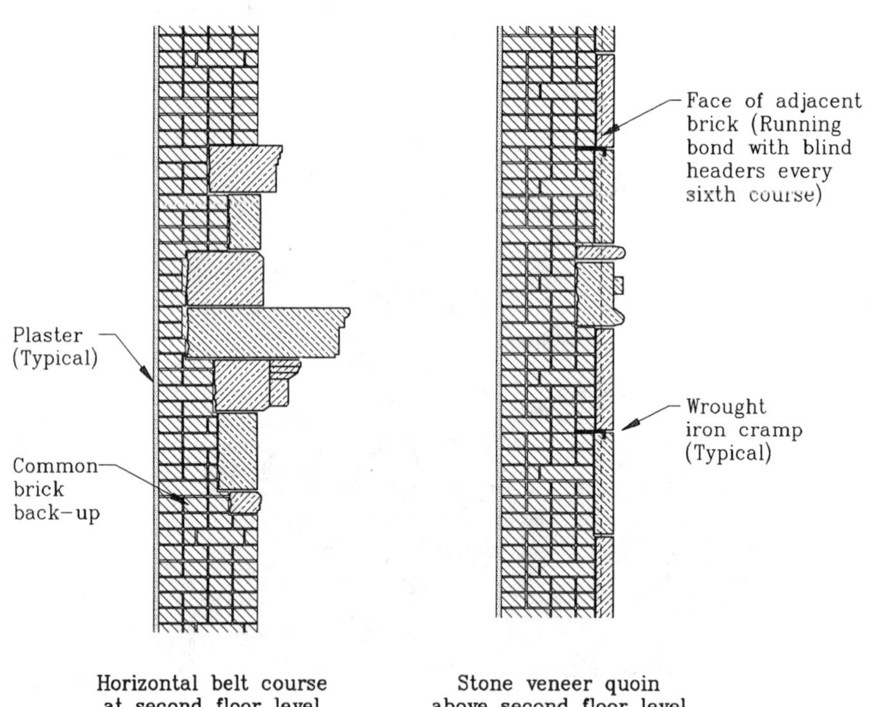

FIG. 2--Typical wall sections.

Metal detectors were used to locate the wrought iron cramps used to support the sandstone veneer. Typical patterns were established showing that cramps were used in every other bed joint of the stone quoins at the corners above the second floor belt course. The use of metal anchorages was very sporadic below this belt course even though many of the pieces would still be considered a veneer. Cramps were also used at the top of the tall window jamb pieces at the 2nd and 3rd floors but never at the sill. This eventually led to bulging of the masonry at several of the 3rd floor window sills because there was no positive anchorage to prevent it. Water infiltrating voids at the 3rd floor sill and freezing gradually pushed the stone masonry out of the wall as shown in Figure 3. Bulges were measured up to 7 cm (2-3/4 in) from what is believed to be the original face of the masonry.

Trial removal and reconstruction of one of the more severely bulged areas exposed some of the wrought iron cramps. All of the anchors were in good condition with only a light patina of corrosion present on them. Anchorage back into the common brick back-up was solid and the anchors were reused to reconstruct the wall section.

Records indicate that the exterior of the Courthouse was sandblasted in the early 1950's in an attempt to clean the masonry. The sandblasting destroyed the exterior skin of the brick exposing the more porous and fissured inner body as shown in Figure 4. Testing with a RILEM uptake tube showed that the brick absorbed water almost as quickly as it could be poured into the tube.

The mortar joints in both the brick and stone masonry were very thin, often 3 mm (1/8 in) or less with several of the units actually touching each other. The majority of the head joints were found to be unfilled with only a thin layer of mortar on the outside face giving the appearance of a full joint. Several of the joints were observed to be completely open and free of any mortar. This condition was particularly prevalent in the horizontal bands of stones at the 2nd floor horizontal belt course and below the 1st floor windows. These conditions coupled with the excessive porosity of the sandblasted brick have allowed extensive water infiltration of the exterior masonry.

OBSERVATIONS AND TESTING

A condition survey was performed on all of the exterior masonry. The work was performed from a hydraulic lift and from ground level. The condition of all of the stones was prioritized to allow contractors to reasonably bid the repair work while ensuring the most critical work would fall within the allotted budget.

Over 1200 of the stones were observed to be exhibiting exfoliation of the outer face of the stone. Thin layers of stone approximately 1.1 mm (0.045 in) were delaminating and falling off as shown in Figure 5. Many of these stones were improperly face-shell bedded or laid such that their natural bedding planes were parallel to the face of the wall. In this configuration there is less resistance between the layers of stone to prevent such delamination. These delaminations did not always follow the natural bedding planes of the stone. There was also a significant number of stones which would normally be thought to be "properly bedded" where delaminations occurred perpendicular to the natural bedding planes. Removal of several stones in a trial repair area revealed some

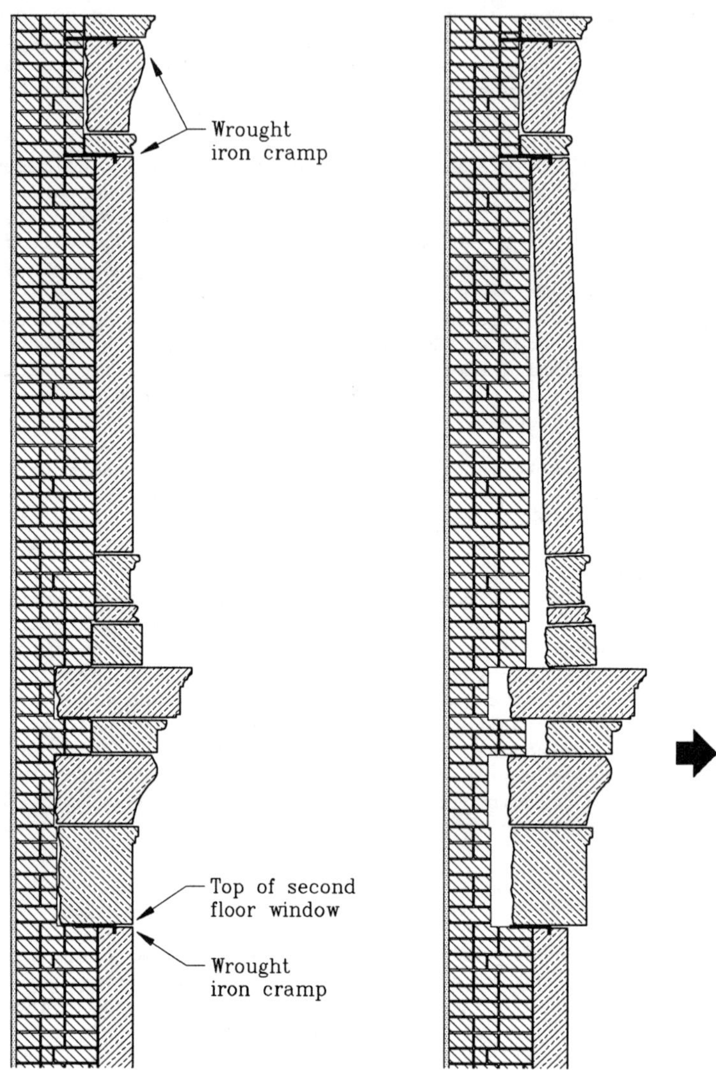

FIG. 3--Schematic illustration of bulging
masonry at third floor window sill.

Fig. 4 -- Sandblasted face brick exposing voids and fissures.

Fig. 5 -- Typical delamination of sandstone units.

of the more severely deteriorated stones were also delaminating on the back side of the stone.

Blind exfoliation of the stone was found to occur in several additional stones by tapping with a wooden mallet. Here the stone was delaminating in much the same manner as described above but the outer layer had not yet begun to fall away giving the appearance that the stone was still sound. Such instances of blind exfoliation were confined to stones which were face-shell bedded.

Samples of the stone were taken at several locations around the building and analyzed. Cores were taken of portions of the horizontal belt course and of the south entrance balcony. These cores were approximately 2 cm (3/4 in) in diameter by 10 cm (4 in) deep. Each was taken vertically from a horizontally bedded stone approximately 20 cm (8 in) thick. Petrographic studies found the primary cementing material of the sandstone to be meager amounts of halloysite clay in which trace amounts of calcite are present. The stone is a fine-grained, finely bedded and finely cross-bedded sandstone. It is finely porous because of an insufficient amount of material to fill interstices between the particles. It has a nominal water absorption of 7 percent. Clay bound sandstones are often associated with durability problems.

Examination of one of the core samples found the exterior surface of the stone to contain several thin incipient delaminations to a depth of approximately 2.5 cm (1 in). The next 5 cm (2 in) of the core were judged sound. The deepest portion of the core corresponding to the interior 2.5 cm (1 in) or middle of the stone also contained thin incipient delaminations. This indicates some of the delaminations may have been present since the stone was first placed in the wall. This particular sample was taken from the edge of the south entrance balcony at the 2nd floor level. White stains appearing to be salts in the stone were observed in this area. Chemical analysis of the sample confirmed chloride has been introduced into the stone. One theory to explain the presence of salts in this location is the possible salting of the balcony slab to prevent ice build-up and leakage over the entrance.

Careful removal of some of the exterior delaminations revealed layers of sand trapped between the delaminations and body of the stone. It appears that the original cementing medium holding together the individual grains of sand had completely been destroyed in these areas leaving only the fine sand particles. Analysis of these exterior delaminations found evidence of fly ash and gypsum masses. The gypsum is believed to be the product of acid rain with lime from the mortar joints. The flyash is a result of older coal fired stoves and furnaces.

A few of the stones had a mottled appearance on portions of their exterior surface. Subsequent investigation found that the blade of a pocketknife could be inserted into these mottled areas and that such areas had been reduced to a powder-like consistency. No pattern of such deterioration was discernible and it was observed to occur in both face-shell bedded and properly bedded stones.

Several patterns of deterioration were identified with the condition survey. The most prevalent pattern was the delamination and often severe erosion of stones beneath horizontal projections. Such delaminations were consistently observed beneath the 2nd floor belt course and beneath sills at all windows. Deterioration patterns similar to those left by wave action on a beach were typically observed in these areas as shown in Figure 6. Such stones were typically oriented

Fig. 6 -- Delaminations form a wave-like pattern in stone beneath horizontal projections.

"properly" with the bedding planes perpendicular to the face of the wall yet the deterioration was often severe with shards of stone weighing up to 260 gm (0.5 lb) loose and in danger of falling. While the more severe deterioration occurred beneath the larger overhangs delaminations were observed beneath ornamental projections as small as 2 cm (3/4 in). It is believed that these horizontal projections contributed to the delamination of the stone. First the horizontal ledges allowed water to seep back into the wall and absorb into the relatively porous sandstone. The water would naturally migrate back out of the stone during dry periods through evaporation. Portions of the stone beneath the projection were naturally protected. Atmospheric pollutants such as fly ash accumulated here over time filling the pores. The surfaces under the projections became relatively impermeable trapping moisture behind them. Subsequent cyclical freezing caused the observed delaminations. These delaminations were more pronounced in areas where they could follow the natural bedding planes of the stone.

   The survey also showed many of the deteriorated stones were face-shell bedded. Entire faces of some of these stones were missing with subsequent layers continuing to delaminate and fall. The exterior layer of many of these stones had changed from the original gray color to more of a light brown color. Petrographic analysis of some of these layers showed ferruginous particles within the sandstone had oxidized to depths of 5 mm (3/16 in). This in conjunction with atmospheric pollutants produced a relatively impermeable crust which trapped water within the stone and with cyclical freezing caused delaminations to occur.

   Another pattern which was evident was the deterioration of many stones at grade. The foundation and basement walls were constructed of rubble limestone. The grade around much of the building was held a few inches below the bottom of the first course of sandstone. However, wherever the grade did come in contact with the sandstone delaminations were observed. Capillary action within the stone appeared to draw moisture up as high as 35 cm (14 in) in some areas. Delaminations occurred in both face-shell and properly bedded stones and again are believed to have been caused by cyclical freezing of the saturated stone.

A similar phenomena was observed at each of the entrances to the building with the most severe deterioration occurring on the south entránce shown in Figure 7. Here the stones were delaminated and deeply eroded as high up as 107 cm (42 in) from the entrance floor slab. White stains were observed on the stone along the edge of the delaminations. Further examination determined these were salts which had wicked upward from de-icing salts thrown out onto the concrete slab and steps. Severe deterioration was also observed along either side of this entrance where snow (presumably mixed with de-icing salts) had been shoveled and piled behind the wingwalls of the entrance stairs. Holes were observed to have eroded in these areas as deep as 7 cm (2-3/4 in).

Research of the literature indicates salts within fine grained stones have caused similar problems in the past [1]. Dissolved salts infiltrate the stone and recrystalize as the stone dries out. This recrystalization occurs in the pores of the stone until compressive stresses are formed eventually causing surface delaminations. While various poultices and methods have been attempted to remove salts from stone none seem to have consistently performed satisfactorily and several are reported to have caused further deterioration of the stone.

Fig. 7 -- Severe deterioration of sandstone at south entrance.

While the predominant form of deterioration observed in the sandstone was delamination of the exterior surface another somewhat prevalent condition observed was cracking. The exterior wall is essentially a monolithic construction of stone and brick on the order of 45 cm (18 in) thick. The interior face is plastered and painted and serves as the interior room finish. When originally constructed the facility was without air conditioning with the exception of fireplaces and steam heat located around the ground floor. The temperature of the exterior wall was close to that of the outside ambient temperature and any thermal gradient through the wall thickness was small. Typical of construction of this time period the exterior walls were built without provision for differential movements.

With the incorporation of air conditioning systems to maintain the interior temperature near 21.1°C (70°F) the wall experienced much greater thermal gradients and movements in general. The result has been cracking of the exterior masonry. Many of these cracks occur both in the stone and brick masonry. Quite often they are associated with an open head joint in the stone masonry and extend vertically or sometimes diagonally up into the brick masonry. Measurements indicate these cracks are typically active, opening and closing in response to temperature.

PREVIOUS MAINTENANCE

It appears that very little maintenance has been performed on the exterior of the Courthouse to date and that often what was performed did more harm than good. Undoubtedly the two most prevalent forms of maintenance have been cleaning of the brick and stone and periodic tuckpointing.

Cleaning of the masonry in the early 1950's was performed by light sandblasting. While considered "state-of-the-art" technology in its day such practice is now shunned by the industry. Sandblasting the brick destroyed the outer protective skin and exposed the much more porous body of the brick. Holes, voids and fissures in the brick were now exposed to the weather. This allowed copious amounts of water to infiltrate the brick and subsequently saturate the stone masonry. The effect of sandblasting on the sandstone itself is not known but presumably it would have removed any loose material and exposed a new layer of stone to the atmosphere accelerating the deterioration.

Both the brick and stone masonry appear to have been tuckpointed at least twice during the life of the building as two different colors of tuckpointing mortar are evident. The tuckpointed mortars are Portland Cement based and much harder than the original lime mortar. With such thin mortar joints the tuckpointing mortar has typically been buttered into the eroded joints so that a skim coat of mortar 3 mm (1/8 in) or less in depth was applied. This harder mortar along the edge of the brick and stone units has resulted in chippage of some of the masonry units.

A number of the stone mortar joints have had a thin layer of sealant tooled over the eroded mortar. In many cases this sealant has failed and peeled back away from the joint where 3-sided adhesion had occurred. This does not appear to have affected the stone itself.

Finally, there are several areas beneath the 2nd floor window sills where the stone has deteriorated and eroded to depths up to 4 cm

(1-1/2 in). Attempts have been made to repair these areas with a cementitious based material. All of the patches have deteriorated and become loose. A number of patches have already fallen. The patching material appears to have trapped water within the stone and accelerated the deterioration as the erosion appears more severe than in similar unpatched areas.

REPAIR RECOMMENDATIONS

The overall rehabilitation plan calls for the Courthouse grounds to have walkways, benches and plantings designed to encourage the public to utilize the space more and get up close to the building. To this end a major thrust of the repair program has been to provide as durable a facade as possible while maintaining the original aesthetics of the building. Experience on this and other buildings has shown that repairs must not be made hastily and must address the causes rather than the symptoms of the problems otherwise the repairs will ultimately become problems themselves. Thus the second major thrust of the program has been to develop a program which utilizes proven methods and emphasizes long term durability so that additional repairs are not necessary within the foreseeable future. The repair program hinges on the following considerations:

- Repair or replace all deteriorated materials such that loose fragments will not be of danger to the public.

- Implement repairs which will not exacerbate or accelerate the deterioration or become problems themselves.

- Utilize tested and proven repair techniques which have shown satisfactory long-term performance.

- Since the presence of water in the masonry is the major cause of the deterioration repairs should minimize the amount of water which can infiltrate the masonry walls in the future.

An extensive review of the literature was made long before the preparation of any repair documents began. Books, articles and papers from around the world from the late 1800's to the present were reviewed to determine what had previously been tried and proven to work in similar circumstances. Outside consultants were brought in to test and offer their experience regarding the various problems at hand. Samples of the various repairs which were eventually decided upon were implemented by an experienced masonry restoration contractor prior to releasing the documents for bidding. This helped to ensure the proposed repair methods were suitable for this building and that they could be performed in a reasonable and cost effective manner.

The delaminating stone proved to be the most difficult problem to address. Consolidation materials and methods were reviewed in detail. Review of the literature shows consolidation of sandstones has been attempted since at least the early 1800's [2]. The general concern appears to be that they often do more harm than good. Some of the references claimed that while consolidation techniques appeared to be successful in the short term that over longer periods of time the stone continued to delaminate and the delaminations were deeper and more severe [3]. Manufacturers of consolidation products were consulted but were unsure how deeply the material could penetrate the stones because of the variation within individual stones and the varied orientation of the bedding planes. Ultimately, while the latest generation of consolidation materials may indeed prove to be useful for such problems it was decided early in the design process not to allow this building to be a "test case."

Others have tried to re-pin delaminating stone with stainless steel dowels and epoxy. As this particular stone is delaminating in such thin layers, down to 1 mm (1/32 in) and less, pressure injection of an epoxy or other similar material to readhere the delaminations was impossible.

A trial repair was made bringing in master stone carvers on site to try to reshape the deteriorated stone in place. The concept was to remove the loose material back to sound substrate and reshape it to blend in with the surrounding work. Attempts were made on various degrees of deteriorated stones without success. It was found that repair of even small areas of fine delaminations caused adjacent "sound" areas to blister and delaminate. Working back and forth across the face of the stone caused progressively deeper delaminations. Reshaping was attempted on both face-shell bedded and properly bedded stones without success. This was attributed to the fine cross-grain beds present in the stones. The reshaping concept was ultimately rejected.

Thus it was determined the delaminated stones could not be "repaired." The delaminations would continue to progress and could not even be reshaped back to sound material. Petrographic analysis indicated that at least some of the stones contained incipient delaminations within them since the time of construction making the stone even more suspect. The decision was made to replace the defective stones. With the aid of the condition survey the stones were prioritized with regard to the apparent extent of deterioration. This was done visually on a sliding scale but as shown earlier may not be a completely reliable indication of the deterioration actually occurring within each stone. The repair plan intends to replace as many of the deteriorated stones as the budget will allow concentrating on replacement of the most severely deteriorated stones first.

Many of the stones to be replaced are thin or roughly 10 cm (4 in) or less in thickness. These will be removed completely and replaced with new stones. Thicker pieces may be cut back with new pieces doweled in and epoxied to the remaining stone substrate. This allows the contractor to perform the bulk of the work without costly underpinning or removal and salvage of additional sound material with the subsequent inherent risk of damage.

Samples of new stone will be physically tested and required to meet ASTM Standard Specification for Quartz-Based Dimension Stone (C 616) criteria. The stone will also be petrographically examined to evaluate its potential durability. An alternate has been included in the contract to replace the deteriorated stones with cast stone but at this point it does not appear to be as cost effective as new sandstone.

The exterior wythe of brick which was damaged by sandblasting will be removed and replaced with new face brick. The original thin 3 mm (1/8 in) wide or less mortar joints will be replaced with nominal 6 mm

(1/4 in) thick joints. This will allow the joints to be filled completely with mortar while accommodating a slightly smaller brick than the original units. The mortar will be tinted slightly from the original white lime mortar to aesthetically de-emphasize the wider joints. All new anchorages for both the brick and stone masonry will be stainless steel. New stainless steel flashings with weeps will be incorporated to limit the amount of water which can infiltrate and saturate the brick and stone masonry.

All of the remaining stone mortar joints will be cut out and caulked to minimize water infiltration. The joints will be cut out with a hand held grinder and hammer and chisel where necessary to minimize damage to the adjacent stone. Severely bulging stonework at the 3rd floor window wills will be removed and replaced. Minor bulges will simply be reanchored.

Cracks in the stonework will be cut out and caulked. The majority of these cracks appear to be "working" and caused by differential thermal movements. The pliable caulking should allow these natural "expansion joints" to continue to function.

All of the remaining stone masonry will be cleaned. The masonry has become soiled with several different types of materials over time. Atmospheric pollutants such as soot and fly ash have darkened most of the stone particularly under overhangs where rain action cannot help clean them. Green organic materials are present in areas on the north elevation. The tops of all horizontal ledges were coated in the 1950's with a material designed to prevent pigeons from roosting. This has proven to be the most difficult material to clean in sample cleaning areas. Sample cleaning areas have identified the optimum methods and materials to clean the masonry without damaging it further. Concentrations of cleaning materials and maximum water pressures will be carefully monitored throughout the work to ensure additional damage does not occur.

SUMMARY

The 114 year old Hamilton County Courthouse is currently undergoing a major rehabilitation effort. The work on the exterior sandstone and brick masonry will be the most extensive since its construction. Improper maintenance including sandblasting in the early 1950's is a major contributor to the present deterioration. The fine-grained clay bound Berea Sandstone is delaminating on all elevations of the building. The two basic patterns of deterioration are beneath horizontal projections and stones which are improperly face-shell bedded. The primary cause of this deterioration is the cyclical freezing of the saturated stone. Atmospheric pollutants and acid rain are also believed to have contributed to the deterioration.

Repair documents have been generated and are in the bidding phase at this time. The proposed repairs have focused on limiting the amount of water infiltration and further deterioration of the masonry. An extensive literature review and investigation has been completed to ensure only tried and proven methods and materials will be utilized so that the proposed repairs are suitable and will not ultimately be the source of additional problems in the future.

REFERENCES

[1]   Gilder, C. B., "Property Owner's Guide to the Maintenance and Repair of Stone Buildings," <u>Technical Series No. 5</u>, The Preservation League of New York State, Inc., 1982, p. 4.

[2]   Schaffer, R. J., "The Weathering of Natural Building Stones," <u>Special Report No. 18</u>, Department of Scientific and Industrial Research, London, 1932, p. 83.

[3]   Caroe, A. D. R., and Caroe, M. B., <u>Stonework: Maintenance and Surface Repair</u>, Church House Publishing, London, 1984.

Author Index

Subject Index